KT-211-057

122 Springer Series in Solid-State Sciences

Edited by Peter Fulde

Springer

Berlin
Heidelberg
New York
Barcelona
Budapest
Hong Kong
London
Milan
Paris
Singapore
Tokyo

Springer Series in Solid-State Sciences

Editors: M. Cardona P. Fulde K. von Klitzing H.-J. Queisser

Managing Editor: H. K.V. Lotsch

Volumes 1–89 are listed at the end of the book

90 **Earlier and Recent Aspects of Superconductivity**
Editors: J. G. Bednorz and K. A. Müller

91 **Electronic Properties of Conjugated Polymers III** Basic Models and Applications
Editors: H. Kuzmany, M. Mehring, and S. Roth

92 **Physics and Engineering Applications of Magnetism** Editors: Y. Ishikawa and N. Miura

93 **Quasicrystals** Editors: T. Fujiwara and T. Ogawa

94 **Electronic Conduction in Oxides**
By N. Tsuda, K. Nasu, A. Yanase, and K. Siratori

95 **Electronic Materials**
A New Era in Materials Science
Editors: J. R. Chelikowsky and A. Franciosi

96 **Electron Liquids** 2nd Edition By A. Isihara

97 **Localization and Confinement of Electrons in Semiconductors**
Editors: F. Kuchar, H. Heinrich, and G. Bauer

98 **Magnetism and the Electronic Structure of Crystals** By V.A. Gubanov, A.I. Liechtenstein, and A.V. Postnikov

99 **Electronic Properties of High-T_c Superconductors and Related Compounds**
Editors: H. Kuzmany, M. Mehring, and J. Fink

100 **Electron Correlations in Molecules and Solids** 3rd Edition By P. Fulde

101 **High Magnetic Fields in Semiconductor Physics III** Quantum Hall Effect, Transport and Optics By G. Landwehr

102 **Conjugated Conducting Polymers**
Editor: H. Kiess

103 **Molecular Dynamics Simulations**
Editor: F. Yonezawa

104 **Products of Random Matrices** in Statistical Physics By A. Crisanti, G. Paladin, and A. Vulpiani

105 **Self-Trapped Excitons**
2nd Edition By K. S. Song and R. T. Williams

106 **Physics of High-Temperature Superconductors**
Editors: S. Maekawa and M. Sato

107 **Electronic Properties of Polymers**
Orientation and Dimensionality of Conjugated Systems Editors: H. Kuzmany, M. Mehring, and S. Roth

108 **Site Symmetry in Crystals**
Theory and Applications 2nd Edition
By R. A. Evarestov and V. P. Smirnov

109 **Transport Phenomena in Mesoscopic Systems** Editors: H. Fukuyama and T. Ando

110 **Superlattices and Other Heterostructures**
Symmetry and Optical Phenomena 2nd Edition
By E. L. Ivchenko and G. E. Pikus

111 **Low-Dimensional Electronic Systems**
New Concepts
Editors: G. Bauer, F. Kuchar, and H. Heinrich

112 **Phonon Scattering in Condensed Matter VII**
Editors: M. Meissner and R. O. Pohl

113 **Electronic Properties of High-T_c Superconductors**
Editors: H. Kuzmany, M. Mehring, and J. Fink

114 **Interatomic Potential and Structural Stability**
Editors: K. Terakura and H. Akai

115 **Ultrafast Spectroscopy of Semiconductors and Semiconductor Nanostructures** By J. Shah

116 **Electron Spectrum of Gapless Semiconductors**
By J. M. Tsidilkovski

117 **Electronic Properties of Fullerenes**
Editors: H. Kuzmany, J. Fink, M. Mehring, and S. Roth

118 **Correlation Effects in Low-Dimensional Electron Systems**
Editors: A. Okiji and N. Kawakami

119 **Spectroscopy of Mott Insulators and Correlated Metals**
Editors: A. Fujimori and Y. Tokura

120 **Optical Properties of III–V Semiconductors**
The Influence of Multi-Valley Band Structures
By H. Kalt

121 **Elementary Processes in Excitations and Reactions on Solid Surfaces**
Editors: A. Okiji, H. Kasai, and K. Makoshi

122 **Theory of Magnetism**
By K. Yosida

123 **Quantum Kinetics in Transport and Optics of Semiconductors** 2nd Edition
By H. Haug and A.-P. Jauho

124 **Relaxations of Excited States and Photo-Induced Structural Phase Transitions**
Editor: K. Nasu

125 **Physics and Chemistry of Transition-Metal Oxides**
Editors: H. Fukuyama and N. Nagaosa

Kei Yosida

Theory of Magnetism

With 47 Figures

Professor Kei Yosida

Science University of Tokyo, Faculty of Science & Technology, Department of Physics, 2641 Yamazaki, Noda-shi Chiba-ken, Japan

Translated by:

Professor Hiroyuki Shiba

Department of Physics
Tokyo Institute of Technology
Tokyo 152, Japan

Professor Kosaku Yamada

Department of Physics
Kyoto University
Kyoto 606-01, Japan

Professor Akio Sakurai

Department of Physics
Kyoto Sangyo University
Kyoto 603, Japan

Series Editors:

Professor Dr., Dres. h. c. Manuel Cardona
Professor Dr., Dres. h. c. Peter Fulde*
Professor Dr., Dres. h. c. Klaus von Klitzing
Professor Dr., Dres. h. c. Hans-Joachim Queisser

Max-Planck-Institut für Festkörperforschung, Heisenbergstrasse 1, D-70569 Stuttgart, Germany
* Max-Planck-Institut für Physik komplexer Systeme, Nöthnitzer Strasse 38, D-01187 Dresden, Germany

Managing Editor:

Dr.-Ing. Helmut K.V. Lotsch

Springer-Verlag, Tiergartenstrasse 17, D-69121 Heidelberg, Germany

---

Title of the original Japanese edition: Jisei (Magnetism) by Kei Yosida

Originally published in Japanese by Iwanami Shoten, Publishers, Tokyo 1991

First Edition 1996
Second Corrected Printing 1998

Library of Congress Cataloging-in-Publication Data.
Yosida, Kei, 1922– [Jisei. English] Theory of magnetism / Kei Yosida.
p. cm. – (Springer series in solid-state sciences; 122) Rev. translation of: Jisei. Includes bibliographical references and index.
ISBN 3-540-60651-3 (hardcover: alk. paper)
1. Magnetism. 2. Magnetic materials. 3. Kondo effect. I. Title. II. Series.
QC753.2.Y6713 1996 538–dc20 96-12857

ISSN 0171-1873
ISBN 3-540-60651-3 Springer-Verlag Berlin Heidelberg New York

Typesetting: Data conversion by A. Leinz, Karlsruhe, using the Springer TEX macro package "cpmono01"
Cover concept: eStudio Calamar Steinen
Cover design: Design & Production, Heidelberg
SPIN: 10677362 54/3144 – 5 4 3 2 1 0 – Printed on acid-free paper

Preface

The electron theory of solids is a large field of physics, which has developed rapidly since the establishment of quantum mechanics. Research into magnetism occupies an important part in this field. In this book, the basic understanding reached in the long history of such research from the viewpoint of the electron theory of solids is explained. This book provides the foundations for the further development of research in this field.

The contents consists of four parts. In Part I, the electronic states of magnetic ions in crystals and the exchange interactions between them are explained. Part II treats problems of ferromagnetism and antiferromagnetism shown by exchange-coupled localized spin systems. In Part III, the magnetic properties of itinerant electron systems, in particular, the problem of strong electron correlation, are discussed in relation to $3d$ electrons in iron group metals. Part IV is allotted to the explanation of the Kondo effect and its related subjects.

Research on the Kondo effect gave rise to many develpoments, particularly in the research into rare-earth intermetallic compounds, including high-density Kondo systems and heavy fermion systems. However, these new developments are not described here, because these subjects are considered to be beyond the scope of this book.

The magnetism of matter originates essentially from the electron correlation due to Coulomb repulsion. The central problems of the present research into magnetism lie in the study of the electron correlation. This book is written, as stated above, with the purpose of providing the necessary bases for research along this line. It is hoped that it will be of some use to younger researchers who intend to further the development of research into magnetism.

Finally, it should be mentioned that the present edition of this book is a revised version of the first one published in Japanese in 1972 by Asakura-Shoten in Tokyo. The revised Japanese version was published in 1991 by Iwanami-Shoten in Tokyo.

This English version was translated from Japanese by three condensed matter theorists, H. Shiba, K. Yamada, and A. Sakurai. In the processes of translation they pointed out some misprints and errors. They also made important comments on the contents. These remarks are all taken care of in this translated edition. The author expresses deep thanks to these three professors for taking on the heavy task of translation and also for their valuable remarks.

Tokyo, January 1996 *Kei Yosida*

Contents

Part II Magnetism of Spin Systems

Part III Magnetism of Metals

Part I

Magnetic Ions in Crystals

1. Electronic States of Free Magnetic Ions

To begin with, we describe, in this chapter, the ground states of free magnetic ions which have incomplete shells. There are three series of magnetic ions, iron-group series, rare-earth series and actinide series. They have $3d$-, $4f$- and $5f$-incomplete shells, respectively. These magnetic ions play an important role in the field of magnetism.

1.1 Magnetic Moments of Electron and Nucleus

The magnetism of materials originates from the magnetic moment of electrons, which are either itinerant or localized. The electron has a characteristic angular momentum $\hbar \boldsymbol{s}$ associated with its spin degree of freedom, and has *spin magnetic moment* proportional to $\boldsymbol{s}$:

$$\boldsymbol{\mu}_{\mathrm{s}} = -g\mu_{\mathrm{B}}\boldsymbol{s} \ , \tag{1.1}$$

where μ_{B} is called the *Bohr magneton*; its magnitude is in cgs-Gauss units

$$\mu_{\mathrm{B}} = \frac{e\hbar}{2mc} = 0.927 \times 10^{-20} \quad \mathrm{emu} \ . \tag{1.2}$$

Here e is the magnitude of the charge of electron, m its mass, c the velocity of light, and $\hbar$ the Planck constant divided by 2π. The minus sign in (1.1) is due to the negative charge of the electron. g is usually called the *g-value*; the magnitude for free electrons is

$$g = 2.0023 \ . \tag{1.3}$$

The spin magnetic moment in (1.1) can be derived from Dirac's relativistic theory of the electron; roughly speaking, it comes from the rotation of the electron. The g-value given by the Dirac equation is 2; the value in (1.3) contains a correction (the anomalous magnetic moment) due to the interaction with the electromagnetic field. The correction Δg to first order in α is given as $\Delta g = \alpha/\pi$, where α is the fine structure constant ($\alpha = e^2/\hbar c \simeq 1/137$).

If we define $\boldsymbol{\sigma}$ as $2\boldsymbol{s}$, the matrices of the x, y and z components of $\boldsymbol{\sigma}$ are expressed as

$$\sigma_x = \begin{pmatrix} 0 & 1 \\ 1 & 0 \end{pmatrix}, \quad \sigma_y = \begin{pmatrix} 0 & -\mathrm{i} \\ \mathrm{i} & 0 \end{pmatrix}, \quad \sigma_z = \begin{pmatrix} 1 & 0 \\ 0 & -1 \end{pmatrix} . \tag{1.4}$$

in the representation where σ_z is diagonal. σ is called the *Pauli spin matrix.* Therefore the eigenvalues of s_z are $+1/2$ and $-1/2$. The following relations hold among the components of σ:

$$\sigma_x^2 = \sigma_y^2 = \sigma_z^2 = 1 , \quad \sigma_x \sigma_y = \mathrm{i}\sigma_z , \dots . \tag{1.5}$$

In addition to the spin magnetic moment, the electron has the *orbital magnetic moment*

$$\boldsymbol{\mu}_\mathrm{o} = -\frac{e}{2c}(\boldsymbol{r} \times \boldsymbol{v}) , \tag{1.6}$$

which arises from the orbital motion of the electron. Here $\boldsymbol{r}$ is the position vector of the electron and $\boldsymbol{v}$ the velocity. This quantity can be written as

$$\boldsymbol{\mu}_\mathrm{o} = -\frac{e}{2mc}(\boldsymbol{r} \times \boldsymbol{p}) = -\mu_\mathrm{B}\boldsymbol{l} , \tag{1.7}$$

where $\boldsymbol{p}$ is the momentum and $\boldsymbol{l}$ is the angular momentum divided by $\hbar$. When an external field $\boldsymbol{H}$ is present, $m\boldsymbol{v}$ is given by

$$m\boldsymbol{v} = \boldsymbol{p} + \frac{e}{c}\boldsymbol{A}(\boldsymbol{r}) ; \tag{1.8}$$

thus the orbital magnetic moment in the magnetic field is

$$\boldsymbol{\mu}_\mathrm{o}(\boldsymbol{H}) = -\mu_\mathrm{B}\boldsymbol{l} - \frac{e^2}{2mc^2}(\boldsymbol{r} \times \boldsymbol{A}) . \tag{1.9}$$

$\boldsymbol{A}$ is the vector potential due to the external magnetic field and the second term in (1.9) is a diamagnetic term. Equation (1.9) remains invariant (i.e., gauge invariant) under the transformation to the vector potential $\boldsymbol{A}'$

$$\boldsymbol{A}' = \boldsymbol{A} + \mathrm{grad}\chi . \tag{1.10}$$

The atomic nuclei in materials also have magnetic moments. Similarly to (1.1), the magnetic moment of a nucleus with nuclear spin $\boldsymbol{I}$ is expressed as

$$\boldsymbol{\mu} = g_\mathrm{N}\mu_\mathrm{N}\boldsymbol{I} , \tag{1.11}$$

where μ_N is given by (1.2), except that the electron mass m is replaced by the proton mass M (i.e., $\mu_\mathrm{N} = (1/1836)\mu_\mathrm{B}$) and is called the *nuclear magneton*. g_N is a quantity corresponding to the g-value and is of order of unity. Therefore, in most cases, the magnetic moments of atomic nuclei can be ignored compared with the magnetic moments of electrons. The magnetic moment of the neutron, which is used in neutron-scattering experiments, has the magnitude $\mu_\mathrm{n} = 1.9132\mu_\mathrm{N}$; it is directed opposite to the spin $\boldsymbol{I}$ (whose magnitude is $1/2$). Incidentally, the magnetic moment of proton is given by $\mu_\mathrm{p} = 2.7929\mu_\mathrm{N}$.

1.2 Ground State of Free Magnetic Ions

Let us consider first the electronic state of a free atom or ion. The electrons in an atom may be regarded, in a first approximation, as moving in the attractive potential due to the atomic nucleus and an average potential (in the Hartree approximation) arising from other electrons. If we assume that the potential has spherical symmetry, the eigenfunction for the electron is expressed in the spherical coordinates (r, θ, φ) as

$$\varphi_{nlm}(r) = R_{nl}(r) Y_l^m(\theta, \varphi) , \tag{1.12}$$

where $R_{nl}(r)$ is the radial part of the wave function; the angular part $Y_l^m(\theta, \varphi)$ is a spherical harmonic defined by

$$\begin{aligned} Y_l^m(\theta, \varphi) &= (-1)^m \Theta_l^m(\theta) \Phi_m(\varphi) \qquad (m \geq 0) , \\ Y_l^{-m}(\theta, \varphi) &= (-1)^m [Y_l^m(\theta, \varphi)]^* \qquad (m > 0) , \end{aligned} \tag{1.13}$$

with

$$\Theta_l^m(\theta) = \left(\frac{2l+1}{2} \frac{(l-|m|)!}{(l+|m|)!} \right)^{1/2} \mathrm{P}_l^{|m|}(\cos\theta) , \tag{1.14}$$

$$\Phi_m(\varphi) = (2\pi)^{-1/2} \mathrm{e}^{\mathrm{i}m\varphi} . \tag{1.15}$$

Here $\mathrm{P}_l^m(\cos\theta)$ is the associated Legendre function defined by

$$\mathrm{P}_l^m(\cos\theta) = \frac{1}{2^l l!} \sin^m\theta \frac{d^{m+l}(-\sin^2\theta)^l}{d(\cos\theta)^{m+l}} . \tag{1.16}$$

The energy eigenvalue ε depends on the principal quantum number n and the orbital angular-momentum (or azimuthal) quantum number l $(l < n)$; it does not depend on the magnetic quantum number m $(-l, -l+1, \ldots, l)$, which corresponds to the z-component of the angular momentum. The quantum number n is a positive integer $(1, 2, 3, \ldots)$. Therefore there are $2(2l+1)$ degenerate states for each ε_{nl}, if we take into account the 2 states related to the spin component; these states are called the *shell*. The electronic state of an atom is such that the energy levels on the shells are successively filled from the bottom according to the Pauli principle. This is called the shell structure of the atom; the states with $l = 0, 1, 2, 3, \ldots$ are denoted as $s, p, d, f, \ldots$, respectively. If the electrons completely fill a shell, the sum of the orbital angular momenta of electrons $\sum_i \boldsymbol{l}_i$ and the sum of the spins $\sum_i \boldsymbol{s}_i$ for the shell vanish so that the filled shell has no magnetic moment. In other words, a nonvanishing magnetic moment of an atom or ion is realized only in the case of an incompletely filled shell. Typical examples of such a situation are transition-metal ions such as iron-group ions and rare-earth ions, in which the $3d$ and $4f$ shells are incompletely filled in iron-group ions and rare-earth ions, respectively. Further, ions belonging to the actinide series have incompletely filled $5f$ shells. Ions having an incompletely filled shell are called the *magnetic ions.*

Let the electron number in the incompletely filled shell of a magnetic ion be n. There are $\begin{bmatrix} 2(2l+1) \\ n \end{bmatrix}$ ways to distribute n electrons among $2(2l+1)$ orbitals and therefore the ground state of the magnetic ion has a degeneracy corresponding to this number. This degeneracy is partially lifted if we take into account the deviation of the electron-electron Coulomb interaction from its average value in the Hartree approximation (which is called the *correlation energy*).

The Coulomb interaction between electrons

$$\mathcal{H}_{\text{Coulomb}} = \sum_{i>j} \frac{e^2}{|\boldsymbol{r}_i - \boldsymbol{r}_j|}$$

is a function only of the distance between two electrons and commutes with the total orbital angular momentum. Thus the total orbital angular momentum $\boldsymbol{L}$ and the total spin $\boldsymbol{S}$

$$\boldsymbol{L} = \sum_i \boldsymbol{l}_i \, , \qquad \boldsymbol{S} = \sum_i \boldsymbol{s}_i$$

are constants of motion. In other words, when the Coulomb interaction is taken into account, the energy is given for each value of the total orbital angular momentum $\boldsymbol{L}$ and the total spin $\boldsymbol{S}$. This state is still $(2L+1)(2S+1)$-fold degenerate with respect to the direction of $\boldsymbol{L}$ and $\boldsymbol{S}$. The state for each value of L and S is called the *multiplet.* The energy difference between two multiplets corresponding to different values of L and S is of order of the magnitude of Coulomb integral, i.e., ~ 10 eV. There is an empirical rule to select the multiplet with the lowest energy from all the possible multiplets, called *Hund's rule.* According to this rule, the state with the largest S has the lowest energy; if there are several states satisfying this criterion, the state with the largest L has the lowest energy among them.

Hund's rule (with some exceptions) has been well confirmed by realistic calculations [1.1]. The first rule can be understood if one notices that the exchange interaction tends to make the electron spins parallel. The second rule may be interpreted as suggesting that the electrons rotating in the same direction can avoid each other effectively so that the cost of the Coulomb interaction is less. The correlation energy due to the Coulomb interaction, which leads to Hund's rule, is called the *Hund coupling.* In Tables 1.1, 2 we show the values of L and S in the ground state for the $3d$ and $4f$ shells, respectively. In spectroscopy, the value of L in the ground state, $L = 0, 1, 2, \ldots$, is denoted by the symbol S, P, D, F, ..., respectively; the degeneracy $2S+1$ of the spin S is denoted with the left superscript.

Table 1.1. The value of λ for iron-group ions [1.2]

ion	n	symbol	λ [cm^{-1}]
Ti^{3+}	$(3d)^1$	^2D	154
V^{3+}	$(3d)^2$	^3F	104
Cr^{3+}	$(3d)^3$	^4F	91
V^{2+}	$(3d)^3$	^4F	56
Cr^{2+}	$(3d)^4$	^5D	58
Mn^{3+}	$(3d)^4$	^5D	88
Mn^{2+}, Fe^{3+}	$(3d)^5$	^6S	–
Fe^{2+}	$(3d)^6$	^5D	−102
Co^{2+}	$(3d)^7$	^4F	−177
Ni^{2+}	$(3d)^8$	^3F	−325
Cu^{2+}	$(3d)^9$	^2D	−829

Table 1.2. The ground state of rare earth ions

n	L	S	J	symbol	g_J
$(4f)^1$	3	1/2	5/2	$^2F_{5/2}$	6/7
$(4f)^2$	5	1	4	$^3H_4$	4/5
$(4f)^3$	6	3/2	9/2	$^4I_{9/2}$	8/11
$(4f)^4$	6	2	4	$^5I_4$	1/5
$(4f)^5$	5	5/2	5/2	$^6H_{5/2}$	2/7
$(4f)^6$	3	3	0	$^7F_0$	0
$(4f)^7$	0	7/2	7/2	$^8S_{7/2}$	2
$(4f)^8$	3	3	6	$^7F_6$	3/2
$(4f)^9$	5	5/2	15/2	$^6H_{15/2}$	4/3
$(4f)^{10}$	6	2	8	$^5I_8$	5/4
$(4f)^{11}$	6	3/2	15/2	$^4I_{15/2}$	6/5
$(4f)^{12}$	5	1	6	$^3H_6$	7/6
$(4f)^{13}$	3	1/2	7/2	$^2F_{7/2}$	8/7

1.3 *LS* Coupling

The energy levels in a multiplet, which is specified by L and S and is $(2L+1)(2S+1)$-fold degenerate, split further if we take into account the magnetic interactions among electrons.

Let us focus on one of the electrons and assume that it is moving with velocity $\boldsymbol{v}$ in an electric field due to the nucleus and the other electrons. In the coordinate system moving with this electron, the nucleus moves with the velocity $-\boldsymbol{v}$. Because of this motion the nucleus (of the charge Ze) produces the following magnetic field at the position of the electron:

$$\boldsymbol{H} = \frac{1}{c} Ze \frac{\boldsymbol{r} \times \boldsymbol{v}}{r^3} = \frac{Ze\hbar}{mc} \frac{1}{r^3} \boldsymbol{l} \tag{1.17}$$

according to the Biot-Savart law; here $\hbar \boldsymbol{l}$ is the orbital angular momentum of the electron. The energy of the spin magnetic moment $-g\mu_{\mathrm{B}} \boldsymbol{s}$ of the electron in this magnetic field is given by

$$\mathcal{H}_{\mathrm{Z}} = \frac{Ze\hbar}{mc} g\mu_{\mathrm{B}} \frac{\boldsymbol{s} \cdot \boldsymbol{l}}{r^3} \ . \tag{1.18}$$

The results in (1.17, 18) have been obtained for the coordinate system which moves with the electron. Since the electron is accelerated by the electric field due to the nucleus, however, the velocity of the coordinate system changes. For this reason, if we consider the problem in a relativistic way, we find that this coordinate system is rotating in the opposite direction relative to the rest system with a frequency which is just one half of the Larmor frequency of the spin precession. Therefore the energy of the spin magnetic moment in the rest system, which takes into account the relativistic correction, is just one half of (1.18). This correction is called the *Thomas correction.* If we use the Dirac equation, the correct expression containing the factor 1/2 can be obtained directly.

Taking this correction into account in (1.18) and summing over all the electrons, we obtain the total spin-orbit interaction

$$\mathcal{H}_{\mathrm{so}} = g\mu_{\mathrm{B}}^2 Z \sum_i \frac{\boldsymbol{s}_i \cdot \boldsymbol{l}_i}{r_i^3} \ . \tag{1.19}$$

In this expression the interaction with the other electrons is not included; the latter can be written as

$$g\mu_{\mathrm{B}} \frac{e}{2mc} \sum_{i \neq j} \left(\boldsymbol{s}_i \cdot \frac{\boldsymbol{r}_{ij} \times (2\boldsymbol{p}_j - \boldsymbol{p}_i)}{r_{ij}^3} \right) \ . \tag{1.20}$$

Here $\boldsymbol{p}_j$ is the momentum of the electron and the absence of the factor 2 in front of $\boldsymbol{p}_i$ is due to the Thomas correction. The interaction between electrons is taken into account, in a first approximation, in the Hartree potential. In this approximation Ze/r^3 in (1.19) should be replaced by $-(1/r)(dU/dr)$, where $U(r)$ is the total potential containing also the Hartree potential. To estimate the magnitude, however, the replacement of Z in (1.19) by an effective charge Z_{eff} is a practical way.

In the ground state of an incompletely filled shell, all the electron spins are parallel to each other, if the electron number n is smaller than $2l + 1$. Therefore, using

$$\boldsymbol{s}_i = \frac{1}{n} \boldsymbol{S} = \frac{1}{2S} \boldsymbol{S} \ , \tag{1.21}$$

we can rewrite the spin-orbit interaction as

$$\mathcal{H}_{\rm so} = \lambda(\boldsymbol{L}\cdot\boldsymbol{S}) = g\mu_{\rm B}^2 Z_{\rm eff}\left\langle\frac{1}{r^3}\right\rangle\frac{1}{2S}(\boldsymbol{L}\cdot\boldsymbol{S}) \ , \tag{1.22}$$

which has the form of a scalar product of $\boldsymbol{L}$ and $\boldsymbol{S}$. Here $\langle 1/r^3\rangle$ represents the average over the electrons in the incompletely filled shell. If n is larger than $2l+1$, the sum of $\boldsymbol{l}_i$ over the electrons having spin parallel to $\boldsymbol{S}$ vanishes and we are left with the contribution from the electrons with antiparallel spin. Therefore we can write the electron spin as

$$\boldsymbol{s}_i = -\frac{1}{10-n}\boldsymbol{S} = -\frac{1}{2S}\boldsymbol{S} \ , \tag{1.23}$$

with which we obtain

$$\mathcal{H}_{\rm so} = \lambda(\boldsymbol{L}\cdot\boldsymbol{S}) = -g\mu_{\rm B}^2 Z_{\rm eff}\left\langle\frac{1}{r^3}\right\rangle\frac{1}{2S}(\boldsymbol{L}\cdot\boldsymbol{S}) \ . \tag{1.24}$$

Equations (1.22, 24) are called the *LS coupling*. The coefficient λ is positive for less than half-filling, and negative for the more than half-filled case. The values of λ for iron-group ions are shown in Table 1.1; these values have been obtained from an analysis of spectra of divalent and trivalent free ions [1.2].

In addition to the spin-orbit interaction mentioned above, the magnetic interaction between electrons includes the interaction between the spin magnetic moments

$$\mathcal{H}_{\rm ss} = (g\mu_{\rm B})^2\sum_{i>j} r_{ij}^{-3}\Big[(\boldsymbol{s}_i\cdot\boldsymbol{s}_j) - 3r_{ij}^{-2}(\boldsymbol{r}_{ij}\cdot\boldsymbol{s}_i)(\boldsymbol{r}_{ij}\cdot\boldsymbol{s}_j)\Big] \ . \tag{1.25}$$

Within the $(2L+1)(2S+1)$-dimensional subspace defined by fixed values of L and S, this interaction has the same matrix elements as

$$\mathcal{H}_{\rm ss} = -\rho\left\langle\frac{1}{r^3}\right\rangle\Big[(\boldsymbol{L}\cdot\boldsymbol{S})^2 + \frac{1}{2}(\boldsymbol{L}\cdot\boldsymbol{S}) - \frac{1}{3}L(L+1)S(S+1)\Big] \ , \tag{1.26}$$

where ρ is a constant. The magnitude of $\rho\langle 1/r^3\rangle$ is smaller than λ by about 2 orders of magnitude and is of order $1{\rm cm}^{-1}$. Thus this interaction can be neglected except for special cases.

The energy levels of a multiplet split due to the LS coupling. To see how this occurs, we rewrite the LS coupling energy as

$$\begin{aligned}\lambda(\boldsymbol{L}\cdot\boldsymbol{S}) &= \frac{1}{2}\lambda\left[(\boldsymbol{L}+\boldsymbol{S})^2 - \boldsymbol{L}^2 - \boldsymbol{S}^2\right]\\ &= \frac{1}{2}\lambda\left[J(J+1) - L(L+1) - S(S+1)\right] \ ,\end{aligned} \tag{1.27}$$

where $\boldsymbol{J}$ is the vector sum of $\boldsymbol{L}$ and $\boldsymbol{S}$, its magnitude being

$$J = L+S,\ L+S-1,\ \dots\ ,|L-S| \ .$$

If we consider the LS coupling in (1.27), $\boldsymbol{L}$ and $\boldsymbol{S}$ are not constants of motion separately, but the total vector $\boldsymbol{J}$ is a contant of motion. Because of this coupling the $(2L+1)(2S+1)$-fold degenerate levels split into $(2J+1)$-fold degenerate $(2S+1)$ or $(2L+1)$ levels for $L > S$ or $S > L$, respectively. In accordance with whether the electron number n in an incompletely filled shell is larger than $2l+1$ or not, the value J of the lowest-energy state is either $J = L+S$ for $\lambda < 0$ or $J = |L-S|$ for $\lambda > 0$, respectively. Since (1.27) gives the energy difference between the states with J and $J+1$ as $\lambda(J+1)$, the energy difference between the ground state and the next excited state for the case $\lambda > 0$ is smaller than that for the case $\lambda < 0$.

If magnetic ions are placed in a crystal, they are subject to the crystal electric field produced by surrounding ionic charges. In iron-group ions the effect of this crystal field is larger than the LS coupling and therefore the electronic states of iron-group ions in crystals cannot be specified by the value of J. On the other hand, in rare-earth ions the effect of the crystal field is smaller than the LS coupling and the ionic state may be regarded as the lowest-energy state among energy levels split by the LS coupling. This difference between the iron group and the rare-earth group is due to the fact that in the former the electrons in the incomplete $3d$ shell are located in an outer region of the core and their orbit is large, while in the latter the electrons in the $(5s)^2(5p)^6$ shell are present outside the incomplete $4f$ shell so that the effective nuclear charge is large and the size of the $4f$ electron orbit is small. As will be explained in Chap. 3, the effect of the crystal field is proportional to a positive power of the radius of the electron orbit like $\langle r^2 \rangle$ or $\langle r^4 \rangle$; in contrast, the parameter λ of the LS coupling is proportional to $\langle 1/r^3 \rangle$ as (1.22, 24) show. For this reason the LS coupling is larger than the crystal field effects if the radius of the electron orbit is small; the situation is reversed if the radius is large.

As mentioned previously, the electronic state of rare-earth ions may be regarded, below room temperature, as in the lowest-energy state which is split off due to the LS coupling. However, when the lowest state has a small value of J due to the positive value of λ, there are exceptions to this rule. If the g-value of the spin is taken as 2, the magnetic moment $\boldsymbol{M}$ is given by

$$\boldsymbol{M} = -\mu_{\mathrm{B}}(2\boldsymbol{S} + \boldsymbol{L}) = -\mu_{\mathrm{B}}(\boldsymbol{J} + \boldsymbol{S}) ;$$

this quantity is not a constant of motion. This situation is explained in Fig. 1.1.

Since $\boldsymbol{J}$ is a constant of motion, we may think that the direction of $\boldsymbol{J}$ is fixed and that $\boldsymbol{L}$ and $\boldsymbol{S}$ rotate about $\boldsymbol{J}$ maintaining the triangular relation depicted in Fig. 1.1. Thus the magnetic moment (for constant $\boldsymbol{J}$) is given by the component of $2\boldsymbol{S} + \boldsymbol{L}$ parallel to $\boldsymbol{J}$. The parallel component of $\boldsymbol{S}$ can be calculated with the angle θ between $\boldsymbol{J}$ and $\boldsymbol{S}$ as

$$\begin{aligned} \boldsymbol{S}_{\parallel} &= |S| \cos\theta \frac{\boldsymbol{J}}{|J|} \\ &= \frac{(\boldsymbol{J} \cdot \boldsymbol{S})}{\boldsymbol{J}^2} \boldsymbol{J} = \frac{1}{2} \frac{1}{\boldsymbol{J}^2} (\boldsymbol{J}^2 - \boldsymbol{L}^2 + \boldsymbol{S}^2) \boldsymbol{J} \\ &= \frac{1}{2J(J+1)} \Big[J(J+1) - L(L+1) + S(S+1) \Big] \boldsymbol{J} . \end{aligned} \tag{1.28}$$

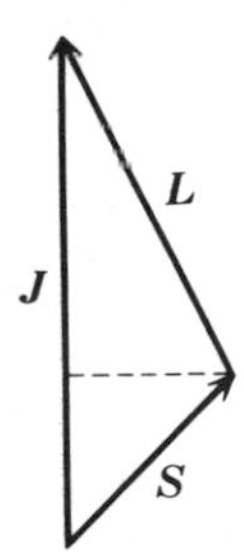

Fig. 1.1. Addition of $\boldsymbol{L}$ and $\boldsymbol{S}$

Using this relation, we obtain for $\boldsymbol{M}$

$$\boldsymbol{M} = -g_J \mu_B \boldsymbol{J} \,, \tag{1.29}$$

$$g_J = 1 + \frac{J(J+1) + S(S+1) - L(L+1)}{2J(J+1)} \,. \tag{1.30}$$

Here g_J is called the *Landé g-value.*

The components of $\boldsymbol{L}$ and $\boldsymbol{S}$ perpendicular to $\boldsymbol{J}$ vanish on average. In the quantum theory this component contributes to off-diagonal elements of $\boldsymbol{L}$ or $\boldsymbol{S}$ among different J's.

The values of J and g_J in the ground state of rare-earth ions are given in Table 1.2.

2. Interactions Between Atomic Nucleus and Electrons

In this chapter an explanation is given on the interactions between electrons and atomic nuclei. These interactions give rise to hyperfine structures in electron spin resonance spectra, relaxation times of nuclear spins in solids and structures in Mössbauer spectra. They are also causes of neutron scatterings by magnetic materials.

2.1 Magnetic Interactions

Among various magnetic interactions between electrons and nucleus, one thinks first of the interaction between two magnetic moments. As is well known, this interaction is given by

$$\frac{1}{r^3}\left[(\boldsymbol{\mu}_{\mathrm{e}} \cdot \boldsymbol{\mu}_{\mathrm{N}}) - 3\frac{1}{r^2}(\boldsymbol{r} \cdot \boldsymbol{\mu}_{\mathrm{e}})(\boldsymbol{r} \cdot \boldsymbol{\mu}_{\mathrm{N}})\right] , \tag{2.1}$$

where $\boldsymbol{\mu}_{\mathrm{e}}$ and $\boldsymbol{\mu}_{\mathrm{N}}$ are the magnetic moments of the electron and the nucleus, respectively. This expression is valid only for $\boldsymbol{r} \neq 0$; in order to discuss s-electrons (having finite probability at $\boldsymbol{r} = 0$), however, we must determine the interaction of two magnetic moments at $\boldsymbol{r} = 0$. That is, the expression (2.1) is incomplete for the interaction at the same point.

Let us approximate, for simplicity, the nucleus as a small sphere and suppose that its magnetic moment is uniformly distributed with density $(1/V)\boldsymbol{\mu}_{\mathrm{N}}$, where V is the volume of the nucleus. The magnetic charge on the surface produces the demagnetization field $-(4\pi/3)(\mu_{\mathrm{N}}/V)$ inside the sphere. The magnetic field which the electron feels within the sphere is $\boldsymbol{B}$, which is given by

$$\boldsymbol{B} = -\frac{4\pi}{3}\frac{\boldsymbol{\mu}_{\mathrm{N}}}{V} + 4\pi\frac{\boldsymbol{\mu}_{\mathrm{N}}}{V} = \frac{8\pi}{3}\frac{\boldsymbol{\mu}_{\mathrm{N}}}{V} . \tag{2.2}$$

The interaction energy between this magnetic field and the magnetic moment of the electron has the form

$$-\boldsymbol{\mu}_{\mathrm{e}} \cdot \frac{8\pi}{3}\frac{\boldsymbol{\mu}_{\mathrm{N}}}{V} \cdot V = -\frac{8\pi}{3}\boldsymbol{\mu}_{\mathrm{e}} \cdot \boldsymbol{\mu}_{\mathrm{N}} .$$

Therefore we conclude that the interaction at $\boldsymbol{r} = 0$ should be

$$-\frac{8\pi}{3}\boldsymbol{\mu}_e \cdot \boldsymbol{\mu}_N \delta(\boldsymbol{r}) \tag{2.3}$$

This interaction is called the *Fermi contact term*; it was first introduced by Fermi to explain the hyperfine structure in spectra.

One should not consider, however, this interaction to be independent of the ordinary interaction between magnetic moments (2.1). To make the situation clear, let us go back to the vector potential. The vector potential of the magnetic field produced by $\boldsymbol{\mu}_N$ can be written as

$$\boldsymbol{A}(\boldsymbol{r}) = \frac{\boldsymbol{\mu}_N \times \boldsymbol{r}}{r^3} . \tag{2.4}$$

Since the magnetic field $\boldsymbol{H}$ is given by curl$\boldsymbol{A}$, we have

$$\begin{aligned} \boldsymbol{H} &= \text{curl}\boldsymbol{A} = \text{curl}\frac{(\boldsymbol{\mu}_N \times \boldsymbol{r})}{r^3} = -\text{curl}\left(\boldsymbol{\mu}_N \times \text{grad}\frac{1}{r}\right) \\ &= \text{curl}\left(\text{grad}\frac{1}{r} \times \boldsymbol{\mu}_N\right) = \text{curl curl}\frac{1}{r}\boldsymbol{\mu}_N . \end{aligned} \tag{2.5}$$

Using the formula curl curl = grad div $- \Delta$ ($\Delta = \nabla^2$: the Laplacian) and the Poisson equation

$$\Delta\frac{1}{r} = -4\pi\delta(\boldsymbol{r}) , \tag{2.6}$$

one can write (2.5) as

$$\begin{aligned} \boldsymbol{H} &= \text{grad}\left(\text{div}\frac{\boldsymbol{\mu}_N}{r}\right) + 4\pi\boldsymbol{\mu}_N\delta(\boldsymbol{r}) \\ &= \left(\text{grad grad} - \frac{1}{3}\Delta\right)\frac{\boldsymbol{\mu}_N}{r} + \frac{8\pi}{3}\boldsymbol{\mu}_N\delta(\boldsymbol{r}) . \end{aligned} \tag{2.7}$$

The first term in this equation is a vector, which is a product of $\boldsymbol{\mu}_N/r$ and the traceless tensor, and there is no singularity at the origin. The singularity at the origin is all contained in the second term. The first term gives the ordinary interaction between magnetic moments (at $\boldsymbol{r} \neq 0$), whereas the second term is the Fermi contact term.

On the other hand, the orbital motion of the electron produces the magnetic field

$$\boldsymbol{H} = \frac{1}{c}\frac{\boldsymbol{r} \times (-e\boldsymbol{v})}{r^3} = -\frac{e\hbar}{mc}\frac{1}{r^3}\boldsymbol{l} \tag{2.8}$$

at the position of nucleus; here $\hbar\boldsymbol{l}$ is the orbital angular momentum of electron. The interaction between this magnetic field and the nuclear magnetic moment takes the form

$$\frac{e\hbar}{mc}\frac{1}{r^3}(\boldsymbol{\mu}_N \cdot \boldsymbol{l}) .$$

Thus the total interaction between the nuclear magnetic moment and the electronic magnetic moment is given by

$$\sum_i g\mu_D\Big\{\Big[\frac{(\boldsymbol{l}_i - \boldsymbol{s}_i)\cdot\boldsymbol{\mu}_N}{r_i^3} + 3\frac{1}{r_i^5}(\boldsymbol{s}_i\cdot\boldsymbol{r}_i)(\boldsymbol{\mu}_N\cdot\boldsymbol{r}_i)\Big] + \frac{8\pi}{3}(\boldsymbol{s}_i\cdot\boldsymbol{\mu}_N)\delta(\boldsymbol{r}_i)\Big\}\,. \quad (2.9)$$

Since the wave function for the s-electron is spherically symmetric, the first term vanishes after taking the average over the wave function and only the δ-function part remains. Other electrons (like p, d, ... electrons) have a vanishing probability at the origin so that only the first contribution remains; the orbital current term is important for these electrons. This part is particularly important in rare-earth ions which have large expectation values of $\boldsymbol{J}$ and $\boldsymbol{L}$. On the other hand, in iron-group ions the total orbital angular momentum $\boldsymbol{L}$ of the ion vanishes, in first approximation, because of the crystal-field effect. For this reason the magnitude of the contribution from the orbital current is of same order to the magnetic dipole-dipole interaction. For iron-group ions the Fermi contact term is an especially important contribution.

Fairly large hyperfine structures of the resonance lines are observed in spin resonance experiments on iron-group ions. The spin of iron-group ions comes from the electrons in the incompletely filled $3d$ shell. Since the $3d$ electrons have zero probability at the nuclear site, there is no direct Fermi contact term. However, because of a strong exchange interaction between the $3d$ electrons and the electrons in inner $1s$, $2s$, $3s$ cores, the Hartree-Fock potential acting on the s-electrons in inner core depends on the spin direction of the s-electrons. Since the inner $1s$, $2s$, $3s$ cores form closed shells, the total spin is zero, but the local electron distributions for up and down spins are different, leading to a local spin polarization. The Fermi contact term is nonvanishing due to this effect and results in an indirect interaction of the form $(\boldsymbol{S}\cdot\boldsymbol{I})$ between the spin $\boldsymbol{S}$ of the $3d$ shell and the nuclear spin $\boldsymbol{I}$. This type of interaction arising from the *core polarization* of the inner s core is considered to be the origin of the hyperfine structure observed in iron-group ions.

2.2 Nuclear Quadrupole Moments

Let us remember that the nucleus is not just a point but has a finite size. If we define the nuclear charge distribution function and the electrostatic potential due to electrons around the nucleus by $\rho(\boldsymbol{r})$ and $V(\boldsymbol{r})$, respectively, the energy of the nucleus in this potential is given by

$$\mathcal{H} = \int \rho(\boldsymbol{r})V(\boldsymbol{r})d\tau\ , \quad (2.10)$$

where $d\tau$ denotes the volume element. Expanding $V(\boldsymbol{r})$ about the origin, we obtain

$$\mathcal{H} = ZeV_0 + \sum_j P_j\Big(\frac{\partial V}{\partial x_j}\Big)_0 + \frac{1}{2}\sum_{jk} Q'_{jk}\Big(\frac{\partial^2 V}{\partial x_j \partial x_k}\Big)_0 + \dots\ . \quad (2.11)$$

Here Ze, P_j and Q'_{jk} are defined by

$$\begin{aligned} Ze &= \int d\tau \rho(\boldsymbol{r}) : && \text{nuclear charge ,} \\ P_j &= \int d\tau \rho(\boldsymbol{r}) x_j : && \text{electric dipole moment ,} \\ Q'_{jk} &= \int d\tau \rho(\boldsymbol{r}) x_j x_k : && \text{electric quadrupole moment ,} \end{aligned}$$

respectively; x_j in the above relations denotes one of x, y and z. The electric dipole moment P_j vanishes if the nuclear charge distribution has inversion symmetry with respect to the origin, as we assume. The first term of (2.11), which is the energy of the nucleus treated as a point charge, is a constant. Omitting this term, we obtain

$$\mathcal{H}_Q = \frac{1}{2} \sum_{jk} Q'_{jk} V_{jk} + \dots \quad , \tag{2.12}$$

where

$$V_{jk} = \left(\frac{\partial^2 V}{\partial x_j \partial x_k} \right)_0 \quad .$$

Equation (2.12) represents the interaction of the electric field gradient and the quadrupole moment. The magnitude of the interaction is similar to that of the interaction of the external magnetic field and the nuclear magnetic moment. Forming from the tensor Q'_{jk} the traceless tensor

$$Q_{jk} = 3Q'_{jk} - \delta_{jk} \sum_i Q'_{ii} \ , \tag{2.13}$$

we can rewrite (2.12) as

$$\mathcal{H}_Q = \frac{1}{6} \sum_{jk} Q_{jk} V_{jk} + \frac{1}{6} \Big(\sum_i Q'_{ii} \Big) \Big(\sum_j V_{jj} \Big) \ . \tag{2.14}$$

$\sum_i Q'_{ii}$ in the second term is independent of the direction of the nuclear spin, as will be shown shortly; therefore we ignore the second term henceforth.

Let us assume that the ground state of the nucleus is nondegenerate except for the degeneracy with respect to the direction of the nuclear spin, and that the excited states are so high that they can be ignored. Then, for the quadrupole moment Q_{jk} only the matrix elements within the ground state need be considered. In this case the state of the nucleus is exclusively determined by the nuclear spin $\boldsymbol{I}$; the quadrupole moment must be proportional to the traceless tensor formed with the nuclear spin $\boldsymbol{I}$. Namely, we have

$$Q_{jk} = C \Big[\frac{3}{2} (I_j I_k + I_k I_j) - \delta_{jk} \boldsymbol{I}^2 \Big] \ , \tag{2.15}$$

where C is a constant. Defining the z component of the nuclear spin $\boldsymbol{I}$ by m, we consider the matrix elements of Q_{jk} within the $2I+1$ levels of $m = -I, \ldots, I$

$$\langle I\ m|Q_{jk}|I\ m'\rangle = C\langle I\ m|\frac{3}{2}(I_jI_k + I_kI_j) - \delta_{jk}\boldsymbol{I}^2|I\ m'\rangle\ . \tag{2.16}$$

If we put the (I, I) component of Q_{zz} as eQ, we obtain

$$\begin{aligned} eQ &= \langle I\ I|Q_{zz}|I\ I\rangle = C\langle I\ I|3I_z^2 - \boldsymbol{I}^2|I\ I\rangle \\ &= CI(2I-1)\ . \end{aligned} \tag{2.17}$$

The quadrupole moment is usually expressed in terms of eQ defined in (2.17) rather than the coefficient C. The quantity eQ defined here is the expectation value of

$$Q_{zz} = 3Q'_{zz} - \sum_i Q'_{ii} = \int d\tau\rho(\boldsymbol{r})(3z^2 - r^2)$$

over the state in which the z component of $\boldsymbol{I}$ is I. In this state the quadrupole moment is uniaxial with respect to the z axis. Further, (2.15) vanishes identically for $I = 1/2$.

For a potential with cubic symmetry in which the principal axes are chosen as the x, y and z axes,

$$V_{xx} = V_{yy} = V_{zz}\ , \qquad V_{xy} = V_{yz} = V_{zx} = 0\ ;$$

therefore there is no quadrupole interaction:

$$\mathcal{H}_Q = \frac{1}{6}\sum_{jk} Q_{jk}V_{jk} = \frac{1}{6}V_{xx}\sum_j Q_{jj} = 0\ .$$

Introducing $I_\pm = I_x \pm \mathrm{i}I_y$, we rewrite $\mathcal{H}_Q$ as

$$\begin{aligned} \langle m'|\mathcal{H}_Q|m\rangle = \frac{eQ}{4I(2I-1)}&\langle m'|(3I_z^2 - \boldsymbol{I}^2)V_0 + (I_+I_z + I_zI_+)V_{-1} \\ &+ (I_-I_z + I_zI_-)V_1 + I_+^2V_{-2} + I_-^2V_2|m\rangle\ , \end{aligned} \tag{2.18}$$

where

$$V_0 = V_{zz}\ , \quad V_{\pm 1} = V_{xz} \pm \mathrm{i}V_{yz}\ , \quad V_{\pm 2} = \frac{1}{2}(V_{xx} - V_{yy}) \pm \mathrm{i}V_{xy}\ . \tag{2.19}$$

Here $\Delta V = 0$ has been used; if one takes the principal axes as the coordinate system, we have $\Delta V = V_{xx} + V_{yy} + V_{zz} = 0$, from which the electric field gradient can be expressed by 2 parameters. Let us choose the z, y and x axes so that

$$|V_{zz}| \geq |V_{yy}| \geq |V_{xx}|\ ,$$

and define eq and η by

$$eq = V_{zz} = \left(\frac{\partial^2 V}{\partial z^2}\right)_0 , \tag{2.20}$$

$$\eta = \frac{V_{xx} - V_{yy}}{V_{zz}} , \quad (0 \leq \eta \leq 1) , \tag{2.21}$$

respectively. η is called the *asymmetry parameter*; it is zero in the case of an axially symmetric potential. For a general potential we can express $\mathcal{H}_Q$ in terms of eq and η as

$$\langle m'|\mathcal{H}_Q|m\rangle = \frac{e^2qQ}{4I(2I-1)}\langle m'|(3I_z^2 - \boldsymbol{I}^2) + \frac{\eta}{2}(I_+^2 + I_-^2)|m\rangle . \tag{2.22}$$

2.3 Isomer Shift

An s electron has a finite charge density $\rho_\mathrm{e} = -e|\psi_\mathrm{s}(0)|^2$ at the position of the nucleus. Let us treat the nucleus as a sphere with radius R_N. Then the interaction between the nuclear charge and the electronic charge depends on the radius. Assuming that the nuclear charge distribution is uniform, we have

$$\rho_\mathrm{N}(r) = \frac{Ze}{(4\pi/3)R_\mathrm{N}^3} .$$

Evaluating the total Coulomb energy between ρ_N and ρ_e, we subtract the corresponding energy for the nucleus whose charge is assumed to be point-like. The difference E is given by a simple calculation as

$$E = -\frac{2}{5}\pi\rho_\mathrm{e}ZeR_\mathrm{N}^2 . \tag{2.23}$$

When the nucleus makes a transition from a metastable excited state (isomer) to the ground state, the interaction energy between the nuclear charge and the s-electron charge changes by

$$\Delta E = Ze^2|\psi_\mathrm{s}(0)|^2\frac{2\pi}{5}(R_\mathrm{is}^2 - R_\mathrm{gr}^2) . \tag{2.24}$$

ΔE depends on the material surrounding the atomic nucleus making such a transition because the value of $|\psi_\mathrm{s}(0)|^2$ in (2.24) changes with the material. The material dependence of ΔE

$$(\Delta E)_\mathrm{d} = Ze^2(|\psi_\mathrm{s}(0)'|^2 - |\psi_\mathrm{s}(0)|^2)\frac{2\pi}{5}(R_\mathrm{is}^2 - R_\mathrm{gr}^2) \tag{2.25}$$

for the transition can be observed by the Mössbauer experiment, which will be described in the next section. Equation (2.24) is called the *isomer shift.*

2.4 Mössbauer Effect

The interaction between the nuclear spin and the electron leads to the hyperfine structure in the Electron Spin Resonance (ESR), as mentioned previously. On the other hand, if it is viewed from the nucleus in connection with Nuclear Magnetic Resonance (NMR) and so on, the interaction is one of the causes of relaxation phenomena in nuclear spins; it also produces an internal field on nuclear spins, shifting the center of the resonance absorption line and sometimes causing the line width. Using this fact, one can obtain information on the electronic state through NMR studies. Similarly, one could also use the Mössbauer effect as a method to explore the electronic state from the nucleus.

Let us consider a phenomenon in which the nucleus makes a transition from an excited state to the ground state via γ-decay, and the emitted γ-ray is then absorbed by another nucleus of the same kind to make a resonant transition from the ground state to the excited state. Let the energy difference between the ground state and the excited state be E_r and the energy of the emitted γ-ray be E_γ. The momentum p of the γ-ray is given by E_γ/c. Because of momentum conservation, the isolated nucleus recoils with momentum $-p$. The kinetic energy R of the nucleus due to this recoil is

$$R = \frac{p^2}{2M} = \frac{E_\gamma^2}{2Mc^2} , \tag{2.26}$$

where M represents the mass of the nucleus; the law of energy conservation

$$E_r = R + E_\gamma \tag{2.27}$$

relates E_γ and E_r. In the following, we shall assume that $Mc^2 \gg E_\gamma$ and approximate E_γ by E_r in (2.26). In order that the second nucleus absorb the γ-ray and make a transition from the ground state to an excited state which is higher by E_r, the γ-ray must have the energy $E_r + R$, because of the recoil energy. Therefore such a transition is possible only when the emitted and absorbed γ-ray has a width larger than $2R$.

If the lifetime of the decay of the excited state is τ, the energy level of the excited state has the width

$$\Gamma = \frac{\hbar}{\tau} \tag{2.28}$$

by the uncertainty principle. Since the energy of the ground state is definite, the energy distribution of the emitted γ-ray has the width given by (2.28); this width is called the *natural width*. However, in a typical γ-decay, E_r is as large as $10^4 \sim 10^6$ eV and the recoil energy is much larger than the natural width. In reality the two nuclei are not at rest, but are in thermal motion. The thermal motion increases further the width in the energy distribution of the emitted and absorbed γ-ray; the associated width of the energy distribution is called the *Doppler broadening*. Let the momentum of the nucleus before emitting the γ-ray be $\boldsymbol{P}_i$. Then the recoil energy R' is

$$R' = \frac{(\boldsymbol{P}_i - \boldsymbol{p})^2}{2M} - \frac{\boldsymbol{P}_i^2}{2M} = R - \frac{\boldsymbol{p}\cdot\boldsymbol{P}_i}{M} ; \tag{2.29}$$

the second term of this equation causes the Doppler broadening. We define $\boldsymbol{P}_i^2/2M$ and the angle between $\boldsymbol{P}_i$ and $\boldsymbol{p}$ to be ε and φ, respectively. The energy E'_γ of the emitted γ-ray is then given by

$$E'_\gamma = E_r - R + D\cos\varphi , \tag{2.30}$$

$$D = 2\sqrt{\varepsilon R} . \tag{2.31}$$

Since $\cos\varphi$ ranges between -1 and 1, the thermal motion of the nucleus gives the energy of the emitted γ-ray a width of order of D. If we take the thermal energy kT as ε, D is much larger than Γ at room temperature and the recoil energy R can become the same order as the width of the γ-ray.

Studying the resonance scattering of the 129 keV γ-ray emitted from ^{191}Ir ($R = 0.05$ eV, $D = 0.1$ eV), Mössbauer discovered that, contrary to expectation, the resonance scattering increases with decreasing temperature and confirmed that the γ-ray emitted or absorbed from the nucleus contains a part which is not affected by the recoil of the nucleus and the Doppler broadening. Such recoilless resonance absorption of the γ-ray is called the *Mössbauer effect*. The isotope ^{57}Fe (which has an incompletely filled $3d$ shell) is usually used to study magnetic materials by means of the Mössbauer effect.

As mentioned before, free nuclei necessarily recoil. However, in the case of a solid where atoms are tightly bound to form a crystalline lattice, the recoil momentum can be accepted, with a finite probability, by the whole crystal. In this case the mass M in (2.26) is replaced by the mass NM of the whole crystal, and therefore the recoil energy is essentially zero. Let us consider this situation classically, following *van Kranendonk* [2.1]. We replace the nucleus emitting the γ-ray by a classical oscillator with the eigenfrequency $\omega_0 = E_r/\hbar$ and express the vector potential of the electromagnetic wave emitted by this oscillator as

$$\boldsymbol{A}(t) = \boldsymbol{A}_0 \exp(\mathrm{i}\omega_0 t) . \tag{2.32}$$

When the frequency changes in time, (2.32) is generalized as follows:

$$\boldsymbol{A}(t) = \boldsymbol{A}_0 \exp\left[\mathrm{i}\int_0^t \omega(t')dt'\right] . \tag{2.33}$$

For simplicity we assume that the nucleus emitting the γ-ray is moving with velocity $v(t) \ll c$ along the direction of the γ-ray (taken as the x axis). The frequency of the emitted electromagnetic wave changes as

$$\omega(t') = \omega_0\left[1 + \frac{v(t')}{c}\right] \tag{2.34}$$

due to the Doppler effect. Subsituting (2.34) into (2.33), we obtain

$$A = A_0 \exp(i\omega_0 t) \exp\Big(\frac{2\pi i x(t)}{\lambda}\Big), \tag{2.35}$$

where λ is the wavelength of the electromagnetic wave with the frequency ω_0. If we assume that the nucleus is oscillating harmonically with angular frequency Ω and amplitude x_0 (i.e., the Einstein model), we can put

$$x(t) = x_0 \sin \Omega t. \tag{2.36}$$

In this case we can write the vector potential A as

$$A = A_0 \exp(i\omega_0 t) \exp\Big(\frac{2\pi i x_0 \sin \Omega t}{\lambda}\Big) . \tag{2.37}$$

Using the formula

$$\exp(iy \sin \theta) = \sum_{n=-\infty}^{\infty} J_n(y) \exp(in\theta) , \tag{2.38}$$

we decompose the second factor of (2.37) into a Fourier series to obtain

$$A = A_0 \sum_{n=-\infty}^{\infty} J_n\Big(\frac{2\pi x_0}{\lambda}\Big) \exp\big[i(\omega_0 + n\Omega)t\big] . \tag{2.39}$$

From this result, we find that the electromagnetic wave emitted from a harmonically oscillating nucleus can be expressed as a superposition of waves having the frequencies $\omega_0, \omega_0 \pm \Omega, \omega_0 \pm 2\Omega, \ldots$ and that the amplitude of each component is given by the Bessel function $J_n(2\pi x_0/\lambda)$. The unshifted wave in this expression corresponds to the Mössbauer effect; the relative intensity is given by

$$f = \frac{|A(n=0)|^2}{|A_0|^2} = J_0^2\Big(\frac{2\pi x_0}{\lambda}\Big) . \tag{2.40}$$

For real crystals one must rather use the Debye model instead of the Einstein model. Then (2.36) is replaced by

$$x(t) = \sum_m x_m \sin \Omega_m t , \tag{2.41}$$

where $x_m \propto N^{-1/2}$. Ω_m is distributed continuously from 0 to the maximum value $k\Theta_D/\hbar$ (Θ_D = Debye temperature), and the total number of modes is equal to $3N$, the number of degrees of freedom of the crystal. Therefore the intensity of the Mössbauer effect is given by

$$f = \prod_{m=1}^{3N} \Big[J_0\Big(\frac{2\pi x_m}{\lambda}\Big)\Big]^2 . \tag{2.42}$$

Expanding J_0 to x_m^2, we obtain f as

$$f = \prod_{m=1}^{3N} \left[1 - \frac{1}{4}\left(\frac{2\pi x_m}{\lambda}\right)^2\right]^2$$
$$= \exp\left(-\frac{2\pi^2}{\lambda^2}\sum_m x_m^2\right) . \tag{2.43}$$

Since x_m^2 is a small quantity proportional to N^{-1}, the result (2.43) is exact due to the definition of the exponential function. The average of x^2 of the nuclear coordinate is given by

$$\langle x^2 \rangle = \frac{1}{2}\sum_m x_m^2 . \tag{2.44}$$

Using this expression for $\langle x^2 \rangle$, we obtain

$$f = \exp\left(-\frac{4\pi^2}{\lambda^2}\langle x^2 \rangle\right) . \tag{2.45}$$

For each Debye mode m

$$\frac{1}{2}x_m^2 = \frac{1}{N}\frac{e_{mx}^2}{M\Omega_m^2}\varepsilon_m \tag{2.46}$$

holds; here e_{mx} is the x component of the polarization vector and ε_m is the average energy of this mode. Let us take for ε the quantized energy including the zero-point oscillation

$$\varepsilon_m = \hbar\Omega_m\left(N_m + \frac{1}{2}\right) , \tag{2.47}$$

$$N_m = \left[\exp\left(\frac{\hbar\Omega_m}{kT}\right) - 1\right]^{-1} . \tag{2.48}$$

Then $\langle x^2 \rangle$ is given by

$$\langle x^2 \rangle = \frac{\hbar}{2M}\frac{1}{N}\sum_m \frac{1}{\Omega_m}\left(1 + \frac{2}{\exp(\hbar\Omega_m/kT) - 1}\right) . \tag{2.49}$$

We assume that Ω_m is independent of the polarization; the summation over m is taken for N independent modes. Approximating the crystal as isotropic and introducing the Debye temperature Θ_D by the maximum frequency $\Omega_D = k\Theta_D/\hbar$, we obtain for the intensity of the Mössbauer line

$$f = \exp\left\{-\frac{3}{2}\frac{R}{k\Theta_D}\left[1 + 4\left(\frac{T}{\Theta_D}\right)^2 \int_0^{\Theta_D/T} \frac{x\,dx}{e^x - 1}\right]\right\} . \tag{2.50}$$

The first term in the exponential function is the contribution from the zero-point oscillation. We find from this result that f decreases with increasing temperature and, in particular, decreases rapidly for $T > \Theta_D$; f is larger for larger Θ_D.

The Mössbauer effect, i.e., the recoilless scattering process, is not a particularly new phenomenon. The Bragg reflection of X rays and the elastic scattering of neutron beams by crystals are both phenomena with recoilless processes. In the latter case the quantity f is nothing but the *Debye-Waller factor*.

The decay processes of ^{57}Fe, which are most frequently used in studies of magnetism, are shown in Fig. 2.1. The 3/2 → 1/2 transition of energy 14.4 keV in this Figure is often used in many Mössbauer experiments. When ^{57}Fe is implanted into ferromagnets or in paramagnets magnetized by the external field, the electron spin in the 3d shell of ^{57}Fe has a finite average $\langle S \rangle$. Consequently, the nuclear spin is subject to an effective internal field because of the interaction between the nuclear magnetic moment and the spin magnetic moment. If the electronic state of ^{57}Fe itself or the arrangement of atoms surrounding ^{57}Fe has symmentry lower than cubic, an interaction proportional to I_z^2 is present between the electric field gradient (whose principal axis is taken as the z axis) and the nuclear quadrupole moment. In this case the energy levels of the excited and ground states in ^{57}Fe are split as shown in Fig. 2.2.

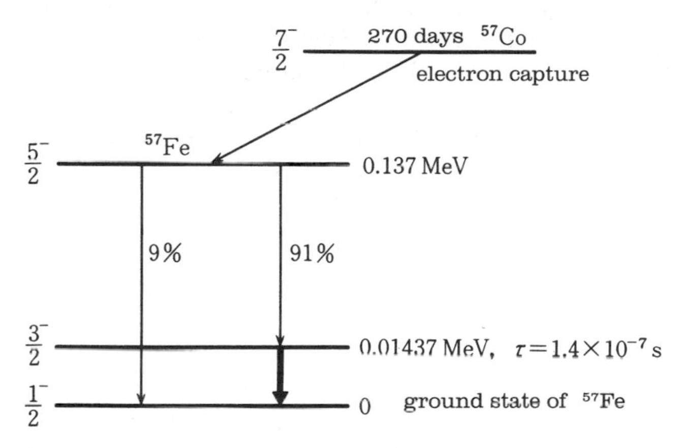

Fig. 2.1. The decay processes of ^{57}Fe

The selection rule for transitions among these levels is $\Delta I_z = 0, \pm 1$; therefore γ-rays with 6 different energies are emitted. Let us suppose that these γ-rays are absorbed by paramagnetic metals with no internal magnetic field such as stainless steel, in which the abundance of ^{57}Fe is about 2%, enough for the Mössbauer-effect experiment. Since no internal field acts on the ^{57}Fe atom in this case, only the γ-ray with E_r can be absorbed. To achieve the absorption, we move the substance containing the emitting nuclei (or the one containing the absorbing nuclei) with a suitable velocity, giving a Doppler

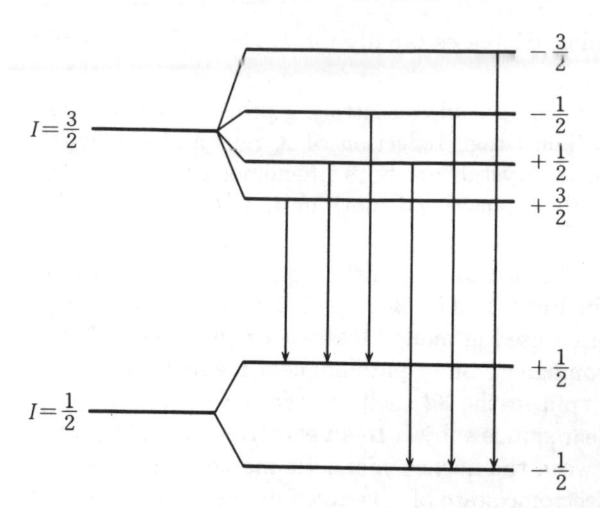

Fig. 2.2. The energy-level splitting of the ground and excited states of ^{57}Fe in the magnetic field and the electric-field gradient. The arrows indicate the possible transitions among energy levels

shift to the emitted or absorbed γ-ray. If the velocity of the absorbing body is v, the energy shift of the γ-ray is 14.4 keV$\times(v/c)$. Thus one can determine the level splitting of the nuclear spin from the velocity needed to achieve the resonance; then the value of the effective field or the magnitude of the electric field gradient can be determined.

The measurement of the internal field by the Mössbauer effect has been carried out for ferromagnetic metals such as Ni, Co and Fe, for many alloys, and for Fe nuclei in ionic crystals. Together with NMR it plays an important role in studies of magnetism.

3. Magnetic Ions in Crystals

Magnetic ions which constitute a crystal lattice are under the influence of surrounding ions and their electronic states are modified from free states. In this chapter, we consider the electronic states of magnetic ions in crystals.

3.1 Ionic Crystals Versus Metals

When atoms are condensed to form a crystal, the electronic state of each atom is modified from the free-atom state because of the interaction with surrounding atoms. The outermost valence electrons of atoms contribute to the cohesive energy of crystals. In metals these electrons move about the crystal, becoming conduction electrons which are described by the Bloch function. On the other hand, in ionic crystals charge transfer occurs between the different types of atoms; as a result positively charged ions and negatively charged ions attract each other by the electrostatic force.

Let us focus our attention on magnetic ions forming a metal or an ionic crystal and represent the orbitals in the incompletely filled shell (3*d* orbitals in iron-group elements or 4*f* orbitals in rare-earth elements), for simplicity, by a nondegenerate localized orbital $\varphi_{\boldsymbol{n}}(\boldsymbol{r})$. Then the Hamiltonian of the system consisting of the magnetic ions can be expressed as

$$\mathcal{H} = \sum_{\boldsymbol{n}\boldsymbol{n}'s} b_{\boldsymbol{n}'-\boldsymbol{n}} a^{\dagger}_{\boldsymbol{n}'s} a_{\boldsymbol{n}s} + U \sum_{\boldsymbol{n}} a^{\dagger}_{\boldsymbol{n}\uparrow} a_{\boldsymbol{n}\uparrow} a^{\dagger}_{\boldsymbol{n}\downarrow} a_{\boldsymbol{n}\downarrow} \ , \tag{3.1}$$

where $\boldsymbol{n}$ and $\boldsymbol{n}'$ represent the lattice points which the magnetic ions occupy. $a^{\dagger}_{\boldsymbol{n}\uparrow}$ is a fermion operator in the second quantization and creates an up-spin electron in the localized orbital at the lattice point $\boldsymbol{n}$; $a_{\boldsymbol{n}\uparrow}$ is the annihilation operator for the same electron. If the localized orbitals $\varphi_{\boldsymbol{n}}(\boldsymbol{r})$ on different lattice points are mutually orthogonal, these operators satisfy the well-known commutation relations for fermion operators:

$$\begin{aligned} a^{\dagger}_{\boldsymbol{n}s} a_{\boldsymbol{n}'s'} + a_{\boldsymbol{n}'s'} a^{\dagger}_{\boldsymbol{n}s} &= \delta_{\boldsymbol{n}\boldsymbol{n}'} \delta_{ss'} \ , \\ a^{\dagger}_{\boldsymbol{n}s} a^{\dagger}_{\boldsymbol{n}'s'} + a^{\dagger}_{\boldsymbol{n}'s'} a^{\dagger}_{\boldsymbol{n}s} &= 0 \ , \\ a_{\boldsymbol{n}s} a_{\boldsymbol{n}'s'} + a_{\boldsymbol{n}'s'} a_{\boldsymbol{n}s} &= 0 \ . \end{aligned} \tag{3.2}$$

In general the atomic wave functions belonging to different ions are not orthogonal to each other; in that case the localized orbitals $\varphi_n(\boldsymbol{r})$ should be interpreted as Wannier functions, which are mutually orthogonal localized orbitals formed with linear combinations of Bloch functions.

The first term in (3.1) represents the kinetic energy; the term transfers an electron at the $\boldsymbol{n}$th site to the $\boldsymbol{n}'$th site without changing the spin direction. $b_{\boldsymbol{n}'-\boldsymbol{n}}$ is the matrix element between $\varphi_{\boldsymbol{n}'}$ and $\varphi_{\boldsymbol{n}}$ of the one-body Hamiltonian including the crystal potential:

$$b_{\boldsymbol{n}'-\boldsymbol{n}} = \int \varphi^*_{\boldsymbol{n}'}(\boldsymbol{r})\mathcal{H}_{\text{cryst}}\varphi_{\boldsymbol{n}}(\boldsymbol{r})d\tau \,. \tag{3.3}$$

The second term represents the Coulomb interaction between two electrons with opposite spins, both located at the $\boldsymbol{n}$th site; U is the Coulomb integral

$$U = \iint |\varphi_{\boldsymbol{n}}(\boldsymbol{r}_1)|^2 \frac{e^2}{r_{12}} |\varphi_{\boldsymbol{n}}(\boldsymbol{r}_2)|^2 d\tau_1 d\tau_2 \,. \tag{3.4}$$

The Pauli principle forbids two electrons with parallel spins to occupy the same localized orbital. If the second term (the Coulomb interaction) is absent, then the Hamiltonian $\mathcal{H}$ in (3.1) can be diagonalized by the Fourier transformation

$$\begin{aligned} a^\dagger_{\boldsymbol{n}s} &= \frac{1}{\sqrt{N}} \sum_{\boldsymbol{k}} \mathrm{e}^{\mathrm{i}\boldsymbol{k}\cdot\boldsymbol{n}} a^\dagger_{\boldsymbol{k}s} \,, \\ a_{\boldsymbol{n}s} &= \frac{1}{\sqrt{N}} \sum_{\boldsymbol{k}} \mathrm{e}^{-\mathrm{i}\boldsymbol{k}\cdot\boldsymbol{n}} a_{\boldsymbol{k}s} \,, \end{aligned} \tag{3.5}$$

as

$$\mathcal{H}_0 = \sum_{\boldsymbol{k}s} \varepsilon_{\boldsymbol{k}} a^\dagger_{\boldsymbol{k}s} a_{\boldsymbol{k}s} \,; \tag{3.6}$$

here $\varepsilon_{\boldsymbol{k}}$, given by

$$\varepsilon_{\boldsymbol{k}} = \sum_{\boldsymbol{n}} b_{\boldsymbol{n}} \mathrm{e}^{\mathrm{i}\boldsymbol{k}\cdot\boldsymbol{n}} \,,$$

is the energy of the state with the wavevector $\boldsymbol{k}$. In this way the electrons localized at each lattice point move about the crystal, becoming the conduction electrons. Conversely, when $b_{\boldsymbol{n}'-\boldsymbol{n}} = b_0\delta_{\boldsymbol{n}'\boldsymbol{n}}$ holds, the electron transfer does not occur at all so that each ion has one localized electron because of the large Coulomb interaction for the case of one electron per site.

When both $b_{\boldsymbol{n}}$ and U are nonzero, the electrons move about the whole crystal and behave as conduction electrons in the case that U is very small compared with $b_{\boldsymbol{n}}$; in this situation the Coulomb repulsive potential U can be treated as a perturbation. In ferromagnetic metals like Fe, Co and Ni, the outer $4s$ electrons are completely in the conduction-electron state and contribute much to the metallic cohesive energy. The $3d$ electrons also move about the metal, playing a role in cohesion. However, the atomic $3d$ orbital is more localized near the center of the atom than the $4s$ orbital so that b_n is small and U is relatively large. As a result, the $3d$ electrons move in the whole crystal, as conduction electrons, but are strongly correlated with each other. It is considered that the s electrons also play an important role in this motion of the $3d$ electrons.

In the opposite case, where b_n is much smaller than U, a state close to the isolated ionic one is expected to be realized. The effect of b_n can be taken into account by perturbation calculation starting from the localized ionic state. Such a state is presumably realized in ionic crystals like MnO and MnF_2, in which magnetic ions are contained as cations. In such crystals the cohesive energy must have a different source. In fact, most of the cohesive energy in ionic crystals originates naturally from the electrostatic force between cations and anions.

As described above, when crystals are formed from free atoms or ions, their electrons either become conduction electrons itinerant in the crystal or remain localized on the atom. The difference between the two is controlled by the relative magnitude of b_n to U in the Hamiltonian (3.1). When the ratio of b_n to U is changed, a transition is expected to occur from one state to the other. This transition (which is presumably discontinuous in 3 and 2 dimensions) is called the *Mott transition* or the Mott problem. It has been studied for a long time, and is one of the most important problems in the electron theory of solids. The Hamiltonian (3.1) is often called the Hubbard Hamiltonian in the literature, although it was used before the work of Hubbard.

3.2 Crystal-Field Effects

We mentioned in the previous section that if the transfer matrix b_n for the electron belonging to an incompletely filled shell of a magnetic ion is much smaller than the Coulomb interaction U, the electron is localized on the ion and the resulting state is close to the free-ion state. In most magnetic ionic crystals or rare-earth metals, the electrons in the incompletely filled shell are in such a localized state. However, the state of the incompletely filled shell is not the same as the free-ion state, but is more or less affected by surrounding ions, notwithstanding its localized nature. Let us call all influences from the surrounding ions the crystal-field effect in a broad sense. The crystal-field effect in the incompletely filled shell of iron-group metals is much larger than in the $4f$ shell of rare-earth metals. The main reason is that the $3d$ shell is outermost in the ions so that it interacts directly with the electrons of surrounding ions. In iron-group elements the crystal-field effect is larger than the LS coupling, whereas the reverse is true in rare-earth elements. Therefore, in the former case the $2L+1$ degenerate energy levels (specified by L) split into several groups under the influence of the crystal field; the LS coupling should then be taken into account for those lowest states split off due to the crystal field. On the other hand, in the latter case, the crystal-field effect must be taken into account for the $2J+1$ degenerate states which belong to the lowest energy split off due to the LS coupling.

The simplest crystal-field effect is the electrostatic effect due to surrounding charges; it is similar to the interaction between the nuclear quadrupole moment and the electric field gradient, which was discussed before. It is the most important crystal-field effect for electrons located at the center of the ion (like the incompletely filled $4f$ shell in rare-earth metals). In contrast, it is not presumably an important effect for the incompletely filled $3d$ shell (which is an outer orbital of the ion); these electrons interact directly with the electrons on the outer closed shell of surrounding anions. Let us consider the effect of this interaction, using the molecular orbital method. First, having in mind an ionic crystal with the NaCl-type structure (like FeO), we assume that the magnetic ion is surrounded octahedrally by O^{2-} anions. For the five independent $3d$ wave functions, we use the following linear combinations of (1.12):

$$
\begin{aligned}
d\varepsilon : \varphi_{xy} &= \frac{1}{\sqrt{2}\mathrm{i}}(\varphi_2 - \varphi_{-2}) \,, \quad \varphi_{yz} = -\frac{1}{\sqrt{2}\mathrm{i}}(\varphi_1 + \varphi_{-1}) \,, \\
\varphi_{zx} &= -\frac{1}{\sqrt{2}}(\varphi_1 - \varphi_{-1}) \,, \\
d\gamma : \varphi_{x^2-y^2} &= \frac{1}{\sqrt{2}}(\varphi_2 + \varphi_{-2}) \,, \quad \varphi_{3z^2-r^2} = \varphi_0 \,.
\end{aligned}
\tag{3.7}
$$

The electron-transfer Hamiltonian such as the first term in (3.1) combines these wave functions with the p orbitals on neighboring O^{2-} ions. As a result, the $3d$ orbitals are mixed with the appropriate combinations of the p orbitals of surrounding anions. We define three p orbitals of O^{2-} as

$$
\begin{aligned}
\varphi_x &= -\frac{1}{\sqrt{2}}(\varphi_1 - \varphi_{-1}) \,, \\
\varphi_y &= -\frac{1}{\sqrt{2}\mathrm{i}}(\varphi_1 + \varphi_{-1}) \,, \\
\varphi_z &= \varphi_0 \,;
\end{aligned}
\tag{3.8}
$$

from symmetry, they mix with the $3d$ orbitals in (3.7) as depicted in Fig. 3.1. As evident from the figure, the $d\varepsilon$ and $d\gamma$ orbitals mix differently with the p orbitals and consequently the $d\varepsilon$ and $d\gamma$ orbitals have different energies. Correspondingly, the p orbitals φ_x, φ_y and φ_z of the O^{2-} ion mix with surrounding d orbitals of the magnetic ions. Since the $2p$ shell is completely filled, those p-d mixed orbitals are filled, with six electrons total. They are the *bonding orbitals* of the p-d mixing, whereas the d-p mixed orbitals derived from the d electrons are orthogonal to the bonding orbitals and are called the *antibonding orbitals*. For the antibonding orbitals, a larger mixing of the

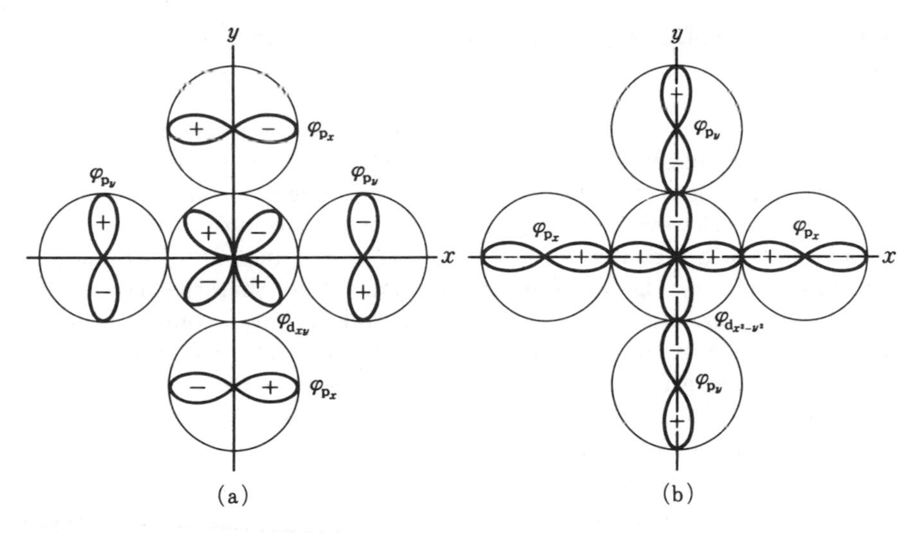

Fig. 3.1. The mixing of the $3d$ orbitals with the $2p$ orbitals of surrounding O^{2-}: (a) $d\varepsilon$ orbital, (b) $d\gamma$ orbital

p orbitals into the d orbitals lead to a higher level. Therefore the $d\gamma$ energy becomes higher, if the energies of the $d\varepsilon$ and $d\gamma$ orbitals are compared. This tendency is the same as for the electrostatic effect, since the $d\varepsilon$ orbital has a larger amplitude in the direction avoiding the negative charge of O^{2-}, while the $d\gamma$ orbital extends along the direction toward the center of the O^{2-} ion.

The energy difference between the $d\varepsilon$ and $d\gamma$ levels is about $10000 \sim 15000$ cm $^{-1}$. Figure 3.2 shows how the energy levels of the 5-fold degenerate d orbitals split in a crystal field of cubic symmetry and also in the case of an additional field of tetragonal symmetry in which the z axis is the principal one.

If there are several d electrons, they successively occupy the energy levels from the bottom, according to Hund's first rule, keeping their spins parallel. In this way one can obtain the ground state in which the crystal-field effect of cubic symmetry is taken into account. Notice, however, that if the crystal-field effect is larger than the intra-atomic exchange interaction, then the $d\gamma$

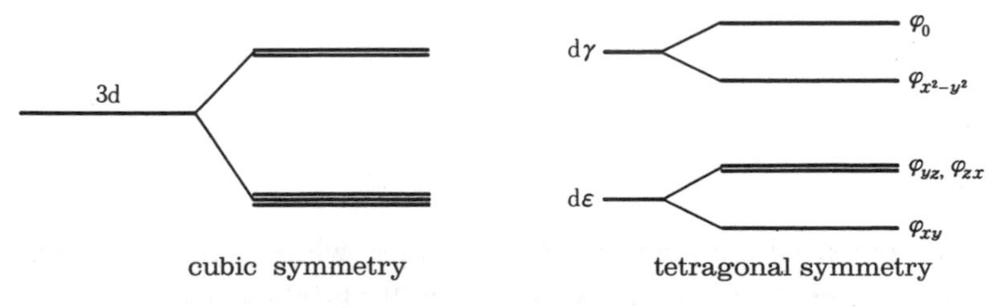

Fig. 3.2. The energy levels of the $3d$ orbitals in a crystal field

orbitals are occupied only after the $d\varepsilon$ orbitals are completely filled, as in complex salts. The ground state for two d electrons is three-fold degenerate, since there are 3 possibilities to choose 2 orbitals from 3 $d\varepsilon$ orbitals. The first excited states correspond to states in which one electron occupies the $d\gamma$ orbital, while the other is in a $d\varepsilon$ orbital; therefore they are 6-fold denenerate. In the second excited state both electrons occupy the $d\gamma$ orbital so that it is nondegenerate. However, this argument does not take into account Hund's second rule, which is related to the correlation energy due to the intra-atomic Coulomb repulsion. The latter energy is usually larger than the crystal-field energy and therefore it is not negligible. If we take it into account, the 6-fold degenerate levels in the first excited state split into two 3-fold degenerate levels, one of which is much higher than the second excited state. If Hund's rule is applied without taking into account the crystal field, the ground state for $(3d)^2$ is ^{3}F. The 3-fold degenerate state at high energy is ^{3}P.

As mentioned before, the electrostatic effect is not so important for $3d$ electrons in the iron group. However, the splitting of the energy levels is determined by the symmetry so that the results obtained in this way hold quite generally if one adjusts the parameters of the electrostatic field.

If we assume that the charge distribution of surrounding ions is spherically symmetric and ignore the overlap of their wave functions with the d electrons at the center, we can treat these ions as point charges. Under this condition the crystal-field potential may be written as

$$V_{\text{cryst}} = \sum_i \frac{e_i}{|\boldsymbol{r} - \boldsymbol{R}_i|} , \tag{3.9}$$

where $\boldsymbol{R}_i$ and e_i are the position vector and charge of the ith ion, respectively. Expressing $\boldsymbol{R}_i$ and $\boldsymbol{r}$ as $(R_i, \theta_i, \varphi_i)$ and (r, θ, φ) in spherical coordinates and expanding (3.9) in spherical harmonics, we obtain

$$V_{\text{cryst}} = \sum_l \sum_{m=-l}^{l} K_{lm} r^l \mathrm{P}_l^{|m|}(\cos\theta)\mathrm{e}^{\mathrm{i}m\varphi} , \tag{3.10}$$

where K_{lm} is given by

$$K_{lm} = \frac{(l-|m|)!}{(l+|m|)!} \sum_i \frac{e_i}{R_i^{l+1}} \mathrm{P}_l^{|m|}(\cos\theta_i)\mathrm{e}^{-\mathrm{i}m\varphi_i} . \tag{3.11}$$

The term for $l = 0$

$$V_0 = \sum_i \frac{e_i}{R_i} \tag{3.12}$$

is the Coulomb potential between two point charges, which is related to the Madelung energy. The terms for $l = 2$ and $l = 4$ are proportional to r^2 and r^4, respectively, and are given by

$$V_2 = \frac{1}{2}K_{20}(3z^2 - r^2) + 6K_{22}(x^2 - y^2) , \tag{3.13}$$

$$V_4 = \frac{1}{8}K_{40}(35z^4 - 30z^2r^2 + 3r^4) + 15K_{42}(7z^2 - r^2)(x^2 - y^2) + 210K_{44}(x^4 - 6x^2y^2 + y^4) . \tag{3.14}$$

It has been assumed here that the configuration of surrounding ions has orthorhombic symmetry.

In the case of hexagonal symmetry, only the K_{20} and K_{40} terms are nonvanishing. (Because of no inversion symmetry, the K_{43} term vanishes.) For cubic symmetry K_{20}, K_{22} and K_{42} vanish; the relation $K_{44} = (1/336)K_{40}$ reduces (3.14) to

$$V_4 = \frac{5}{2}K_{40}\left(x^4 + y^4 + z^4 - \frac{3}{5}r^4\right) . \tag{3.15}$$

The potential higher than fourth order is not necessary for d electrons.

Once the electrostatic crystal-field potential V_{cryst} is obtained, the potential energy for n electrons in the incompletely filled shell is given by

$$\mathcal{H}_{\text{cryst}} = -e\sum_i V_{\text{cryst}}(\boldsymbol{r}_i) . \tag{3.16}$$

In order to determine the energy eigenvalue for this crystal field, one forms first $2L+1$ wave functions for the n-electron system determined by Hund's rule, calculates matrix elements among these states and determines the eigenvalues and eigenfunctions. However, as far as the matrix elements within the $2L+1$ dimensional subspace are concerned, $\mathcal{H}_{\text{cryst}}$ is equivalent to a polynomial in the total angular momentum $\boldsymbol{L}$. The operators equivalent to the terms in (3.13–15) are given by

$$\sum(3z^2 - r^2) = \alpha\langle r^2\rangle[3L_z^2 - L(L+1)] , \tag{3.17}$$

$$\sum(35z^4 - 30z^2r^2 + 3r^4) = \beta\langle r^4\rangle[35L_z^4 - 30L(L+1)L_z^2 + 25L_z^2 - 6L(L+1) + 3L^2(L+1)^2] , \tag{3.18}$$

$$\sum(x^4-6x^2y^2+y^4) = \frac{1}{2}\sum[(x+\mathrm{i}y)^4+(x-\mathrm{i}y)^4] = \frac{1}{2}\beta\langle r^4\rangle(L_+^4+L_-^4) , \tag{3.19}$$

$$\sum\left(x^4 + y^4 + z^4 - \frac{3}{5}r^4\right) = \beta\langle r^4\rangle\left[L_x^4 + L_y^4 + L_z^4 - \frac{1}{5}L(L+1)(3L^2 + 3L - 1)\right] , \tag{3.20}$$

where $\langle r^2\rangle$ and $\langle r^4\rangle$ represent averages over the radial wave function [3.1].

The K_{22} and K_{42} terms can be easily obtained from (3.17, 18), respectively. The coefficients α and β can be determined by calculating one matrix element and comparing both sides. The results are as follows:

$$\alpha = \mp \frac{2(2l+1-4S)}{(2l-1)(2l+3)(2L-1)}, \tag{3.21}$$

$$\beta = \mp \frac{3(2l+1-4S)[-7(l-2S)(l-2S+1)+3(l-1)(l+2)]}{(2l-3)(2l-1)(2l+3)(2l+5)(L-1)(2L-1)(2L-3)}, \tag{3.22}$$

where the sign $\mp$ is chosen according to $n \lessgtr 2l+1$, n being $3d$ electron number.

The expression of the crystal-field Hamiltonian in terms of a polynomial in the angular momentum operator $\boldsymbol{L}$ allows one to evaluate its eigenvalue relatively easily, if one uses the properties of $\boldsymbol{L}$ (the commutation relations and so on). For instance, the eigenvalues and eigenfunctions for the operator equivalent of the cubic crystal field are given by

energy eigenvalue   **eigenfunction**

$\underline{L=2}$

$$\frac{18}{5}\beta\langle r^4\rangle \qquad \left.\begin{array}{l} \psi_0 \propto \frac{1}{2}(3z^2-r^2) \\ \frac{1}{\sqrt{2}}(\psi_2+\psi_{-2}) \propto \frac{\sqrt{3}}{2}(x^2-y^2) \end{array}\right\} \; : \Gamma_3$$

$$-\frac{12}{5}\beta\langle r^4\rangle \qquad \left.\begin{array}{l} \frac{\mathrm{i}}{\sqrt{2}}(\psi_{-1}+\psi_1) \propto yz \\ \frac{1}{\sqrt{2}}(\psi_{-1}-\psi_1) \propto zx \\ \frac{\mathrm{i}}{\sqrt{2}}(\psi_{-2}-\psi_2) \propto xy \end{array}\right\} \; : \Gamma_5 \tag{3.23a}$$

$\underline{L=3}$

$$18\beta\langle r^4\rangle \qquad \left.\begin{array}{l} -\frac{1}{\sqrt{2}}\left[\sqrt{\frac{3}{8}}(\psi_{-1}-\psi_1)+\sqrt{\frac{5}{8}}(\psi_3-\psi_{-3})\right] \propto x(5x^2-3r^2) \\ -\frac{\mathrm{i}}{\sqrt{2}}\left[\sqrt{\frac{3}{8}}(\psi_{-1}+\psi_1)+\sqrt{\frac{5}{8}}(\psi_3+\psi_{-3})\right] \propto y(5y^2-3r^2) \\ \psi_0 \propto z(5z^2-3r^2) \end{array}\right\} \; : \Gamma_4$$

$$-6\beta\langle r^4\rangle \qquad \left.\begin{array}{l} -\frac{1}{\sqrt{2}}\left[\sqrt{\frac{5}{8}}(\psi_{-1}-\psi_1)-\sqrt{\frac{3}{8}}(\psi_3-\psi_{-3})\right] \propto x(y^2-z^2) \\ \frac{\mathrm{i}}{\sqrt{2}}\left[\sqrt{\frac{5}{8}}(\psi_{-1}+\psi_1)-\sqrt{\frac{3}{8}}(\psi_3+\psi_{-3})\right] \propto y(z^2-x^2) \\ \frac{1}{\sqrt{2}}(\psi_2+\psi_{-2}) \propto z(x^2-y^2) \end{array}\right\} \; : \Gamma_5$$

$$-36\beta\langle r^4\rangle \qquad \frac{\mathrm{i}}{\sqrt{2}}(\psi_{-2}-\psi_2) \propto xyz \qquad : \Gamma_2 \tag{3.23b}$$

Here ψ_M represents the eigenfunction for the z-component M of the angular momentum $\boldsymbol{L}$. When the operator equivalent is used, it is not necessary to consider the wave function for the $(3d)^n$-electron system. From this result we see that the energy levels of iron-group ions split under a cubic crystal field as shown in Fig. 3.3. The sign of the crystal field is the same as that due to anions placed at the vertices of an octahedron.

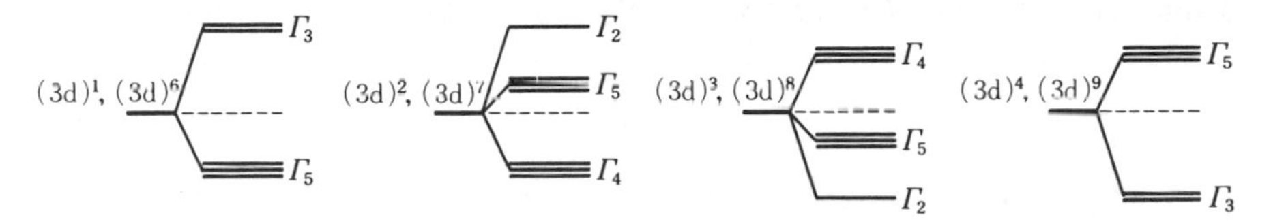

Fig. 3.3. The energy levels of iron-group ions in a cubic crystal field

If the ground-state subspace is formed with $2J+1$ dimensional states ($\boldsymbol{J}$ is the sum of the orbital angular momentum and spin angular momentum) as for rare-earth ions, the electrostatic crystal-field Hamiltonian within this subspace is equivalent to a polynomial in the components J_x, J_y and J_z of the total angular momentum. Let us assume that the crystal-field potential is expanded as

$$\begin{aligned} V_{\text{cryst}} &= A_{20}(3z^2 - r^2) + A_{40}(35z^4 - 30z^2r^2 + 3r^4) \\ &\quad + A_{60}(231z^6 - 315z^4r^2 + 105z^2r^4 - 5r^6) \\ &\quad + A_{66}(x^6 - 15x^4y^2 + 15x^2y^4 - y^6) \,. \end{aligned} \tag{3.24}$$

Having rare-earth metals in mind, we have assumed here that the crystal field has hexagonal symmetry; the 6-fold axis is taken as the z-axis. For f electrons the crystal-field terms higher than 6th order do not contribute to the splitting of energy levels because of the orthogonality of the Legendre functions. The operators equivalent to the terms in (3.24) are as follows:

$$A_{20}\sum(3z^2 - r^2) = \alpha A_{20}\langle r^2\rangle[3J_z^2 - J(J+1)] \,, \tag{3.25}$$

$$\begin{aligned} A_{40}\sum(35z^4 - 30z^2r^2 + 3r^4) &= \beta A_{40}\langle r^4\rangle[35J_z^4 - 30J(J+1)J_z^2 \\ &\quad + 3J^2(J+1)^2 + 25J_z^2 - 6J(J+1)] \,, \end{aligned} \tag{3.26}$$

$$\begin{aligned} A_{60}\sum(231z^6 - 315z^4r^2 + 105z^2r^4 - 5r^6) &= \gamma A_{60}\langle r^6\rangle[231J_z^6 \\ &\quad - 315J(J+1)J_z^4 + 105J^2(J+1)^2J_z^2 - 5J^3(J+1)^3 + 735J_z^4 \\ &\quad - 525J(J+1)J_z^2 + 40J^2(J+1)^2 + 294J_z^2 - 60J(J+1)] \,, \end{aligned} \tag{3.27}$$

$$\begin{aligned} A_{66}\frac{1}{2}\sum\left[(x+\mathrm{i}y)^6 + (x-\mathrm{i}y)^6\right] &= \gamma A_{66}\langle r^6\rangle \\ &\quad \times \frac{1}{2}\left[(J_x+\mathrm{i}J_y)^6 + (J_x-\mathrm{i}J_y)^6\right] \,, \end{aligned} \tag{3.28}$$

where α, β and γ are coefficients depending on J, L, S and l; they are given by

$$\alpha = -\frac{2}{45} \,, \quad \beta = \frac{2}{45\cdot 11} \,, \quad \gamma = -\frac{4}{9\cdot 13\cdot 33} \tag{3.29}$$

for one f electron having no spin. In Table 3.1 we show the values of α, β and γ for the ground state of rare-earth ions [3.2].

Table 3.1. The values for α, β and γ in the ground state of rare earth ions [3.2]

ion			α	β	γ
Ce^{3+}	$(4f)^1$	$^2F_{5/2}$	$-2/5{\cdot}7$	$2/5{\cdot}7{\cdot}9$	0
Pr^{3+}	$(4f)^2$	$^3H_4$	$-13{\cdot}2^2/11{\cdot}5^2{\cdot}3^2$	$-2^2/11^2{\cdot}5{\cdot}3^2$	$17{\cdot}2^4/13{\cdot}11^2{\cdot}7{\cdot}5{\cdot}3^4$
Nd^{3+}	$(4f)^3$	$^4I_{9/2}$	$-7/11^2{\cdot}3^2$	$-2^3{\cdot}17/13{\cdot}11^3{\cdot}3^3$	$-19{\cdot}17{\cdot}5/13^2{\cdot}11^3{\cdot}7{\cdot}3^3$
Pm^{3+}	$(4f)^4$	$^5I_4$	$2{\cdot}7/11^2{\cdot}5{\cdot}3$	$8{\cdot}7{\cdot}17/13{\cdot}11^3{\cdot}5{\cdot}3^3$	$19{\cdot}17{\cdot}2^3/13^2{\cdot}11^2{\cdot}7{\cdot}3^3$
Sm^{3+}	$(4f)^5$	$^6H_{5/2}$	$13/7{\cdot}5{\cdot}3^2$	$2{\cdot}13/11{\cdot}7{\cdot}5{\cdot}3^3$	0
Eu^{3+}	$(4f)^6$	$^7F_0$	0	0	0
Gd^{3+}	$(4f)^7$	$^8S_{7/2}$	0	0	0
Tb^{3+}	$(4f)^8$	$^7F_6$	$-1/11{\cdot}3^2$	$2/11^2{\cdot}5{\cdot}3^3$	$-1/13{\cdot}11^2{\cdot}7{\cdot}3^4$
Dy^{3+}	$(4f)^9$	$^6H_{15/2}$	$-2/7{\cdot}5{\cdot}3^2$	$-8/13{\cdot}11{\cdot}7{\cdot}5{\cdot}3^3$	$2^2/13^2{\cdot}11^2{\cdot}7{\cdot}3^3$
Ho^{3+}	$(4f)^{10}$	$^5I_8$	$-1/5^2{\cdot}3^2{\cdot}2$	$-1/13{\cdot}11{\cdot}7{\cdot}5{\cdot}3{\cdot}2$	$-5/13^2{\cdot}11^2{\cdot}7{\cdot}3^3$
Er^{3+}	$(4f)^{11}$	$^4I_{15/2}$	$2^2/7{\cdot}5^2{\cdot}3^2$	$2/13{\cdot}11{\cdot}7{\cdot}5{\cdot}3^2$	$2^3/13^2{\cdot}11^2{\cdot}7{\cdot}3^3$
Tm^{3+}	$(4f)^{12}$	$^3H_6$	$1/11{\cdot}3^2$	$2^3/11^2{\cdot}5{\cdot}3^4$	$-5/13{\cdot}11^2{\cdot}7{\cdot}3^4$
Yb^{3+}	$(4f)^{13}$	$^2F_{7/2}$	$2/7{\cdot}3^2$	$-2/11{\cdot}7{\cdot}5{\cdot}3$	$2^2/13{\cdot}11{\cdot}7{\cdot}3^3$

3.3 Effective Hamiltonians Describing One-Ion Anisotropy

The magnetic moment of rare-earth ions is proportional to the total angular momentum $\boldsymbol{J}$. Therefore the Hamiltonians in (3.25–28) give the anisotropy energy for the magnetic moment of the ion. In iron-group ions, on the other hand, the crystal-field splitting of the $2L+1$ degenerate levels is much larger than both kT (T: temperature) and the LS coupling, so that we have to consider the ground state due to the crystal field. In most ionic crystals of iron-group compounds, magnetic ions are located at the center of an octahedron formed with anions. In this case, the energy levels split due to the cubic field in the way shown in Fig. 3.3, from which we see that ions with $(3d)^3$ or $(3d)^8$ (corresponding to Cr^{3+}, Ni^{2+} etc.) are nondegenerate in the ground state. Even when ground-state degeneracy is present in the cubic field, it is usually lifted by a crystal field of lower symmetry. In most cases the effect of the lower-symmetry crystal field is smaller than that of the cubic field, but is still larger than the LS coupling. In such a case one considers simply the nondegenerate ground state.

Since the crystal-field Hamiltonian is given as a real function, its eigenfunctions can be expressed also with real functions. On the other hand, the operator of the total angular momentum $\boldsymbol{L}$ is pure imaginary; thus the expectation value of $\boldsymbol{L}$ over a real eigenfunction is pure imaginary. Since $\boldsymbol{L}$ is a Hermitian operator, the diagonal matrix element must be real. From this we see that the expectation value of the angular momentum over a nondegenerate eigenstate must be zero: namely we obtain for a nondegenerate ground state $|0\rangle$

$$\langle 0|\boldsymbol{L}|0\rangle = 0 \ . \tag{3.30}$$

This means that the orbital angular momentum is quenched in a nondegenerate ground state, which is realized by the crystal-field splitting. This is called the *quenching of the orbital angular momentum.* The quenched orbital angular momentum is partially restored by the LS coupling, as shown below.

Let E_n and $|n\rangle$ be the energy level and the corresponding eigenfunction due to the crystal-field splitting. The function $|n\rangle$ may be regarded as the orbital part of the eigenfunction of a real many-electron system or as the eigenfunction of the Hamiltonian written with equivalent operators. In both cases one can assume that the eigenfunction including the spin is given as a product of the orbital and spin parts. At this stage the orbital state of the ion is the ground state $|0\rangle$, in which the orbital angular momentum is quenched, and the spin $\boldsymbol{S}$ is completely free with $(2S+1)$ fold degeneracy; in this case the magnetic moment of the ion is given exclusively by the spin. The free spin couples to the lattice only when we take into account the LS coupling.

Let us treat the LS coupling and the Zeeman energy,

$$V = \lambda \boldsymbol{L} \cdot \boldsymbol{S} + \mu_{\mathrm{B}} \boldsymbol{H} \cdot (2\boldsymbol{S} + \boldsymbol{L}) \ , \tag{3.31}$$

as a perturbation. Since the spin wave function is independent of the orbital part, the spin $\boldsymbol{S}$ is left as an operator in this perturbation calculation. Because of the quenching of $\boldsymbol{L}$, first-order perturbation theory leads to

$$\Delta E^{(1)} = 2\mu_{\mathrm{B}} \boldsymbol{H} \cdot \boldsymbol{S} \ . \tag{3.32}$$

Introducing $\Lambda_{\mu\nu}$ defined by

$$\Lambda_{\mu\nu} = \sum_n \frac{\langle 0|L_\mu|n\rangle\langle n|L_\nu|0\rangle}{E_n - E_0} \ , \tag{3.33}$$

we obtain the second-order energy as

$$\Delta E^{(2)} = -\sum_{\mu\nu} \left[\lambda^2 \Lambda_{\mu\nu} S_\mu S_\nu + 2\lambda\mu_{\mathrm{B}} \Lambda_{\mu\nu} H_\mu S_\nu + \mu_{\mathrm{B}}^2 \Lambda_{\mu\nu} H_\mu H_\nu\right] \ , \tag{3.34}$$

where μ and ν represent x, y or z. Adding $\Delta E^{(1)}$ and $\Delta E^{(2)}$, we have

$$\mathcal{H}_{\mathrm{S}} = \sum_{\mu\nu} \left[2\mu_{\mathrm{B}} H_\mu (\delta_{\mu\nu} - \lambda\Lambda_{\mu\nu}) S_\nu - \lambda^2 S_\mu \Lambda_{\mu\nu} S_\nu - \mu_{\mathrm{B}}^2 H_\mu \Lambda_{\mu\nu} H_\nu\right] \tag{3.35}$$

as the effective Hamiltonian for a nondegenerate ground state split off by the crystal field. The first term represents an effective Zeeman energy, which means that the g value has been replaced by the g tensor

$$g_{\mu\nu} = 2(\delta_{\mu\nu} - \lambda\Lambda_{\mu\nu}) \ . \tag{3.36}$$

Here the additional tensor $-2\lambda\Lambda_{\mu\nu}$ in the induced orbital moment, which arises from the mixing with high-energy orbital states due to the LS coupling and is expressed as a change of the magnetic moment accompanied with the spin $\boldsymbol{S}$.

The second term is the spin Hamiltonian in a narrow sense or the *anisotropy spin Hamiltonian*, which represents the anisotropy energy for the spin direction. Let us take the principal axes of the crystal as x, y and z axes and express the components of Λ as Λ_x, Λ_y and Λ_z. Then the anisotropy spin Hamiltonian can be written as

$$\begin{aligned}\mathcal{H} = -\lambda^2\Big\{&\frac{1}{3}(\Lambda_x+\Lambda_y+\Lambda_z)S(S+1)\\&+\frac{1}{3}\Big[\Lambda_z-\frac{1}{2}(\Lambda_x+\Lambda_y)\Big][3S_z^2-S(S+1)]\\&+\frac{1}{2}(\Lambda_x-\Lambda_y)(S_x^2-S_y^2)\Big\}\,. \qquad (3.37)\end{aligned}$$

The anisotropy Hamiltonian lifts the $(2S+1)$-fold degeneracy of the spin. Omitting the constant term, we obtain from (3.37)

$$\mathcal{H} = DS_z^2 + E(S_x^2 - S_y^2)\,. \qquad (3.38)$$

For integer S the first term in this Hamiltonian splits the spin energy levels into doubly degenerate S levels $S_z = \pm S, \pm(S-1), \dots, \pm 1$ and a non-degenerate one with $S_z = 0$; for half-odd integer S it leads to doubly degenerate $S+1/2$ levels with $S_z = \pm S, \pm(S-1), \dots, \pm 1/2$. The second term has finite matrix elements between states with $\Delta S_z = \pm 2$. Therefore, for integer S, the doubly degenerate levels $S_z = \pm M$, which are produced by the first term, are split by the second term; as a result the $(2S+1)$-fold degeneracy is lifted by the anisotropy Hamiltonian. However, for half-odd integer S the difference $2M$ between $S_z = \pm M$ of the doubly degenerate levels is an odd integer so that there is no matrix element between these states. Consequently, the double degeneracy due to the first term remains. The case of half-odd integer S corresponds to a system with an odd number of electrons; for this case the crystal field cannot lift the degeneracy completely, leaving double degeneracy. This is called the *Kramers theorem*; the doubly degenerate levels remaining are called a Kramers doublet.

The Kramers theorem is a general result which can be derived when the Hamiltonian for the electron system is invariant in time reversal. Under time reversal the orbital and spin angular momenta change signs; therefore the Kramers degeneracy is lifted first by the Zeeman energy (which changes sign under time reversal).

The third term in (3.35) is not related to the LS coupling; it comes rather from the second-order perturbation of the Zeeman energy for the orbital angular momentum. This gives a temperature independent (anisotropic) paramagnetic susceptibility, which is called the *Van Vleck orbital paramagnetism*. The Van Vleck orbital paramagnetism gives a non-negligible contribution when the energy of the excited states is not too high. In the case of transition metals, iron-group metals in particular, excited states are present continuously from the Fermi energy; therefore a large orbital paramagnetism is expected as pointed out by *Kubo* and *Obata* [3.3]. In vanadium metal the paramagnetic susceptibility hardly changes between the normal and superconducting states, suggesting that most of the paramagnetism in this metal originates from the orbital paramagnetism [3.4].

In the case of cubic symmetry, the change of the effective g value and the orbital paramagnetism are finite and isotropic, since the anisotropy Hamiltonian in second order is merely a constant. In this case the anisotropy shows up in fourth-order perturbation; the anisotropy Hamiltonian is usually written as

$$\mathcal{H}_{\mathrm{S}} = \frac{a}{6}\left\{S_x^4 + S_y^4 + S_z^4 - \frac{1}{5}S(S+1)[3S(S+1)-1]\right\}. \tag{3.39}$$

The second-order terms are constant for $S = 1/2$ even when the crystal field has a low symmetry. Similarly (3.39) vanishes for $S < 2$. In such a situation the anisotropy Hamiltonian comes from the anisotropic interaction. The anisotropy Hamiltonian in (3.39) becomes important in Mn^{2+} and Fe^{3+}, which have $S = 5/2$. For these ions $L = 0$, and thus the Hamiltonian of (3.39) can be derived from the perturbation energy between different multiplets in which the perturbation is the sum of the ls coupling for each electron. The origin of the magnetic anisotropy in ferrites containing Cu, Ni or Mn is considered to be such anisotropy Hamiltonian of Fe^{3+} ions [3.5].

3.4 Jahn-Teller Effect

In the previous section we considered the case in which the orbital state of magnetic ions is nondegenerate in the ground state. In that case the $(2S+1)$-fold degeneracy in the spin state is lifted by the anisotropy spin Hamiltonian, which arises due to the LS coupling: a crystal field with sufficiently low symmetry completely lifts the degeneracy in the system containing an even number of electrons, while the Kramers doublet remains in a system of an odd number of electrons. The degeneracy present in the Kramers doublet is removed by the external magnetic field or, in its absence, by the interaction between the spins of the ions.

When the orbital ground state is degenerate, the splitting of the electronic energy levels of the ion, including the spin state, can be different from the above. For a cubic crystal field, the d orbitals split into $d\gamma$ (Γ_3) and $d\varepsilon$ (Γ_5) levels, which are doubly and triply degenerate, respectively. If we consider the same problem for the n-electron system, a triply degenerate state Γ_5 or Γ_4 becomes the ground state for $(3d)^1$, $(3d)^6$, $(3d)^2$, and $(3d)^7$. On the other hand, the doubly degenerate state Γ_3 is the ground state for $(3d)^4$ and $(3d)^9$. Finite matrix elements of the angular momentum $\boldsymbol{L}$ exist among the three states in Γ_5 and Γ_4; therefore these states are connected with each other via the LS coupling. However, no matrix element of $\boldsymbol{L}$ is present between the two states of Γ_3 so that the two states are not directly connected by the LS coupling.

In general, the orbital states of electrons couple strongly to the lattice; the splitting of the energy levels due to the crystal field is a realization of this effect. This splitting of energy levels is an effect of fixed lattice points. If we regard the positions of the lattice points as dynamical variables, they are naturally subject to a reaction due to the electronic states of the magnetic ions. In other words, magnetic ions always accompany distortions of the lattice of surrounding ions. Those distortions tend to lift the remaining degeneracy, if the energy levels of the orbital state due to the fixed crystal field are still degenerate. The lifting of the orbital degeneracy by the interaction between the lattice distortion and the orbital state is called the *Jahn-Teller effect*. In this case the surrounding ionic configuration is distorted so as to lower the symmetry of the crystal field. The essential point of the Jahn-Teller effect is that the energy increase due to the lattice distortion is proportional to the second power of the distortion, whereas the energy level splitting of the orbital states due to the distortion is linearly proportional to the distortion.

This effect occurs quite generally. For instance, let us consider a metal which takes a new structure, of lower symmetry, due to a small distortion of the lattice. In this case the lowering of the symmetry produces new Brillouin zone boundaries; if the latter are close to the Fermi surface, the energy of conduction electrons is lowered because of the energy gap (which is proportional to the distortion). In fact, this is believed to be realized in some semimetals. It is also possible to realize the lowering of the symmetry by a spin configuration instead of the lattice distortion: the spin ordered states in rare-earth metals and metallic chromium may be regarded as examples of this effect. The Jahn-Teller interaction between the orbitals and the lattice is also a cause of the spin-lattice relaxation in paramagnetic salts.

Let us assume that a magnetic ion is surrounded by anions in an octahedral way; the displacement of each ion from the octahedral position is given by (X_i, Y_i, Z_i). We expand the electrostatic potential due to these ions in terms of these ionic coordinates and the electron coordinates x, y and z. We keep the terms to first order in X_i, Y_i, Z_i and to second order in x, y, z. (The terms to first order in the latter do not affect the $3d$ electrons.) Defining the index i for the anions as in Fig. 3.4, we obtain the change of the potential energy as

$$\begin{aligned}\Delta V = &-\frac{1}{2}a\left[x^2(X_1 - X_4) + y^2(Y_2 - Y_5) + z^2(Z_3 - Z_6)\right]\\&+\frac{1}{2}b\left[yz(Y_3 + Z_2 - Y_6 - Z_5) + zx(Z_1 + X_3 - Z_4 - X_6)\right.\\&\left.\quad +xy(X_2 + Y_1 - X_5 - Y_4)\right] , \qquad (3.40)\end{aligned}$$

where a and b are positive constants. If we separate the displacements into two parts, whose directions are symmetric and antisymmetric with respect

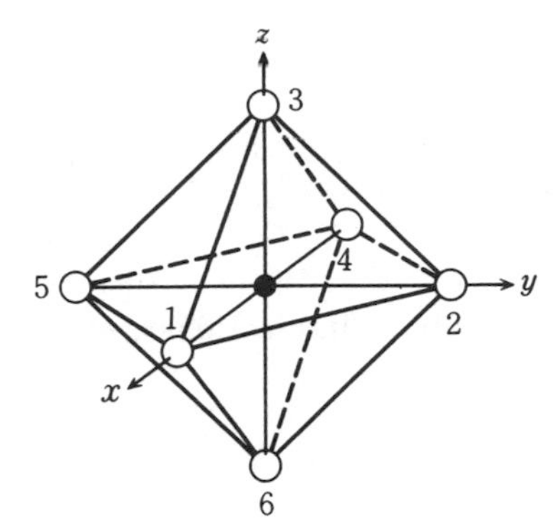

Fig. 3.4. The octahedron formed with anions surrounding a magnetic ion. The open circles represent anions

to inversion through the origin, only the symmetric part remains. The displacements symmetric under inversion are then decomposed into the normal modes of the octahedron. As shown in Fig. 3.5, there are 6 such independent normal modes. Among these modes q_2 and q_3 (which have the same symmetry as $d\gamma$) have the same eigenfrequency. The modes q_4, q_5, q_6 have the same symmetry as $d\varepsilon$ and are degenerate with each other. If we express (3.40) in terms of these coordinates, we obtain

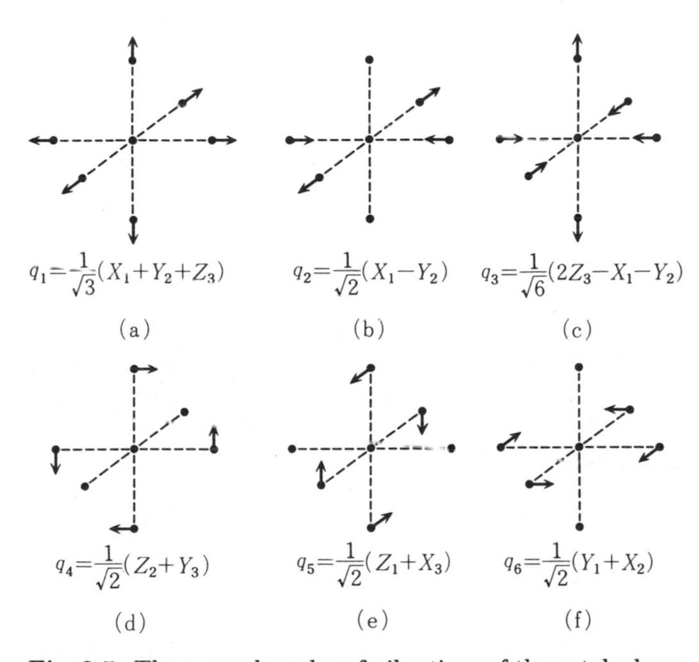

Fig. 3.5. The normal modes of vibrations of the octahedron

$$\Delta V = V_1 + V_2 ,$$

$$V_1 = -\sqrt{\frac{2}{3}} a \Big[\frac{\sqrt{3}}{2}(x^2 - y^2) q_2 + \frac{1}{2}(2z^2 - x^2 - y^2) q_3 \Big] , \tag{3.41}$$

$$V_2 = \sqrt{\frac{2}{3}} b \Big[\sqrt{3} yz q_4 + \sqrt{3} zx q_5 + \sqrt{3} xy q_6 \Big] , \tag{3.42}$$

where the isotropic part has been omitted. The above interactions can be written, by using the operator equivalent, as

$$V_1 = -\sqrt{\frac{2}{3}} a\alpha\langle r^2\rangle \Big\{ \frac{\sqrt{3}}{2}(L_x^2 - L_y^2) q_2 + \frac{1}{2}[3L_z^2 - L(L+1)] q_3 \Big\} , \tag{3.43}$$

$$V_2 = \sqrt{\frac{2}{3}} b\alpha\langle r^2\rangle \Big[\frac{\sqrt{3}}{2}(L_y L_z + L_z L_y) q_4 + \frac{\sqrt{3}}{2}(L_z L_x + L_x L_z) q_5 + \frac{\sqrt{3}}{2}(L_x L_y + L_y L_x) q_6 \Big] , \tag{3.44}$$

where α is a constant given by (3.21).

Let us first consider Γ_3 for $L = 2$ as degenerate states; this applies to the ground state of Cu^{2+}, Mn^{3+} and Cr^{2+}. It is easy to evaluate the matrix elements of (3.43, 44) with respect to the two eigenfunctions belonging to Γ_3, $(1/\sqrt{2})(\psi_2 + \psi_{-2})$ and ψ_0, the result being

$$V_1 \to -3\sqrt{\frac{2}{3}} a\alpha\langle r^2\rangle \left[q_2 \begin{pmatrix} 0 & 1 \\ 1 & 0 \end{pmatrix} + q_3 \begin{pmatrix} 1 & 0 \\ 0 & -1 \end{pmatrix} \right] ,$$

$$V_2 \to 0 .$$

Therefore the Hamiltonian describing the energy change due to the distortion is

$$W = -A(\sigma_x q_2 + \sigma_z q_3) + \frac{1}{2} M\omega^2 (q_2^2 + q_3^2) . \tag{3.45}$$

Here σ_x and σ_z are 2×2 Pauli spin matrices defined by (1.4). The second term is the increase of the potential energy due to the distortion, ω and M being the angular frequency of the normal mode and the mass of the anion, respectively. The value of A for $d\gamma$ is

$$A = -\frac{2}{7}\sqrt{\frac{2}{3}} a\langle r^2\rangle \tag{3.46}$$

by using $\alpha = -2/21$. If we introduce polar coordinates in the q_3–q_2 plane and express q_3 and q_2 as

$$q_3 = q\cos\theta , \quad q_2 = q\sin\theta , \tag{3.47}$$

(3.45) can be written as

$$W = -Aq(\sigma_x \sin\theta + \sigma_z \cos\theta) + \frac{1}{2} M\omega^2 q^2 . \tag{3.48}$$

The first term is the same as the Zeeman energy for a magnetic field Aq/μ_B applied along the direction making an angle θ to the z axis in the zx plane. Thus the eigenvalues of (3.48) are given by

$$E = \mp Aq + \frac{1}{2}M\omega^2 q^2 , \tag{3.49}$$

which are independent of θ; if the quantization axis for σ is the z axis, the eigenfunctions are

$$\varphi_{+1} = \begin{pmatrix} \cos\frac{\theta}{2} \\ \sin\frac{\theta}{2} \end{pmatrix} , \quad \varphi_{-1} = \begin{pmatrix} -\sin\frac{\theta}{2} \\ \cos\frac{\theta}{2} \end{pmatrix} . \tag{3.50}$$

Here the values ± 1 in $\varphi_{\pm 1}$ denote the component of σ along the θ direction. The distortion q is obtained by minimizing (3.49), as

$$q_0 = \frac{|A|}{M\omega^2} , \tag{3.51}$$

the corresponding energy being

$$E_{\min} = -\frac{A^2}{2M\omega^2} . \tag{3.52}$$

The energy lowering $E_{\min}$ has been obtained without taking into account the vibration of the anions, including the zero-point motion. If the kinetic energy due to the vibration of the anions is so large that its amplitude exceeds the distortion in (3.51), then the calculation of the energy lowering does not make sense, and the whole problem has to be treated dynamically including the kinetic-energy term of the anions in the Hamiltonian for the distortion, (3.48).

If the vibrational energy is smaller than (3.52), however, the distortion given by (3.51) is stable. The energy lowering due to this static Jahn-Teller effect does not depend on the direction of the distortion; in other words any combination of the distortions q_2 and q_3 leads to the same energy.

The degeneracy of the energy with respect to the direction of the distortion is lifted on taking into account the third-order anharmonic term in X_i (as pointed out by *Öpic* and *Pryce* [3.6]) and higher-order coupling terms between the orbitals and the lattice, namely second-order in X_i and second-order in x (suggested by *Liehr* and *Ballhausen* [3.7]). The anharmonic term in the distortion of the octahedron can be expressed as

$$\begin{aligned} W' &= c_1(X_1^3 + Y_2^3 + Z_3^3) + 3c_2 X_1 Y_2 Z_3 \\ &= \frac{1}{\sqrt{6}}(c_1 + c_2) q_3 (q_3^2 - 3q_2^2) = A_3 q^3 \cos 3\theta , \end{aligned} \tag{3.53}$$

while the higher-order interaction is given by

$$\begin{aligned} W' &= B_3[(q_3^2 - q_2^2)\langle\sigma_z\rangle - 2q_2 q_3\langle\sigma_x\rangle] \\ &= \pm B_3 q^2 \cos 3\theta . \end{aligned} \tag{3.54}$$

The second relation in (3.53) can be derived by expressing X_1, Y_2, Z_3 in terms of q_1, q_2, q_3 and putting $q_1 = 0$. The sign $\pm$ in (3.54) is chosen for $\mp$ sign in front of A in (3.49). Both (3.53, 54) give a three-fold anisotropy in the q_3-q_2 plane. The case $\theta = 0$ corresponds to a distortion of q_3, i.e., an elongation along the z axis and a contraction along the x and y axes, whereas the case $\theta = \pi$ means the opposite distortion, see Fig. 3.5(c). On the other hand, the distortion for $\theta = 2\pi/3$ is

$$q = q_3 \cos\frac{2\pi}{3} + q_2 \sin\frac{2\pi}{3} = \frac{1}{\sqrt{6}}(2X_1 - Y_2 - Z_3) , \tag{3.55}$$

which gives a tetragonal distortion with an elongation along the x axis; the distortion for $\theta = 4\pi/3$ is tetragonal with elongation along the y axis. Therefore, the distortions for $\theta = 0, 2\pi/3, 4\pi/3$ are tetragonal, with elongation axis z, x, and y, respectively. Since they are all distortions of the otherwise perfect octahedron, they are equivalent to each other. The anisotropy energies in (3.53, 54) determine the sign of the tetragonal distortion, namely whether the material elongates or contracts along the axis. According to Öpic and Pryce, the coefficient A_3 in the anharmonic term is negative in general. Then an elongation along the c axis occurs due to the anharmonic term. Which of the two orbital states is chosen depends on the sign of A. The higher-order interaction is considered to be generally smaller than the anharmonic term.

In the above we discussed a cluster of one magnetic ion and its surrounding octahedral anions. When the magnetic ions form a crystalline lattice together with the anions, an anion next to a magnetic ion is at the same time one of the anions around a neighboring magnetic ion. Therefore the distortions of the octahedra surrounding the magnetic ions must be mutually consistent. One solution we can think of is a uniform distortion; this occurs actually in $CuFe_2O_4$ [3.8] and Mn_3O_4 [3.9], in which the Cu^{2+} ion or the Mn^{3+} ion is located at the center of the octahedron formed by the anions. The crystal distortions from cubic to tetragonal symmetry in these materials are due to the Jahn-Teller effect, $(c-a)/a$ being 6% and 15%, respectively.

Such lattice distortions decrease with increasing temperature and the cubic symmetry is restored above a cetain temperature (360°C in $CuFe_2O_4$ and 1170°C in Mn_3O_4). For a uniform distortion, the value of q is the same for all the magnetic ions; therefore the energy in (3.49), due to the distortion, is given by

$$W = -Aq\sigma_z + \frac{1}{2}M\omega^2 q^2 \tag{3.56}$$

for each ion. Remembering that the eigenvalue of σ_z is $\pm$ and taking the thermal average $\langle\sigma_z\rangle$, we obtain for the distortion $\bar{q}$ minimizing (3.56)

$$\bar{q} = \frac{A}{M\omega^2}\langle\sigma_z\rangle = q_0 \tanh\frac{|A|\bar{q}}{kT} . \tag{3.57}$$

The solution $\bar{q}$ of (3.57) decreases gradually from q_0 at $T = 0$ with increasing temperature, and then drops rapidly near $T_c - |\Lambda|q_0/k$; above T_c the crystal lattice recovers the original cubic symmetry. If we take into account the anisotropy energy in (3.53, 54), this transition becomes first order [3.10].

It is possible to realize a consistent distortion of the octahedra surrounding magnetic ions not only by a uniform lattice distortion, but also by changing the sign of the distortion from one magnetic ion to another. The latter case is seen in MnF_3 and $KCuF_3$. Figure 3.6 shows the ReO_3-type crystal structure of MnF_3 [3.11]. In this structure, each Mn^{3+} ion has neighboring F^- ions along the x, y, z axes at distances of l, s, m $(l > m > s)$, respectively. This crystal structure can be obtained from the perfect ReO_3 lattice in the following way: first the distortion $q_2 = (1/\sqrt{2})(l - s)$ of the octahedron formed with F^- ions is added in a staggered way along the three directions, and then the distortion $q_3 = (1/\sqrt{6})(2m - l - s)$ is superposed on it.

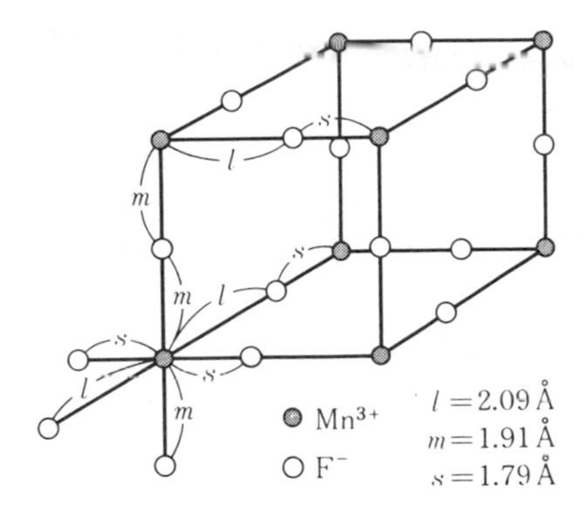

Fig. 3.6. The crystal structure of MnF_3

As shown in Fig. 3.7, there are three possible ways to introduce to the ReO_3-type lattice the distortion q_2 or q_3 of the octahedron, in a consistent manner. Kanamori explained why the structure of MnF_3 is stabilized among these three possibilities [3.10]. If we use the q vector in the q_3-q_2 plane, the states depicted in Figs. 3.7(a, b) can be expressed with two q vectors which are parallel and antiparallel to the q_2 axis. On the other hand, the state (c) can be represented by the q vectors parallel and antiparallel to the q_3 axis. Since the anisotropy energy arising from the anharmonicity gives the lowest energy for $\theta = 0, \pm 2\pi/3$, the total energy in the states (a) and (b) is lowered by tilting the direction of q a little to the direction $\pm 2\pi/3$ (as shown in Fig. 3.8). By this effect, a uniform distortion with $-q_3$ is added to the distortion (a) or (b). Defining the tilting angle by ϕ, we obtain

$$\tan\phi = \frac{|q_3|}{|q_2|} = \frac{1}{\sqrt{3}}\frac{l + s - 2m}{l - s} = 0.115\ , \quad \phi = 6°35' \tag{3.58}$$

on using observed values.

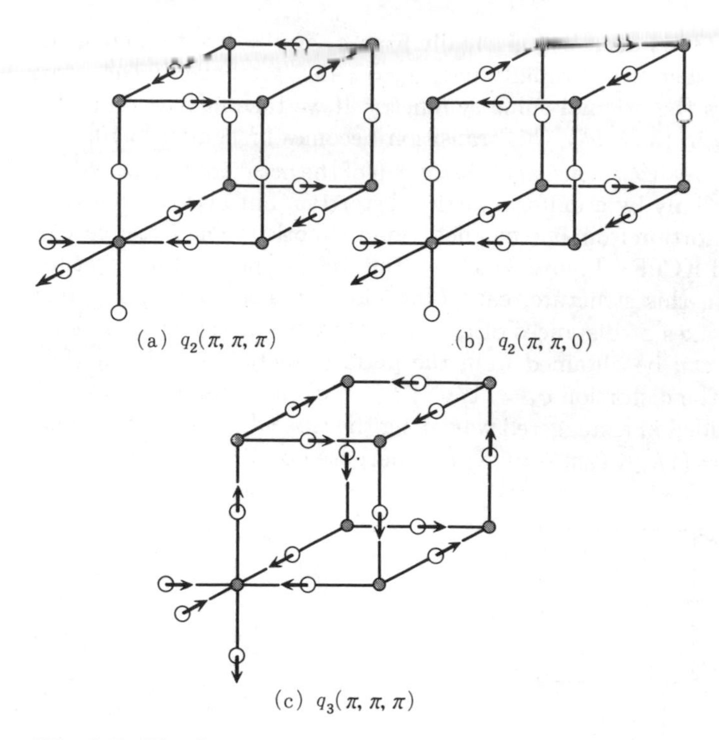

Fig. 3.7. The distortions of the octahedron in the ReO_3-type lattice

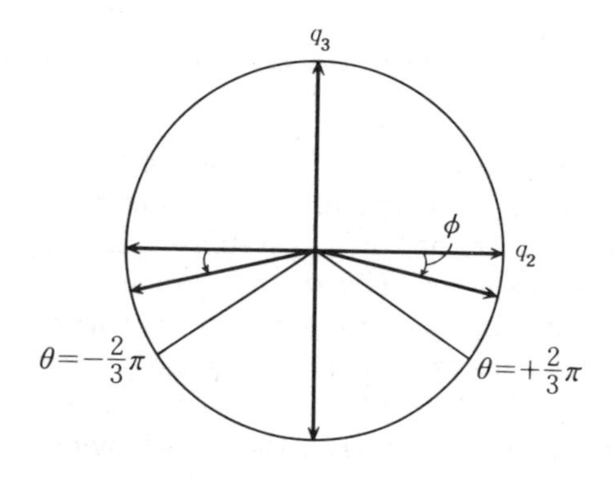

Fig. 3.8. The distortion of the octahedron one expects when the anisotropy energy is taken into accout

$KCuF_3$ has a perovskite-type structure; if the K^+ ions are removed, the lattice formed by the Cu^{2+} and F^- ions is of the ReO_3-type. From an X-ray analysis, *Okazaki* and *Suemune* [3.12] and *Okazaki* [3.13] found the values

$$l = 2.25\text{Å}\ , \quad m = 1.96\text{Å}\ , \quad s = 1.89\text{Å}\ . \tag{3.59}$$

From these we obtain $\tan\phi \simeq 0.35$, a much larger value than for MnF_3. This consideration does not tell us which of the two structures in Figs. 3.7(a, b) is lower in energy. In fact, the energy difference is fairly small and both structures in Figs. 3.7(a, b) are known to exist in $KCuF_3$. The orbital basis function of Cu^{2+} or Mn^{3+} for this type of distortion is given from (3.50) by

$$\begin{aligned} \varphi_{+1}^{\pm} &= \cos\left[\frac{1}{2}\left(\frac{\pi}{2}+\phi\right)\right]\varphi_{x^2-y^2} \pm \sin\left[\frac{1}{2}\left(\frac{\pi}{2}+\phi\right)\right]\varphi_{3z^2-r^2}\ , \quad (A>0) \\ \varphi_{-1}^{\pm} &= \mp\sin\left[\frac{1}{2}\left(\frac{\pi}{2}+\phi\right)\right]\varphi_{x^2-y^2} + \cos\left[\frac{1}{2}\left(\frac{\pi}{2}+\phi\right)\right]\varphi_{3z^2-r^2}\ , (A<0) \end{aligned} \tag{3.60}$$

where the suffix $\pm$ denotes the sign of q_2. For Cu^{2+} and Mn^{3+}, $\alpha = 2/21$ so that A is considered to be positive. Let the angular part of $(1/\sqrt{2})(\varphi_2+\varphi_{-2})$ and φ_0 be

$$\frac{1}{\sqrt{2}}(\varphi_2+\varphi_{-2}) = \frac{\sqrt{3}}{2}(x^2-y^2)\ , \quad \varphi_0 = \frac{1}{2}(3z^2-r^2)\ ; \tag{3.61}$$

then using the value of ϕ for $KCuF_3$, we obtain

$$\begin{aligned} \varphi_{+1}^{+} &= 0.092x^2 - 0.908y^2 + 0.816z^2\ , \\ \varphi_{+1}^{-} &= 0.908x^2 - 0.092y^2 - 0.816z^2\ , \end{aligned} \tag{3.62}$$

for the angular part of the orbital basis function for the two distortions. φ_{+1}^{+} represents the orbital elongated along the y, z direction and flat along the x direction; φ_{+1}^{-} can be obtained from φ_{+1}^{+} by interchanging x and y. Therefore, in this crystal the $3d$ orbitals can form chains, which are connected along the z direction, but disconnected in the x-y plane. In fact, a typical one-dimensional Heisenberg spin system, which is exchange-coupled along the chain, is realized in this substance, according to the study of *Hirakawa* and *Kurogi* [3.14].

Once the energy levels of the orbital ground state Γ_3 are split by the Jahn-Teller effect, the anisotropy spin Hamiltonian for such a magnetic ion can be calculated by the procedure in the last section, in which the LS coupling is taken as a perturbation. In the case of Cu^{2+} having $S = 1/2$, however, the anisotropy energy does not arise from this process, but from the anisotropic interaction between magnetic ions.

3.5 Case with Unquenched Orbital Angular Momentum

Let us consider next the case in which the ground state is either Γ_5 or Γ_4 split by the cubic crystal field. This applies to Fe^{2+} and Co^{2+} located at the center of an octahedron of anions. Finite matrix elements of the orbital angular momentum are present within the subspace of triply degenerate Γ_5 and Γ_4. The matrix elements of the components L_x, L_y, L_z of the orbital angular momentum are calculated, for $\Gamma_5(L=2)$ in (3.23a) and for $\Gamma_4(L=3)$ in (3.23b), as

$$(L_z) = \rho \begin{pmatrix} 0 & \mathrm{i} & 0 \\ -\mathrm{i} & 0 & 0 \\ 0 & 0 & 0 \end{pmatrix}, \quad (L_x) = \rho \begin{pmatrix} 0 & 0 & 0 \\ 0 & 0 & \mathrm{i} \\ 0 & -\mathrm{i} & 0 \end{pmatrix},$$

$$(L_y) = \rho \begin{pmatrix} 0 & 0 & -\mathrm{i} \\ 0 & 0 & 0 \\ \mathrm{i} & 0 & 0 \end{pmatrix}, \tag{3.63}$$

where the value of ρ is 1 for Γ_5 and 3/2 for Γ_4. Let us remember that the angular momentum $\boldsymbol{l}$ with the magnitude 1 has, except for the reversed sign, the same matrix elements among three eigenfunctions, $\varphi_x = -(1/\sqrt{2})(\varphi_1 - \varphi_{-1})$, $\varphi_y = -(1/\sqrt{2}\mathrm{i})(\varphi_1 + \varphi_{-1})$, $\varphi_z = \varphi_0$, as in (3.63). Therefore (3.63) is equivalent to $-\rho\boldsymbol{l}$ with angular momentum $l = 1$; this means that the LS coupling in the Γ_5 or Γ_4 subspace is equivalent to $-\rho\lambda\boldsymbol{l}\cdot\boldsymbol{S}$.

The lattice-orbital Jahn-Teller interaction in (3.43, 44) can be expressed, with the fictitious angular momentum $\boldsymbol{l}$ of magnitude 1, as

$$\begin{aligned} V = &-\sqrt{\frac{2}{3}}a\tau\alpha\langle r^2\rangle\Big\{\frac{\sqrt{3}}{2}(l_x^2 - l_y^2)q_2 + \frac{1}{2}[3l_z^2 - l(l+1)]q_3\Big\} \\ &+\sqrt{\frac{2}{3}}b\tau\alpha\langle r^2\rangle\Big[\frac{\sqrt{3}}{2}(l_yl_z + l_zl_y)q_4 + \frac{\sqrt{3}}{2}(l_zl_x + l_xl_z)q_5 \\ &\qquad + \frac{\sqrt{3}}{2}(l_xl_y + l_yl_x)q_6\Big] . \end{aligned} \tag{3.64}$$

Here τ is a constant different from ρ: $\tau = -3$ for $\Gamma_5(L = 2)$ and $\tau = 6$ for $\Gamma_4(L = 3)$. If we assume for simplicity that q_4, q_5 and q_6 are all equal, the effective Hamiltonian for the magnetic ion including the LS coupling can be expressed as

$$\begin{aligned} \mathcal{H} = &-\rho\lambda\boldsymbol{l}\cdot\boldsymbol{S} + D\big[l_z^2 - \frac{1}{3}l(l+1)\big] + E(l_x^2 - l_y^2) \\ &+ \frac{1}{2}F(l_yl_z + l_zl_y + l_zl_x + l_xl_z + l_xl_y + l_yl_x) . \end{aligned} \tag{3.65}$$

Here the second and third terms are the effective Hamiltonian of the crystal field for tetragonal and orthorhombic symmetries, respectively, and the fourth term is that of a trigonal crystal field with the trigonal axis along [111].

The distortion is determined by minimizing the sum of the energy eigenvalue of (3.65) and the increase of the potential energy due to the distortion. If the Jahn-Teller coupling is larger than the LS coupling, such a distortion becomes nearly independent of the LS-coupling constant $\rho\lambda$. This case applies to $CuCr_2O_4$, which has the spinel structure, in which the oxygen ions form a face-centered-cubic lattice and the interstitial sites are occupied by Cu^{2+} and Cr^{3+}. There are two interstitial sites: the center of the octahedron and the center of the tetrahedron formed by the O^{2-} ions. In $CuCr_2O_4$, the Cu^{2+} ions occupy the center of the tetrahedron. The cubic crystal field due to the tetrahedral anions has sign opposite to the octahedral one; therefore Γ_5 becomes the ground state. In this substance a tetragonal distortion ($c/a < 1$) is present even above the Curie temperature where the spins of Cu^{2+} ions are disordered. In this case, $a < 0, \alpha > 0$ and $\tau < 0$ hold; thus, if q_3 is negative, we obtain $D > 0$, resulting in a nondegenerate ground state. Therefore we have a contraction along the c axis. A detailed account of this problem is given in the papers of *Kanamori* [3.10], and *Dunitz* and *Orgel* [3.15].

It is exceptional that the Jahn-Teller coupling is much larger than the LS coupling as in $CuCr_2O_4$; in most cases the LS coupling is larger than the Jahn-Teller coupling. In the latter case, the fictitious angular momentum $\boldsymbol{l}$ should be coupled first with $\boldsymbol{S}$ and the sum

$$\boldsymbol{j} = \boldsymbol{l} + \boldsymbol{S} \tag{3.66}$$

becomes a good quantum number. Since $\lambda < 0$ for Fe^{2+} and Co^{2+}, the state with the smallest $\boldsymbol{j}$ is the ground state; the magnetic moment of the ion is, as in the case of rare-earth ions, proportional to $\boldsymbol{j}$. If the direction of $\boldsymbol{j}$ is forced to align ferromagnetically or antiferromagnetically either by a strong external field or by the exchange interaction between the two $\boldsymbol{j}$'s of neighboring magnetic ions, the Jahn-Teller coupling together with aligned moments tends to distort the crystal. The origin of the large magnetostriction in the NaCl-type antiferromagnets CoO and FeO is believed due to this effect [3.16].

In some cases, the octahedron of anions surrounding a magnetic ion is distorted by the symmetry of the crystal, or the crystal field due to the ions beyond the nearest neighbors has low symmetry. In such a case, since the low-symmetry crystal field is much smaller than the cubic field, the effective Hamiltonian for the magnetic ion is given by (3.65), in which the parameters D, E, F are determined by the low-symmetry crystal field. There are two possibilities in this case: in the first, the crystal field is larger than the LS coupling, and in the other the two are comparable. Typical examples of the former case are the Co^{2+} and Fe^{2+} ions, which are surrounded by the octahedron of O^{2-} ions, in the spinel-structure ferrite $CoFe_2O_4$, the magnetite Fe_3O_4, and CoF_2. The Fe^{2+} ions in Fe_3O_4 are subject, in addition to the cubic field, to the trigonal crystal field along [111] and other equivalent directions. The ground state due to this crystal field is $l_\zeta = 0$ and nondegenerate; therefore the treatment for the nondegenerate ground state applies to this case. However, for the Co^{2+} ions in the same-type material $CoFe_2O_4$ the sign of α is the same, but the sign of τ in (3.64) is opposite; therefore the doublet $l_\zeta = \pm 1$ becomes lower. Because of the LS coupling we have $S_\zeta = -3/2$ for $l_\zeta = 1$ and $S_\zeta = 3/2$ for $l_\zeta = -1$. As a result, a large energy is needed to tilt the spin from the ζ axis. This is the reason why the Co-ferrite has a large magnetic anisotropy compared with other ferrites [3.17, 18].

An example of the case in which the LS coupling and the low-symmetry crystal field are comparable is found in Co-Tutton salt $K_2Co(SO_4)_2{\cdot}6H_2O$. In this case the Hamiltonian must be treated without any approximation, the result being 6 Kramers doublets for Co^{2+}. If the lowest state is described by a fictitious spin with magnitude 1/2, the effective g value deviates appreciably from 2 because of a contribution from the orbital magnetic moment to the total magnetic moment [3.19].

4. Exchange Interactions

In ionic crystals which have magnetic ions as cations, the spin moments of the ions order at low temperatures antiparallel or parallel to each other; typical examples are MnO and MnF_2. These ordered states are called antiferromagnetic or ferromagnetic, respectively. The best known ferromagnets are the metals Fe, Co and Ni. The energy levels of the $3d$-electrons which give rise to the spin magnetic moments in these metals form a band, like those of the conduction electrons. As a result, a Fermi surface exists in the momentum distribution of the $3d$-electrons, which take part in the metallic cohesion. Because of this fact, we cannot treat the $3d$-electrons simply as localized spins. Therefore we omit here the iron-group ferromagnets from our consideration; we discuss them in another chapter.

In this chapter we consider the interactions acting between the spins of the magnetic ions in the ionic crystals in which electrons are localized. These interactions are electric in their origin and usually much stronger than the magnetic-dipole interaction.

4.1 Direct Exchange Interaction

Now it is well known that the antiferromagnetism or ferromagnetism in ionic crystals originates from the interaction between spins of magnetic ions; the interaction is written as

$$-2J\boldsymbol{S}_i \cdot \boldsymbol{S}_j \; . \tag{4.1}$$

An interaction of this type is called the *exchange interaction*; the coupling constant J is the *exchange integral*. The exchange interaction originates from a quantum exchange term of the Coulomb interaction between d electrons on neighboring ions so that it has the coupling strength of the Coulomb interaction, as discussed by *Heisenberg* [4.1] in 1928.

Denoting the wave function of the d-electron as $\psi(\boldsymbol{r})$, we can write the expectation value of the Coulomb interaction e^2/r_{12} as

$$\mathcal{H}_{\text{Coulomb}} = \frac{1}{2}\iint \psi^*(\boldsymbol{r}_1)\psi^*(\boldsymbol{r}_2)\frac{e^2}{r_{12}}\psi(\boldsymbol{r}_2)\psi(\boldsymbol{r}_1)d\tau_1 d\tau_2 \; . \tag{4.2}$$

Here the integral over $d\tau$ means both the integral with respect to the position of the d electron and the sum over the spin state ($s_z = \pm 1/2$). If we regard $\psi(\boldsymbol{r})$ as a Fermi field operator for the electron, this term gives an interaction part in the second-quantized Hamiltonian. In this case, $\psi^*(\boldsymbol{r})$ becomes the Fermi operator $\psi^\dagger(\boldsymbol{r})$ which creates an electron at position $\boldsymbol{r}$, and $\psi(\boldsymbol{r})$ becomes the annihilation operator. We expand the field $\psi(\boldsymbol{r})$ in terms of the orthogonal wave functions $\phi_{nm}(\boldsymbol{r})v_s$ localized at magnetic ions as

$$\psi(\boldsymbol{r}) = \sum_{nms} a_{nms}\phi_{nm}(\boldsymbol{r})v_s \ . \tag{4.3}$$

Here the $\phi_{nm}(\boldsymbol{r})$ are orthogonal to each other: ϕ_{nm} means the d-orbital function which occupies the nth lattice point and has the component of angular momentum m. The spin eigenfunction v_s represents the up- or down-spin state and has the following two components:

$$\begin{aligned} v_\uparrow\left(\frac{1}{2}\right) &= 1 \ , \qquad v_\downarrow\left(\frac{1}{2}\right) = 0 \ , \\ v_\uparrow\left(-\frac{1}{2}\right) &= 0 \ , \qquad v_\downarrow\left(-\frac{1}{2}\right) = 1 \ . \end{aligned} \tag{4.4}$$

Substituting (4.3) into (4.2) and performing the integration, we obtain

$$\begin{aligned} \mathcal{H}_{\text{Coulomb}} &= \frac{1}{2}\sum_{nms}\left\langle \boldsymbol{n}_1 m_1, \boldsymbol{n}_2 m_2 \left| \frac{e^2}{r_{12}} \right| \boldsymbol{n}_3 m_3, \boldsymbol{n}_4 m_4 \right\rangle \\ &\quad \times a^+_{\boldsymbol{n}_1 m_1 s_1} a^+_{\boldsymbol{n}_2 m_2 s_2} a_{\boldsymbol{n}_3 m_3 s_2} a_{\boldsymbol{n}_4 m_4 s_1} \ . \end{aligned} \tag{4.5}$$

Here $a^\dagger_{nms}$ and a_{nms} are Fermi operators which satisfy the commutation relations (3.2). Thus (4.2, 5) are the Hamiltonian of the Coulomb interaction in the second-quantized form.

The term in which all $\boldsymbol{n}$ in (4.5) are equal corresponds to the Coulomb interaction between d-electrons belonging to the same magnetic ion; this term is the Hund coupling. For simplicity, we assume nondegenerate d-orbitals and represent the orbitals by a single function; since $a_{n\uparrow}a_{n\uparrow}$ and $a^\dagger_{n\uparrow}a^\dagger_{n\uparrow}$ are zero, the term thus obtained gives the second term in (3.1). In the first approximation it can be considered that electrons in ionic crystals are localized at each ion. Assuming that one electron is localized on each magnetic ion, let us consider the expectation value of (4.5) in the ground state. If we assume a nondegenerate orbital, then a finite expectation value is obtained only when $\boldsymbol{n}_1 = \boldsymbol{n}_4$ and $\boldsymbol{n}_2 = \boldsymbol{n}_3$ or $\boldsymbol{n}_1 = \boldsymbol{n}_3$ and $\boldsymbol{n}_2 = \boldsymbol{n}_4$ are satisfied. The first term is

$$\left\langle \boldsymbol{n}_1\boldsymbol{n}_2 \left| \frac{e^2}{r_{12}} \right| \boldsymbol{n}_2\boldsymbol{n}_1 \right\rangle \sum_{s_1 s_2} a^\dagger_{\boldsymbol{n}_1 s_1} a_{\boldsymbol{n}_1 s_1} a^\dagger_{\boldsymbol{n}_2 s_2} a_{\boldsymbol{n}_2 s_2} \ , \tag{4.6}$$

and the second term is

$$-\left\langle n_1n_2\left|\frac{e^2}{r_{12}}\right|n_1n_2\right\rangle\sum_{s_1s_2}a^+_{n_1s_1}a_{n_1s_2}a^\dagger_{n_2s_2}a_{n_2s_1} \ . \tag{4.7}$$

The first term gives the usual Coulomb interaction between two electrons localized at $\boldsymbol{n}_1$ and $\boldsymbol{n}_2$; the second term is a quantum effect due to the property of Fermi operators. The diagonal part of this term with respect to the spin suffix ($s_1 = s_2$) lowers the energy between two electrons with parallel spins by $\langle \boldsymbol{n}_1\boldsymbol{n}_2|e^2/r_{12}|\boldsymbol{n}_1\boldsymbol{n}_2\rangle$ compared with that between two electrons with antiparallel spins. This integral is called the exchange integral and usually written as $J_{n_1n_2}$, since it corresponds to the term with the second indices $\boldsymbol{n}_1$ and $\boldsymbol{n}_2$ exchanged in the first Coulomb term. We rewrite the exchange term (4.7) as

$$\begin{aligned}-J_{n_1n_2}\Big[&\frac{1}{2}(a^\dagger_{n_1\uparrow}a_{n_1\uparrow}+a^\dagger_{n_1\downarrow}a_{n_1\downarrow})(a^\dagger_{n_2\uparrow}a_{n_2\uparrow}+a^\dagger_{n_2\downarrow}a_{n_2\downarrow})\\ &+\frac{1}{2}(a^\dagger_{n_1\uparrow}a_{n_1\uparrow}-a^\dagger_{n_1\downarrow}a_{n_1\downarrow})(a^\dagger_{n_2\uparrow}a_{n_2\uparrow}-a^\dagger_{n_2\downarrow}a_{n_2\downarrow})\\ &+a^\dagger_{n_1\uparrow}a_{n_1\downarrow}a^+_{n_2\downarrow}a_{n_2\uparrow}+a^\dagger_{n_1\downarrow}a_{n_1\uparrow}a^\dagger_{n_2\uparrow}a_{n_2\downarrow}\Big]\end{aligned} \tag{4.8}$$

and use the following relations between the spin operators s_z, s_x, s_y and the Fermi operators $a^\dagger_{n\uparrow}, a^\dagger_{n\downarrow}, a_{n\uparrow}, a_{n\downarrow}$:

$$\begin{aligned}s_{nz} &= \frac{1}{2}(a^\dagger_{n\uparrow}a_{n\uparrow}-a^\dagger_{n\downarrow}a_{n\downarrow}) \ ,\\ s_{n+} &= s_{nx}+\mathrm{i}s_{ny}=a^\dagger_{n\uparrow}a_{n\downarrow} \ , \quad s_{n-}=s_{nx}-\mathrm{i}s_{ny}=a^\dagger_{n\downarrow}a_{n\uparrow} \ .\end{aligned} \tag{4.9}$$

Then, noting that the electron numbers at sites $\boldsymbol{n}_1$ and $\boldsymbol{n}_2$ are unity, we can write the average of the Coulomb energy in the ground state between the two electrons at sites $\boldsymbol{n}_1$ and $\boldsymbol{n}_2$ as

$$K_{n_1n_2}-J_{n_1n_2}\left(\frac{1}{2}+2\boldsymbol{s}_{n_1}\cdot\boldsymbol{s}_{n_2}\right) , \tag{4.10}$$

where $K_{n_1n_2}$ is the Coulomb integral in (4.6). The spin function $\frac{1}{2}(1+4\boldsymbol{s}_{n_1}\cdot\boldsymbol{s}_{n_2})$ in (4.10) is the operator P^σ_{12} which exchanges two spin variables. The exchange interaction in the second term is called the *direct exchange interaction*, which Heisenberg supposed to be the origin of the ferromagnetism, as stated above. Heisenberg discussed the exchange interaction on the basis of the Heitler-London model, which will be described later. In the Heitler-London model the discussion is more complicated, since the atomic orbitals are used as the localized d-orbitals. However, there are very few materials (CrO_2 and $CrBr_3$ are such examples) in which the ferromagnetism arises from the direct exchange interaction proposed by Heisenberg.

The integral $J_{n_1n_2}$ is always positive, as we show below. In the expression

$$J_{n_1n_2}=\int\phi^*_{n_1}(\boldsymbol{r}_1)\phi^*_{n_2}(\boldsymbol{r}_2)\frac{e^2}{r_{12}}\phi_{n_1}(\boldsymbol{r}_2)\phi_{n_2}(\boldsymbol{r}_1)d\tau_1d\tau_2 \ , \tag{4.11}$$

we expand $1/r$ in a Fourier series as

$$\frac{e^2}{r_{12}} = \frac{1}{V}\sum_k \frac{4\pi e^2}{k^2} \mathrm{e}^{\mathrm{i}\boldsymbol{k}\cdot(\boldsymbol{r}_1-\boldsymbol{r}_2)} , \tag{4.12}$$

and rewrite the exchange integral as

$$J_{n_1 n_2} = \frac{1}{V}\sum_k \frac{4\pi e^2}{k^2} \int \phi^*_{n_1}(\boldsymbol{r}_1)\phi_{n_2}(\boldsymbol{r}_1)\mathrm{e}^{\mathrm{i}\boldsymbol{k}\cdot\boldsymbol{r}_1} d\tau_1 \times \int \phi^*_{n_2}(\boldsymbol{r}_2)\phi_{n_1}(\boldsymbol{r}_2)\mathrm{e}^{-\mathrm{i}\boldsymbol{k}\cdot\boldsymbol{r}_2} d\tau_2 . \tag{4.13}$$

which is clearly positive.

In the above we treated $\phi_{n_1}(\boldsymbol{r})$ and $\phi_{n_2}(\boldsymbol{r})$ as localized orbitals orthogonal to each other. This means that we assumed for $\phi_{n_1}(\boldsymbol{r})$ and $\phi_{n_2}(\boldsymbol{r})$ Wannier functions, which are given by linear combinations of the eigenfunctions of the Hamiltonian possessing a crystal potential only, i.e., the Bloch functions. We discuss later the Heitler-London model in which nonorthogonal orbitals of free atoms are used; the treatment using the set of orthogonal functions is very convenient for understanding the physical meaning, while it may be not convenient for explicit calculations.

4.2 Exchange Interaction Due to the Second-Order Perturbation – Superexchange Interaction

The direct exchange interaction was derived as the expectation value of the Coulomb interaction in the ground state of the localized electron system, but spin interactions of the same type arise from the second-order perturbation. We assume single d-orbitals, retain only the Coulomb interaction between the electrons on the same ion, (which is dominant), and omit the interactions on different ions. By this procedure we obtain the Hubbard Hamiltonian given by (3.1) as that in the localized d-electron system. As stated before, it is the repulsive Coulomb interaction U on the same ions (given by the second term) which hinders the itinerant motion of d-electrons in the crystal and keeps them in localized states. When this Coulomb interaction is large compared to the transfer matrix element $b_{\boldsymbol{n}}$ between electrons at neighboring sites, the system becomes insulating. Therefore, in this case, the Coulomb energy U is included in the unperturbed Hamiltonian and the kinetic energy given by the first term (except b_0) is treated as a perturbation.

In the perturbation expansion for the energy, the first-order term is the sum of Nb_0 and (4.10) obtained in the previous section, since one d-electron exists at each ion in the ground state. On the other hand, the second-order term is given by the process where an electron at ion $\boldsymbol{n}$ transfers to a neighboring ion $\boldsymbol{n}'$ and after that one of the two electrons at $\boldsymbol{n}'$ returns to the original ion $\boldsymbol{n}$. By this second-order process, d-electrons situated at $\boldsymbol{n}$ and $\boldsymbol{n}'$ exchange positions with each other. However, only when the spins at $\boldsymbol{n}$ and $\boldsymbol{n}'$ are antiparallel to each other, this type of processes is possible; the process for the parallel spins is prohibited by the Pauli principle. The second-order term for this process is given by

$$\Delta E_2 = -\sum_{\substack{nn'\\ ss'}} \frac{|b_{n'-n}|^2}{U} a^\dagger_{ns'} a_{n's'} a^\dagger_{n's} a_{ns} \ . \tag{4.14}$$

We rewrite the summation over the spins as follows :

$$\begin{aligned}
&\sum_{ss'} a^\dagger_{ns'} a_{n's'} a^\dagger_{n's} a_{ns} \\
&\quad = (a^\dagger_{n\uparrow} a_{n'\uparrow} + a^\dagger_{n\downarrow} a_{n'\downarrow})(a^\dagger_{n'\uparrow} a_{n\uparrow} + a^\dagger_{n'\downarrow} a_{n\downarrow}) \\
&\quad = (a^\dagger_{n\uparrow} a_{n\uparrow} + a^\dagger_{n\downarrow} a_{n\downarrow}) - \frac{1}{2}(a^\dagger_{n\uparrow} a_{n\uparrow} + a^\dagger_{n\downarrow} a_{n\downarrow})(a^\dagger_{n'\uparrow} a_{n'\uparrow} + a^\dagger_{n'\downarrow} a_{n'\downarrow}) \\
&\qquad - \frac{1}{2}(a^\dagger_{n\uparrow} a_{n\uparrow} - a^\dagger_{n\downarrow} a_{n\downarrow})(a^\dagger_{n'\uparrow} a_{n'\uparrow} - a^\dagger_{n'\downarrow} a_{n'\downarrow}) - a^\dagger_{n\uparrow} a_{n\downarrow} a^\dagger_{n'\downarrow} a_{n'\uparrow} \\
&\qquad - a^\dagger_{n\downarrow} a_{n\uparrow} a^\dagger_{n'\uparrow} a_{n'\downarrow} \ .
\end{aligned} \tag{4.15}$$

Then, using the relation (4.9) and $a^\dagger_{n\uparrow} a_{n\uparrow} + a^\dagger_{n\downarrow} a_{n\downarrow} = 1$, we obtain from (4.14)

$$\Delta E_2 = -\sum_{nn'} \frac{|b_{n'-n}|^2}{U} \left(\frac{1}{2} - 2\boldsymbol{s}_n \cdot \boldsymbol{s}_{n'} \right) . \tag{4.16}$$

This quantity vanishes for $\boldsymbol{s}_n$ and $\boldsymbol{s}_{n'}$ parallel to each other, as it should. This energy stabilizes the antiparallel spin pair. This type of exchange interaction between localized spins was called the *kinetic exchange* by *Anderson* [4.2]. On the other hand, the direct exchange interaction is called the *potential exchange.*

In fourth-order, the following interaction arises:

$$\Delta E_4 \to -\frac{|b_{n'-n}|^4}{U^3} (\boldsymbol{s}_n \cdot \boldsymbol{s}_{n'})^2 \ . \tag{4.17}$$

This term, which is smaller than (4.16) by the factor $(|b_{n'-n}|/U)^2$, is called *biquadratic exchange*; the existence of this term has been confirmed in materials such as MnO [4.3]. Among the other interactions in fourth-order exists the four-spin exchange interaction which arises from the process $1 \to 2 \to 3 \to 4 \to 1$:

$$K[(\boldsymbol{s}_1 \cdot \boldsymbol{s}_2)(\boldsymbol{s}_3 \cdot \boldsymbol{s}_4) + (\boldsymbol{s}_1 \cdot \boldsymbol{s}_4)(\boldsymbol{s}_3 \cdot \boldsymbol{s}_2) - (\boldsymbol{s}_1 \cdot \boldsymbol{s}_3)(\boldsymbol{s}_2 \cdot \boldsymbol{s}_4)] \ . \tag{4.18}$$

Such higher-order exchange interactions play an important role in the spin structure of antiferromagnets which are supposed to be insulating, but near the transition point to a metallic state, like NiS_2 [4.4, 5]. Moreover, the special antiferromagnetic spin structure observed below 1mK in the nuclear spin system of solid ^{3}He in the body centered cubic lattice arises from the 4-spin exchange interaction (4.18).

The exchange interaction given by (4.16, 10) can be extended to the cases possessing the general number of electrons; for pair $(\boldsymbol{n}_1, \boldsymbol{n}_2)$ we obtain

$$E_{\text{direct}} = -2 \sum_{mm'} J_{\boldsymbol{n}_1\boldsymbol{n}_2}^{mm'} \boldsymbol{s}_{\boldsymbol{n}_1}^m \cdot \boldsymbol{s}_{\boldsymbol{n}_2}^{m'} , \tag{4.19}$$

$$E_{\text{kin}} = +4 \sum_{mm'} \frac{|b_{\boldsymbol{n}_1\boldsymbol{n}_2}^{mm'}|^2}{U} \boldsymbol{s}_{\boldsymbol{n}_1}^m \cdot \boldsymbol{s}_{\boldsymbol{n}_2}^{m'} . \tag{4.20}$$

Here $b_{\boldsymbol{n}'\boldsymbol{n}}^{m'm}$ is the transfer matrix element due to the crystal potential between $\phi_{\boldsymbol{n}'m'}(\boldsymbol{r})$ and $\phi_{\boldsymbol{n}m}(\boldsymbol{r})$ and $J_{\boldsymbol{n}'\boldsymbol{n}}^{m'm}$ is the exchange integral between them; they are given respectively by

$$b_{\boldsymbol{n}'\boldsymbol{n}}^{m'm} = \int \phi_{m'}^*(\boldsymbol{r} - \boldsymbol{n}')\mathcal{H}_{\text{cryst}}\phi_m(\boldsymbol{r} - \boldsymbol{n})d\tau , \tag{4.21}$$

$$J_{\boldsymbol{n}'\boldsymbol{n}}^{m'm} = \int \phi_{m'}^*(\boldsymbol{r}_1 - \boldsymbol{n}')\phi_m^*(\boldsymbol{r}_2 - \boldsymbol{n})\frac{e^2}{r_{12}}\phi_m(\boldsymbol{r}_1 - \boldsymbol{n})\phi_{m'}(\boldsymbol{r}_2 - \boldsymbol{n}')d\tau_1 d\tau_2 . \tag{4.22}$$

The Coulomb interaction on a single ion having n d-electrons is given by the term in which all $\boldsymbol{n}$ are equal in (4.5). Retaining the largest Coulomb terms among these, which is given by putting $m_1 = m_4$ and $m_2 = m_3$, and ignoring the orbital dependence of the Coulomb integral, we approximate them by the same value U. In this case, the energy denominator of the second-order perturbation is the excitation energy to the state in which $n-1$ d-electrons are present on the ion $\boldsymbol{n}$ and $n+1$ d-electrons are on the ion $\boldsymbol{n}'$; it is given by

$$\Delta E = U\left[\frac{n(n+1)}{2} + \frac{(n-1)(n-2)}{2} - 2\frac{n(n-1)}{2}\right] = U . \tag{4.23}$$

This value is the same as that in the case with one electron per ion. When the total spin of the magnetic ion is determined as $\boldsymbol{S}$ by Hund's first rule, we put

$$\boldsymbol{s}^m = \frac{1}{2S}\boldsymbol{S} , \tag{4.24}$$

in (4.19, 20). Then we can write the exchange interaction in the form

$$E_{\text{exch}} = -2J_{\boldsymbol{n}_1\boldsymbol{n}_2}\boldsymbol{S}_{\boldsymbol{n}_1} \cdot \boldsymbol{S}_{\boldsymbol{n}_2} , \tag{4.25}$$

where the effective exchange integral is given by

$$J_{\boldsymbol{n}_1\boldsymbol{n}_2} = -\frac{1}{(2S)^2} \sum_{mm'} \left(\frac{2|b_{\boldsymbol{n}_1\boldsymbol{n}_2}^{mm'}|^2}{U} - J_{\boldsymbol{n}_1\boldsymbol{n}_2}^{mm'}\right) . \tag{4.26}$$

Here the summation over m and m' is taken for the occupied orbitals, when less-than-half orbitals in the incomplete shell are occupied; when more-than-half orbitals are occupied, the summation is taken over the unoccupied orbitals by considering that the unoccupied orbitals possess the spin given by (4.24). In this procedure, the orbitals to be summed over should be chosen properly, according to the ground state of the magnetic ion. For the case with degenerate orbitals in the ground state, the calculation becomes complicated. The cases where the ground state has no orbital degeneracy are $d^3(Cr^{3+}), d^5(Mn^{2+})$ and $d^8(Ni^{2+})$. For d^3, we choose $d\varepsilon$ orbitals (xy, yz and zx) as m; for d^5 $d\gamma$ orbitals, ($x^2 - y^2$ and $3z^2 - r^2$) should be added to these, and for d^8, two $d\gamma$ orbitals ($x^2 - y^2$ and $3z^2 - r^2$) should be chosen [4.2]. In the iron-group oxides and fluorides, generally the antiferromagnetic states are realized; it means that in these materials the first term in (4.26) (the superexchange interaction) dominates the second term (the direct exchange interaction). It is rather difficult, however, to evaluate quantitatively the terms for real magnetic materials.

One important feature of the superexchange interaction is that it vanishes identically in some cases, since the transfer matrix $b_{n_1 n_2}^{mm'}$ depends not only on the distance between the magnetic ions n_1 and n_2 but also the symmetry of the orbitals m and m'. Because of this fact, the superexchange interaction depends on the relative positions of two magnetic ions with respect to the anion situated between them. The values of the effective exchange interaction are different in the following two cases: in one case two magnetic ions are located on the line at the opposite positions with respect to anion, in another case the two directions connecting two magnetic ions and the anion are perpendicular to each other. In general, when two magnetic ions and the anion between them are on the same line, i.e., in the 180° coupling, the absolute value of the effective exchange interaction is large.

The exchange interaction between two magnetic ions in magnetic compounds arises via anions situated between the ions and is called the superexchange interaction. It was proposed by *Kramers* [4.6] in 1934 and details of the mechanism were studied by many researchers: *Anderson* [4.7], *Goodenough* [4.8], *Kanamori* [4.9], *Wollan* et al. [4.10], J. Yamashita, J. Kondo and J.C. Slater. Goodenough, Kanamori and Wollan introduced into the theory the symmetry of d-orbitals. Subsequently, *Anderson* [4.2] discussed the superexchange interaction on the basis of the Mott insulator; he simplified and unified various mechanisms giving rise to the superexchange interaction into two types: the direct exchange interaction and the kinetic exchange due to electron transfer. The description in this section is based on the viewpoint by Anderson.

4.3 Anisotropic Exchange Interaction

The spin Hamiltonian of one-ion anisotropy, which was discussed in Sect. 3.3, is the energy which arises from the fact that the orbital moments induced by the LS coupling depend on the direction of the spin $\boldsymbol{S}$ with respect to the crystal axes. The same effect adds to the isotropic exchange interaction an anisotropic part. Since the one-ion anisotropy Hamiltonian vanishes identically for $S = 1/2$ ions such as Cu^{2+}, the anisotropic exchange interaction plays an important role in the magnetic anisotropy of antiferromagnetic cuprates.

Let us consider two magnetic ions, in which there is no orbital degeneracy in the ground state and the orbital moments vanish. We treat the sum of the LS coupling terms on the two magnetic ions and the exchange interaction V_{exch} as the perturbation Hamiltonian

$$\mathcal{H}' = \lambda(\boldsymbol{L}_1 \cdot \boldsymbol{S}_1) + \lambda(\boldsymbol{L}_2 \cdot \boldsymbol{S}_2) + V_{\text{exch}} \ . \tag{4.27}$$

We consider the following third-order process: first the $\boldsymbol{L}_1 \cdot \boldsymbol{S}_1$ term excites the ion 1 to the state ψ_{n_1}, then the exchange interaction acts between the ion 1 (in the excited state) and the ion 2, and finally the $\boldsymbol{L}_1 \cdot \boldsymbol{S}_1$ term returns the ion 1 to its ground state. By this process the following energy is derived:

$$\Delta E_3 = -\sum_{\mu\nu} \left[S_{1\mu} \Gamma^{(1)}_{\mu\nu} (\boldsymbol{S}_1 \cdot \boldsymbol{S}_2) S_{1\nu} + S_{2\mu} \Gamma^{(2)}_{\mu\nu} (\boldsymbol{S}_1 \cdot \boldsymbol{S}_2) S_{2\nu} \right] , \tag{4.28}$$

$$\Gamma^{(1)}_{\mu\nu} = 2\lambda^2 \sum_{n_1 n_1'} \frac{\langle g_1 | L_\mu | n_1 \rangle J(n_1 g_2, n_1' g_2) \langle n_1' | L_\nu | g_1 \rangle}{(E_{n_1} - E_{g_1})(E_{n_1'} - E_{g_1})} \ . \tag{4.29}$$

$J(n_1 g_2, n_1' g_2)$ is the exchange integral between ψ_{n_1} for the ion 1 and the ground state for the ion 2; however, since the off-diagonal terms for the states of the ion 1 are present generally, we include them in (4.29). Since $\Gamma^{(1,2)}_{\mu\nu}$, in contrast to $\Lambda_{\mu\nu}$ in (3.33), contains the factor $J(n_1 g_2, n_1 g_2)$ of the exchange interaction, the anisotropy for a spin pair exists generally, even when the crystal field has cubic symmetry. If $S = 1/2$ is assumed in (4.28), the spin Hamiltonian can be written as

$$\mathcal{H}_{\text{aniso}} = -\frac{1}{4} \sum_{\mu\nu} \sum_{i=1,2} \left[(\Gamma^{(i)}_{\mu\nu} + \Gamma^{(i)}_{\nu\mu}) - \delta_{\mu\nu} (\Gamma^{(i)}_{xx} + \Gamma^{(i)}_{yy} + \Gamma^{(i)}_{zz}) \right] S_{1\mu} S_{2\nu} \, . \tag{4.30}$$

This term can be considered as an anisotropic exchange interaction and also as a generalization of the usual magnetic-dipole interaction; thus it is called the *pseudo-dipole interaction.* If we assume that $J(n_1 g_2, n_1 g_2)$ is of the same order as the exchange integral in the ground state, the order of the interaction is estimated as

$$\mathcal{H}_{\text{aniso}} \sim \frac{\lambda^2 J}{(\Delta E)^2} \sim (\Delta g)^2 J \ . \tag{4.31}$$

Here Δg is the shift of the g-value from 2 ($\sim$0.2) and ΔE is the energy splitting ($10^4 \sim 10^3 \mathrm{cm}^{-1}$). In $CuCl_2 \cdot 2H_2O$, the anisotropic exchange interaction and the magnetic-dipole interaction contribute to the magnetic anisotropy [4.11]. On the other hand, in CoF_2 and $CoCl_2 \cdot 6H_2O$ the energy splittings between the Kramers doublets are large enough for us to retain only the lowest Kramers doublet [4.12]. This doublet can be described by a fictitious spin of magnitude 1/2. The exchange interaction for this case is obtained by projecting the original isotropic exchange interaction onto the two-dimensional subspace. The projected exchange interaction can be written as an anisotropic exchange interaction in terms of fictitious spins and gives rise to the magnetic anisotropy.

In the above discussion of the exchange interaction, we considered mainly its diagonal elements with respect to the orbital states, and included the off-diagonal terms only in the terms given by (4.29). If we include all the off-diagonal terms, then processes other than those described above are possible. For example, there is a process in which two ions are simultaneously excited by the exchange interaction $J(g_1g_2, n_1n_2)$ and then the LS coupling returns each ion to its ground state. However, it is considered that the term in (4.28) which includes the diagonal elements is the dominant one.

Taking into account the off-diagonal terms of the exchange interaction with respect to the orbital states, we obtain, as the second-order term in the perturbation,

$$\begin{aligned}\mathcal{H}_{\mathrm{DM}} = & \\ -\lambda\Bigg(\sum_{n_1} & \frac{\langle g_1|\boldsymbol{L}_1\cdot\boldsymbol{S}_1|n_1\rangle\langle n_1g_2|V_{\mathrm{exch}}|g_1g_2\rangle+\langle g_1g_2|V_{\mathrm{exch}}|n_1g_2\rangle\langle n_1|\boldsymbol{L}_1\cdot\boldsymbol{S}_1|g_1\rangle}{E_{n_1}-E_{g_1}} \\ & + \text{ term with 1 and 2 exchanged}\Bigg) \; . \qquad (4.32)\end{aligned}$$

Here we have used once the exchange interaction and the LS coupling. This can be rewritten as

$$\begin{aligned}\mathcal{H}_{\mathrm{DM}} = 2\lambda\sum_{\mu}\Bigg\{ & \sum_{n_1}\frac{J(n_1g_2, g_1g_2)\langle g_1|L_{1\mu}|n_1\rangle[S_{1\mu}, (\boldsymbol{S}_1\cdot\boldsymbol{S}_2)]}{E_{n_1}-E_{g_1}} \\ & +\sum_{n_2}\frac{J(g_1n_2, g_1g_2)\langle g_2|L_{2\mu}|n_2\rangle[S_{2\mu}, (\boldsymbol{S}_1\cdot\boldsymbol{S}_2)]}{E_{n_2}-E_{g_2}}\Bigg\} \; , \qquad (4.33)\end{aligned}$$

where μ represents x, y and z; the fact that the matrix element of $\boldsymbol{L}_{1,2}$ is pure imaginary has been used. Using the relation

$$[\boldsymbol{S}_1, (\boldsymbol{S}_1\cdot\boldsymbol{S}_2)] = -\mathrm{i}\boldsymbol{S}_1\times\boldsymbol{S}_2 \; , \qquad (4.34)$$

we can write (4.33) as an antisymmetric interaction between the two spins:

$$\mathcal{H}_{\mathrm{DM}} = \boldsymbol{D} \cdot (\boldsymbol{S}_1 \times \boldsymbol{S}_2) \,, \tag{4.35}$$

$$\boldsymbol{D} = -2\mathrm{i}\lambda \left[\sum_{n_1} \frac{\langle g_1 | \boldsymbol{L}_1 | n_1 \rangle}{E_{n_1} - E_{g_1}} J(n_1 g_2, g_1 g_2) - \sum_{n_2} \frac{\langle g_2 | \boldsymbol{L}_2 | n_2 \rangle}{E_{n_2} - E_{g_2}} J(g_1 n_2, g_1 g_2) \right] . \tag{4.36}$$

The existence of this kind of antisymmetric interaction was pointed out by *Dzyaloshinsky* [4.13] on the basis of symmetry analysis of the crystal structure and the microscopic derivation of the interaction was done by *Moriya* [4.14]. Therefore this interaction is called the *antisymmetric exchange interaction of Dzyaloshinsky-Moriya.* The vector $\boldsymbol{D}$ in (4.36) does not vanish in the general case, though it clearly vanishes when the crystal field surrounding the magnetic ion has inversion symmetry with respect to the center between the two magnetic ions. The rules concerning the direction of the vector $\boldsymbol{D}$ (as obtained by Moriya) are as follows (we denote AB as the line segment connecting the two spins, and C as the midpoint of AB):

(i) case with mirror plane perpendicular to AB through C:
 $\boldsymbol{D} \perp \mathrm{AB}$
(ii) case with mirror plane including AB:
 $\boldsymbol{D} \perp$ mirror plane
(iii) case with two-fold rotation axis perpendicular to AB through C:
 $\boldsymbol{D} \perp$ two-fold rotation axis
(iv) case with n-fold rotation ($n > 2$) axis along AB:
 $\boldsymbol{D} \parallel \mathrm{AB}$

and others. In case (i), if we take AB to be the z-axis and C at the origin, then S_1 and S_2 transform as $S_{1x} \rightleftarrows S_{2x}$, $S_{1y} \rightleftarrows S_{2y}$ and $S_{1z} \rightleftarrows -S_{2z}$; the x and y components in (4.35) are invariant and the z-component changes the sign. Therefore D_z vanishes. In case (ii), if we assign the mirror plane including AB to the xz-plane and take mirror reflection with respect to this plane, the transformation is such as $S_{1,2x} \rightarrow S_{1,2x}$, $S_{1,2y} \rightarrow -S_{1,2y}$ and $S_{1,2z} \rightarrow S_{1,2z}$; the y component is invariant and the x and z components change signs. Thus $D_x = D_z = 0$ and only the D_y component remains. In case (iii), taking the two-fold rotation axis as the x-axis and rotating by 180° about it, we obtain the transformation: $S_{1z} \rightleftarrows -S_{2z}$, $S_{1x} \rightleftarrows S_{2x}$ and $S_{1y} \rightleftarrows -S_{2y}$; the x component changes the sign and the y and z components are invariant. Thus $D_x = 0$ and $\boldsymbol{D}$ is in the yz-plane. Similar considerations apply to case (iv). As shown here, the four conditions are the rules to determine the direction of $\boldsymbol{D}$ when (4.36) is put aside and only the interaction given by (4.35) is assumed to be present. The direction of $\boldsymbol{D}$ is uniquely determined in cases (ii) and (iv) but in cases (i) and (iii) it is not; to determine the direction of $\boldsymbol{D}$ in the latter cases, we need to inspect (4.36) in detail.

As an example, let us consider α-Fe_2O_3, an antiferromagnet with rhombohedral structure; the positions of the cations and anions in the unit cell and the directions of the spins in the ordered state are shown in Fig. 4.1.

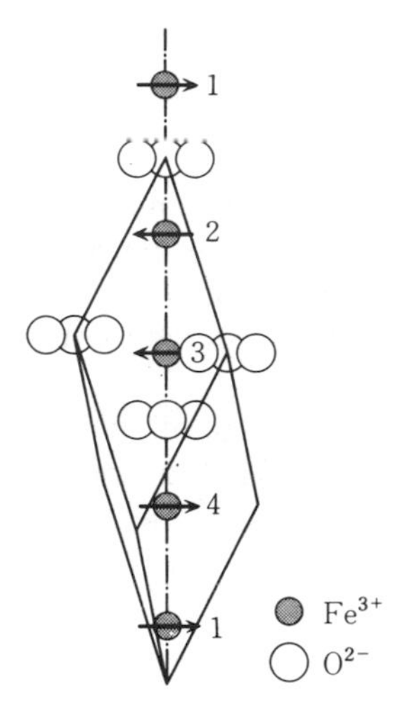

Fig. 4.1. Crystal structure and spin ordering of α-Fe_2O_3

In the temperature range below the Néel temperature T_N =950K and above the Morin temperature T_M =260K, the Fe^{3+} spins are aligned perpendicular to the c-axis; below T_M they are parallel to the c-axis. Since the midpoint between Fe^{3+} ions 1 and 4 or 2 and 3 is a center of inversion symmetry as shown in Fig. 4.1, $\boldsymbol{D}$ vanishes between these Fe^{3+} pairs. On the other hand, $\boldsymbol{D}$ does not vanish between Fe^{3+} pairs 1 and 2, 3 and 4 or 1 and 3 or 2 and 4; since the c-axis has threefold rotation symmetry, $\boldsymbol{D}$ for these pairs is parallel to the c-axis. Therefore for this spin ordering the antisymmetric interaction given by $\boldsymbol{D}_z(\boldsymbol{S}_1 \times \boldsymbol{S}_2)_z$ acts between spins on the sublattices of $\pm$ spins. By this interaction, the spins of sublattices aligned antiparallel to each other due to the exchange interaction cant by a small angle θ and possess a ferromagnetic component perpendicular to the spin axis of the antiferromagnet. The weak ferromagnetism which is observed in α-Fe_2O_3 in the temperature range between T_N and T_M and lies in the direction perpendicular to the c-axis originates from this mechanism. The canting angle θ is given by

$$\theta = \frac{D_z}{2|J|} \sim \frac{\lambda}{\Delta E} \sim \Delta g \; . \tag{4.37}$$

The same weak ferromagnetism parasitic on the antiferromagnetism is observed in $MnCO_3$, $CoCO_3$ [4.15, 16] and $RFeO_3$, $RCrO_3$ containing the rare-earth metal ions for R [4.17, 18]. The weak ferromagnetism parasitic on the antiferromagnetism occurs also when the easy axes of the one-ion anisotropy energy is orthogonal to two sublattices with $\pm$ spins. Examples are NiF_2 [4.19] and $KMnF_3$ [4.20].

4.4 Direct Exchange Interaction in Heitler-London Theory

The direct exchange interaction was derived originally in the *Heitler-London* theory [4.21] which explained the binding energy of the hydrogen molecule; Heisenberg developed the theory of ferromagnetism, keeping in mind the direct exchange interaction of Heitler-London as an exchange interaction between spins on two neighboring atoms.

Let us consider two hydrogen atoms a and b, and denote the atomic orbitals of isolated atoms as $\phi_a(\boldsymbol{r})$ and $\phi_b(\boldsymbol{r})$, respectively. The Heitler-London model assumes, as the wave function in the zeroth approximation, the linear combination of the wave functions for isolated atoms:

$$\psi_{s,a}(\boldsymbol{r}_1,\boldsymbol{r}_2)=2^{-1/2}(1\pm|\langle a|b\rangle|^2)^{-1/2}[\phi_a(\boldsymbol{r}_1)\phi_b(\boldsymbol{r}_2)\pm\phi_a(\boldsymbol{r}_2)\phi_b(\boldsymbol{r}_1)]\,. \tag{4.38}$$

Here s and a mean symmetric and antisymmetric with respect to interchange of the electron coordinates $\boldsymbol{r}_1$ and $\boldsymbol{r}_2$; on the right-hand side of (4.38) we take the $+$ sign for s and the $-$ sign for a. For the triplet state with parallel spins, whose spin function is symmetric with respect to spin interchange, the orbital wave function should be the antisymmetric function of (4.38), because of the Pauli principle. On the other hand, for the singlet state with antiparallel spins where spin wave function is antisymmetric with respect to spin interchange, the symmetric wave function of (4.38) should be taken as the orbital wave function. The prefactor in (4.38) is the normalization factor; $\langle a|b\rangle$ is the overlap integral between ϕ_a and ϕ_b:

$$\langle a|b\rangle=\int\phi_a^*(\boldsymbol{r})\phi_b(\boldsymbol{r})d\tau\,. \tag{4.39}$$

The Hamiltonian $\mathcal{H}$ of the hydrogen molecule is given by the sum of those for the isolated atoms:

$$\mathcal{H}_0=-\frac{\hbar^2}{2m}\nabla_1^2-\frac{\hbar^2}{2m}\nabla_2^2-\frac{e^2}{r_{a1}}-\frac{e^2}{r_{b2}}\,, \tag{4.40}$$

and the additional term:

$$\Delta H=\frac{e^2}{r_{12}}+\frac{e^2}{r_{ab}}-\frac{e^2}{r_{b1}}-\frac{e^2}{r_{a2}}\,. \tag{4.41}$$

Therefore the expectation values of the Hamiltonian with respect to the symmetric and antisymmetric functions are calculated as

$$\begin{aligned}\langle H\rangle_{s,a}&=\frac{\langle ab|H|ab\rangle\pm\langle ab|H|ba\rangle}{1\pm|\langle a|b\rangle|^2}\\&=2\varepsilon_{\mathrm{H}}+\frac{\langle ab|\Delta H|ab\rangle\pm\langle ab|\Delta H|ba\rangle}{1\pm|\langle a|b\rangle|^2}\,.\end{aligned} \tag{4.42}$$

Equation (4.42) gives the energies of the spin-singlet and spin-triplet states; $\varepsilon_{\rm H}$ is the energy of the $1s$ ground state for the hydrogen atom. The direct exchange integral is given by half the energy difference between the two states.

In terms of the distance r between electron and nucleus and the Bohr radius $a_{\rm H}, \phi_a(r)$ is written as $\exp(-r/a_{\rm H})$. Therefore all three integrals in the second term of the second equation of (4.42) are expressed as products of powers of R and $\exp(-2R/a_{\rm H})$ in the limit of large distance R between the hydrogen atoms. Thus the main part of energy difference between the singlet and triplet states in the limit $R \to \infty$ is given by

$$\langle H \rangle_s - \langle H \rangle_a \to 2\langle ab|\Delta H|ba\rangle \, . \tag{4.43}$$

The integral on the right-hand side is found in *Sugiura*'s famous paper [4.22] on the energy of the hydrogen molecule. Using his result and taking the leading part in the limit $R \to \infty$, we obtain

$$\langle H \rangle_s - \langle H \rangle_a \to \left[-\frac{56}{45} + \frac{4}{15}\gamma + \frac{4}{15}\log\left(\frac{R}{a_{\rm H}}\right)\right]\left(\frac{R}{a_{\rm H}}\right)^3 \exp\left(-\frac{2R}{a_{\rm H}}\right) . \tag{4.44}$$

Here we use atomic units, in which the unit of energy is 27.2 eV; γ is the Euler constant ($\gamma \simeq 0.577$). From (4.44), we see that the energy difference between the singlet and triplet states is negative up to moderately large values of R but changes to positive for large values of R ($R > 50a_{\rm H}$) due to the log term.

On the other hand, there is a general mathematical theorem that the lowest energy eigenfunction of a differential operator of Sturm-Liouville type has no node; therefore $\langle H \rangle_s$ should be always lower than $\langle H \rangle_a$. Thus the Heitler-London model does not give the correct asymptotic form in the limit $R \to \infty$. *Herring* [4.23] pointed out this fact and found a way to obtain the correct asymptotic form. According to Herring, the log term arises from the exchange integral of e^2/r_{12},

$$\int \phi_a(\boldsymbol{r}_1)\phi_b(\boldsymbol{r}_2)\frac{e^2}{r_{12}}\phi_a(\boldsymbol{r}_2)\phi_b(\boldsymbol{r}_1)d\tau_1 d\tau_2 \, . \tag{4.45}$$

This is the self-energy of exchange charge given by $\phi_a(\boldsymbol{r})\phi_b(\boldsymbol{r})$. The equal density surface of the exchange density is given by $|\boldsymbol{r} - \boldsymbol{R}_a| + |\boldsymbol{r} - \boldsymbol{R}_b| =$ constant, namely, the surface of a spheroid with foci at $\boldsymbol{R}_a$ and $\boldsymbol{R}_b$. For $R = |\boldsymbol{R}_a - \boldsymbol{R}_b| \gg a_{\rm H}$, the overlap charge is localized almost in the long cigar-shaped region with length R and diameter $(R \cdot a_{\rm H})^{1/2}$. This charge distribution gives the logarithmic term.

The matrix element $\langle ab|e^2/r_{12}|ba\rangle$ in the Heitler-London model is an approximate form of

$$\left\langle \phi \left| \frac{e^2}{r_{12}} \right| {\rm P}_{12}\phi \right\rangle , \tag{4.46}$$

where ϕ is the sum of the exact functions for the singlet and triplet states ϕ_s and ϕ_a, respectively, and $\rm P_{12}$ is the permutation operator which interchanges the coordinates of electrons 1 and 2. The exact matrix element is expected not to give the logarithmic term. In the correct wave function ϕ, the two electrons keep away from each other to reduce the contribution of e^2/r_{12}. In other words, $\phi_a\phi_b$ in the Heitler-London approximation ignores the electron correlation effect and gives the unreasonable result in the limit $R \to \infty$. As mentioned before, the direct exchange interaction proportional to the exchange integral of e^2/r_{12} acts in Mott insulators, in addition to the superexchange interaction. One has to be cautious of the above point also for this case. When we consider the direct exchange interaction on the basis of the Heitler-London model, we must keep Herring's remark in mind.

Next, let us extend the Heitler-London model for the hydrogen molecule to the crystal lattice system and denote the wave function as $\psi(\boldsymbol{r}_1, \boldsymbol{r}_2, \ldots \boldsymbol{r}_N)$; here the electron with coordinate $\boldsymbol{r}_i$ is assumed to be localized at the lattice point $\boldsymbol{R}_i$. The energy of the localized state is denoted as E_0. In the Heitler-London model, the ground state is approximated as

$$\psi = \prod_{i=1}^{N} \phi_i(\boldsymbol{r}_i) \, . \tag{4.47}$$

Here electron spins are ignored and electrons are supposed to be distinguishable. Since the total Hamiltonian is symmetric in the electron coordinates, $\mathrm{P}^{(r)}\psi$, which is generated by the permutation operation $\mathrm{P}^{(r)}$ with respect to the coordinate of the electron or of the atom, is an independent wave function belonging to the same energy E_0. Therefore we obtain $N!$ degenerate wave functions by permutation of the N electrons.

Let us consider the wave function ψ and the wave function $\mathrm{P}^{(r)}\psi$ obtained by applying the permutation operator $\mathrm{P}^{(r)}$ to ψ. These two wave functions are linked by the Hamiltonian $\mathcal{H}$; the symmetric and antisymmetric functions are eigenfunctions in the presence of the perturbation. The energy splitting is $2J_\mathrm{P}$ with J_P given by

$$J_\mathrm{P} = \int \ldots \int \psi^*(\boldsymbol{r}_1, \boldsymbol{r}_2, \ldots, \boldsymbol{r}_N)(H - E_0)\mathrm{P}^{(r)}\psi(\boldsymbol{r}_1, \boldsymbol{r}_2, \ldots, \boldsymbol{r}_N) \\ \times d\tau_1 d\tau_2 \ldots d\tau_N \, . \tag{4.48}$$

Thus the effective Hamiltonian in this subspace can be written as

$$\mathcal{H}_{\mathrm{P}^{(r)}} = E_0 + J_\mathrm{P}[\mathrm{P}^{(r)} + (\mathrm{P}^{(r)})^{-1}] \, . \tag{4.49}$$

Taking all the permutations into account, we can extend the effective Hamiltonian to the total effective Hamiltonian:

$$\mathcal{H} = E_0 + \sum_\mathrm{P} J_\mathrm{P}\mathrm{P}^{(r)} \, . \tag{4.50}$$

Here J_P is the direct exchange integral given by (4.48); more precisely, the sum of the exact symmetric and antisymmetric wave function must be used as the ψ in (4.48). In this case J_P is always negative because of the Sturm-Liouville theorem.

In the above discussion, the spin degree of freedom for each electron was not taken into account. Now we restore indistinguishability of electrons and the spin degree of freedom. Because of the Pauli principle, we confine ourselves to the function antisymmetric with respect to the permutation of electron coordinates and spins; then only the spin degrees of freedom are left as the freedom of electrons. Thus the Hamiltonian can be written in terms of the operators in the spin-space. Using the permutation operator P^σ for electron spins, we can write the permutation operator P with respect to coordinate and spin, as

$$\mathrm{P} = \mathrm{P}^\sigma \mathrm{P}^{(r)} = \delta_\mathrm{P} \ , \tag{4.51}$$

where δ_P is 1 for an even permutation and -1 for an odd one. By substituting the relation (4.51) into (4.50), the Hamiltonian is written in terms of the permutation P^σ with respect to spin variables as

$$\mathcal{H} = E_0 + \sum_{\mathrm{P}\neq\mathrm{E}} \delta_\mathrm{P} J_\mathrm{P} \mathrm{P}^\sigma \ , \tag{4.52}$$

where we have used $J_\mathrm{P} = J_{\mathrm{P}^{-1}}$. This result gives the effective Hamiltonian for the N spin system composed of electrons, which are localized at each lattice point. In the sum over P in the second term, the identical permutation E is excluded.

Now we restrict ourselves to the permutation P^σ_{12} between two electrons belonging to nearest-neighbor atoms. Using the relation

$$\mathrm{P}^\sigma_{12} = \frac{1}{2}(1 + \boldsymbol{\sigma}_1 \cdot \boldsymbol{\sigma}_2) \ , \tag{4.53}$$

we can write (4.52) as

$$\mathcal{H} = E_0 - \frac{1}{2} J \sum_{\langle i,j \rangle} (1 + \boldsymbol{\sigma}_i \cdot \boldsymbol{\sigma}_j) \ . \tag{4.54}$$

Thus we have derived the Heisenberg Hamiltonian. Here $\boldsymbol{\sigma}$ is the Pauli spin matrix (twice the spin $\boldsymbol{s}$), and $\langle i,j \rangle$ means the sum over spin pairs located at nearest-neighbor sites. Since J is negative, this spin interaction is antiferromagnetic.

The next simplest exchange interaction involves the cyclic permutation of electrons on three neighboring sites. Let us consider the body-centered-cubic lattice. If 1-2 and 2-3 are assumed to be nearest-neighbor atoms, then 3-1 is a next nearest neighbor pair. Such permutation (123) can be written in terms of the operator P^σ_{12} as

$$\begin{aligned}\mathrm{P}^\sigma_{(123)} &= \mathrm{P}^\sigma_{31} \mathrm{P}^\sigma_{32} = \frac{1}{4}(1 + \boldsymbol{\sigma}_3 \cdot \boldsymbol{\sigma}_1)(1 + \boldsymbol{\sigma}_3 \cdot \boldsymbol{\sigma}_2) \\ &= \frac{1}{4}\left[1 + (\boldsymbol{\sigma}_1 \cdot \boldsymbol{\sigma}_3) + (\boldsymbol{\sigma}_2 \cdot \boldsymbol{\sigma}_3) + (\boldsymbol{\sigma}_1 \cdot \boldsymbol{\sigma}_2) + \mathrm{i}\boldsymbol{\sigma}_3 \cdot (\boldsymbol{\sigma}_1 \times \boldsymbol{\sigma}_2)\right] \ . \end{aligned} \tag{4.55}$$

If we add to this permutation the inverse permutation $\mathrm{P}^{\sigma}_{(321)}$ with the same J_{P}, the last term of (4.55) is cancelled. Thus as the spin Hamiltonian we obtain the sum of two-spin exchange interactions. This gives a ferromagnetic exchange interaction since $\mathrm{P}^{\sigma}_{(123)}$ is an even permutation; the sign in front of J_{P} is positive and J_{P} itself is negative.

New types of spin interactions appear from the four-site cyclic permutation; among them the cyclic permutation between nearest-neighbor atoms gives rather large J_{P}. The spin operator $\mathrm{P}^{\sigma}_{(1234)}$ is calculated as

$$
\begin{aligned}
&\mathrm{P}^{\sigma}_{(1234)} \\
&= \mathrm{P}^{\sigma}_{14}\mathrm{P}^{\sigma}_{13}\mathrm{P}^{\sigma}_{12} \\
&= \frac{1}{8}(1+\boldsymbol{\sigma}_1\cdot\boldsymbol{\sigma}_4)(1+\boldsymbol{\sigma}_1\cdot\boldsymbol{\sigma}_3)(1+\boldsymbol{\sigma}_1\cdot\boldsymbol{\sigma}_2) \\
&= \frac{1}{8}+\frac{1}{8}\left[(\boldsymbol{\sigma}_1\cdot\boldsymbol{\sigma}_2)+(\boldsymbol{\sigma}_1\cdot\boldsymbol{\sigma}_3)+(\boldsymbol{\sigma}_1\cdot\boldsymbol{\sigma}_4)+(\boldsymbol{\sigma}_3\cdot\boldsymbol{\sigma}_4)+(\boldsymbol{\sigma}_2\cdot\boldsymbol{\sigma}_4)+(\boldsymbol{\sigma}_2\cdot\boldsymbol{\sigma}_3)\right] \\
&\quad +\frac{\mathrm{i}}{8}\left[\boldsymbol{\sigma}_1\cdot(\boldsymbol{\sigma}_4\times\boldsymbol{\sigma}_3)+\boldsymbol{\sigma}_1\cdot(\boldsymbol{\sigma}_4\times\boldsymbol{\sigma}_2)+\boldsymbol{\sigma}_1\cdot(\boldsymbol{\sigma}_3\times\boldsymbol{\sigma}_2)+\boldsymbol{\sigma}_2\cdot(\boldsymbol{\sigma}_4\times\boldsymbol{\sigma}_3)\right] \\
&\quad +\frac{1}{8}\left[(\boldsymbol{\sigma}_1\cdot\boldsymbol{\sigma}_2)(\boldsymbol{\sigma}_3\cdot\boldsymbol{\sigma}_4)+(\boldsymbol{\sigma}_1\cdot\boldsymbol{\sigma}_4)(\boldsymbol{\sigma}_2\cdot\boldsymbol{\sigma}_3)-(\boldsymbol{\sigma}_1\cdot\boldsymbol{\sigma}_3)(\boldsymbol{\sigma}_2\cdot\boldsymbol{\sigma}_4)\right]. \qquad (4.56)
\end{aligned}
$$

After adding the terms due to the inverse permutation, the imaginary terms cancel with each other; we are left with only the sum of two-body and four-body interactions. In this case the permutation is odd so that the exchange interaction is antiferromagnetic.

The exchange interaction between spins on the basis of the Heitler-London model, which we described above, can be applied to the nuclear spin system of solid helium 3. Here the ^{3}He atoms (fermions) play the role of the electrons and the nuclear spins of ^{3}He play the role of spins. In this case the three-body and four-body cyclic permutations are more important than the two-body permutation. This is because the He atom (unlike the electron) possesses a hard core due to the closed shell [4.24].

Although the consideration of the exchange interaction between electrons on the basis of the Heitler-London model may look different from that of the Mott insulator, it should be noted that actually both of them discuss the same phenomenon (the exchange mechanism of electrons) from different viewpoints.

Part II

Magnetism of Spin Systems

5. Molecular-Field Theory

In this chapter we show that the most stable spin structures of classical spins coupled with exchange interactions are the helical spin structures. Ferromagnetic and antiferromagnetic spin structures are regarded as special cases of the helical structures. Helically ordered spin configurations are destroyed by the thermal motion of spins and changed to disordered states as the temperature increases. The phenomena of such phase tansitions are discussed on the basis of the molecular-field approximation.

5.1 Heisenberg Model

Let us assume that the state of a magnetic ion is specified by an angular momentum (measured in units of $\hbar$), which is either the spin angular momentum $\boldsymbol{S}$ in the case of iron-group ions or the total angular momentum $\boldsymbol{J}$ of orbital and spin angular momenta in the case of rare-earth ions. Henceforth, we write this angular momentum as $\boldsymbol{S}$ and call it a spin. The $(2S+1)$-fold degeneracy related to the orientation of $\boldsymbol{S}$ is lifted, except for a Kramers doublet, by a one-ion Hamiltonian. If the splitting of the energy levels due to the one-ion Hamiltonian is much smaller than the exchange interaction between ions, the degeneracy is lifted by the latter. Even when the one-ion Hamiltonian is much larger than the exchange interaction, the degeneracy of the Kramers doublet is removed by the exchange interaction. In such a situation the Hamiltonian of the spin system is primarily given by the exchange interaction

$$\mathcal{H} = -2 \sum_{\langle n,m \rangle} J(\boldsymbol{R}_m - \boldsymbol{R}_n) \boldsymbol{S}_m \cdot \boldsymbol{S}_n \ . \tag{5.1}$$

In this case the one-ion Hamiltonian and anisotropic interactions can be treated as small perturbations. The Hamiltonian (5.1), which was used by Heisenberg to discuss ferromagnets, is called the *Heisenberg Hamiltonian* or *Heisenberg model*. In (5.1), $\boldsymbol{R}_m$ and $\boldsymbol{R}_n$ represent the position vectors of the lattice points and the summation is taken over $\boldsymbol{n}, \boldsymbol{m}$ pairs.

The Heisenberg Hamiltonian has been widely studied to discuss ferromagnetism and antiferromagnetism. At the same time, together with the *Ising model*, it has been one of the standard models for cooperative phenomena and phase transitions.

The Ising model is described by the Hamiltonian

$$\mathcal{H} = -2 \sum_{\langle n,m \rangle} J_{nm} \mu_n \mu_m \ , \tag{5.2}$$

where μ_n is a variable taking the value $+1$ or -1. This Hamiltonian can be regarded as a limiting case of (5.1), in which $\boldsymbol{S}$ is assumed to be 1/2 and the coefficient of the transverse component $S_{nx}S_{mx} + S_{ny}S_{my}$ is small. In other words, one may regard it as the limit of large anisotropy along the z axis. On the other hand, the opposite limit of a large transverse component, i.e.,

$$\mathcal{H} = -2 \sum_{\langle n,m \rangle} J_{nm} (S_{nx}S_{mx} + S_{ny}S_{my}) \ , \tag{5.3}$$

is called the *XY model.*

In statistical mechanics, the Ising and XY models are often used instead of the Heisenberg model because of their greater simplicity. In addition, they are sometimes reasonable approximations for real spin systems. For instance, a system may be well represented by the Ising model, if the level scheme due to the one-ion Hamiltonian is such that the lowest levels correspond to $J_z = \pm J$ and the higher levels for $J_z = \pm(J-1)$ are far above these, compared with the exchange interaction.

5.2 Spin Configuration in the Ground State

In general, it is not possible to determine the eigenvalues and eigenfunctions of the Heisenberg Hamiltonian, since it is a many-body problem. Even the problem of determining the ground state is extremely difficult except in the case that all the exchange interactions $J(\boldsymbol{R}_n - \boldsymbol{R}_m)$ are positive. To simplify the problem, therefore, let us assume that $\boldsymbol{S}_n$ is a classical vector. Under this assumption the ground state corresponds to the configuration of the vectors $\boldsymbol{S}_n$ which minimizes the energy (5.1). In most three-dimensional cases the ordered configuration is hardly modified by quantum effects.

Let us consider a system in which the spins sit on a Bravais lattice with one spin per unit cell. Moreover, we assume that the interaction is symmetric:

$$J(\boldsymbol{R}_m - \boldsymbol{R}_n) = J(\boldsymbol{R}_n - \boldsymbol{R}_m) \ . \tag{5.4}$$

We now construct a linear combination of the $\boldsymbol{S}_n$ as

$$\boldsymbol{S}_q = N^{-1/2} \sum_n \boldsymbol{S}_n \mathrm{e}^{-\mathrm{i}\boldsymbol{q}\cdot\boldsymbol{R}_n} \ . \tag{5.5}$$

The lattice vectors $\boldsymbol{R}_n$ can be expressed as

$$\boldsymbol{R}_n = n_1\boldsymbol{a}_1 + n_2\boldsymbol{a}_2 + n_3\boldsymbol{a}_3 \quad (n_i : \text{integer}) \tag{5.6}$$

in terms of the primitive vectors $\boldsymbol{a}_1$, $\boldsymbol{a}_2$, and $\boldsymbol{a}_3$; using these, we construct basis vectors $\boldsymbol{b}_1$, $\boldsymbol{b}_2$, and $\boldsymbol{b}_3$ defined by

$$b_1 = \frac{2\pi(a_2 \times a_3)}{a_1 \cdot (a_2 \times a_3)}\,, \quad b_2 = \frac{2\pi(a_3 \times a_1)}{a_2 \cdot (a_3 \times a_1)}\,, \quad b_3 = \frac{2\pi(a_1 \times a_2)}{a_3 \cdot (a_1 \times a_2)}\,. \tag{5.7}$$

The vectors given by

$$K_n = n_1 b_1 + n_2 b_2 + n_3 b_3 \qquad (n_i : \text{integer}) \tag{5.8}$$

are called reciprocal-lattice vectors; the lattice formed with them is called the reciprocal lattice. If we express the vector q as

$$q = q_1 b_1 + q_2 b_2 + q_3 b_3\,,$$

then the component q_i has the value

$$q_i = \frac{n_i}{N_i} \qquad (n_i : \text{integer})\,. \tag{5.9}$$

Here N_i is the number of unit cells in the a_i direction, and the total number of lattice points is equal to $N_1 N_2 N_3$. The number of independent q vectors given by (5.9) is N; usually these vectors are chosen to lie within the first Brillouin zone, which is the region surrounded by planes bisecting the vectors connecting the origin and neighboring reciprocal lattice points.

According to (5.5), the Fourier coefficients S_q satisfy

$$S_q^* = S_{-q}\,. \tag{5.10}$$

Further, the relations

$$N^{-1}\sum_n \mathrm{e}^{\mathrm{i}q\cdot R_n} = \delta_{q,0}\,, \qquad N^{-1}\sum_q \mathrm{e}^{\mathrm{i}q\cdot R_n} = \delta_{n,0} \tag{5.11}$$

hold from (5.9). With these relations S_n can be expressed as the Fourier series

$$S_n = N^{-1/2}\sum_q S_q \mathrm{e}^{\mathrm{i}q\cdot R_n}\,, \tag{5.12}$$

which is the inverse of (5.5).

Substituting (5.12) into (5.1), we obtain for the Heisenberg Hamiltonian

$$\mathcal{H} = -\sum_q J(q) S_q \cdot S_{-q}\,, \tag{5.13}$$

which has the form of a sum of squares of Fourier components S_q. Here $J(q)$ is a real quantity defined by

$$J(q) = \sum_n J(R_n)\mathrm{e}^{-\mathrm{i}q\cdot R_n} = J(-q) = J(q)^*\,. \tag{5.14}$$

If the magnitude of each spin S_n is S, its square is given by

$$S^2 = S_n \cdot S_n = N^{-1}\sum_{qq'} S_q \cdot S_{-q'}\mathrm{e}^{\mathrm{i}(q-q')\cdot R_n}\,. \tag{5.15}$$

This has to be satisfied by all $\boldsymbol{S}_n$ independent of $\boldsymbol{R}_n$. Instead of applying these N relations, we look for $\boldsymbol{S}_q$, which minimize the energy (5.13) under the relaxed condition that the sum of (5.15) over $\boldsymbol{R}_n$ satisfies

$$NS^2 = \sum_q \boldsymbol{S}_q \cdot \boldsymbol{S}_{-q} \,. \tag{5.16}$$

Clearly, the solution is such that all $\boldsymbol{S}_q$ should vanish, except for $\boldsymbol{S}_Q$ and $\boldsymbol{S}_{-Q}$; here $\pm\boldsymbol{Q}$ are the values of $\boldsymbol{q}$ corresponding to the maximum of $J(\boldsymbol{q})$, and we assume that $\boldsymbol{Q} \neq 0$. For such $\boldsymbol{S}_q$, the condition (5.15) takes the form

$$S^2 = N^{-1}\left(2\boldsymbol{S}_Q \cdot \boldsymbol{S}_{-Q} + \boldsymbol{S}_Q \cdot \boldsymbol{S}_Q \mathrm{e}^{2\mathrm{i}\boldsymbol{Q}\cdot\boldsymbol{R}_n} + \boldsymbol{S}_{-Q} \cdot \boldsymbol{S}_{-Q} \mathrm{e}^{-2\mathrm{i}\boldsymbol{Q}\cdot\boldsymbol{R}_n}\right) . \tag{5.17}$$

In order that the right-hand side be independent of $\boldsymbol{R}_n$, the relations

$$\boldsymbol{S}_Q \cdot \boldsymbol{S}_Q = 0 \,, \quad \boldsymbol{S}_Q \cdot \boldsymbol{S}_{-Q} \neq 0 \tag{5.18}$$

should hold. Introducing real vectors $\boldsymbol{R}_Q$, $\boldsymbol{I}_Q$ to express $\boldsymbol{S}_Q$ as

$$\boldsymbol{S}_Q = \boldsymbol{R}_Q + \mathrm{i}\boldsymbol{I}_Q \,, \quad \boldsymbol{S}_{-Q} = \boldsymbol{R}_Q - \mathrm{i}\boldsymbol{I}_Q \,, \tag{5.19}$$

we obtain

$$\boldsymbol{S}_Q \cdot \boldsymbol{S}_{-Q} = \boldsymbol{R}_Q^2 + \boldsymbol{I}_Q^2 \,, \tag{5.20}$$

$$\boldsymbol{S}_Q \cdot \boldsymbol{S}_Q = \boldsymbol{R}_Q^2 - \boldsymbol{I}_Q^2 + 2\mathrm{i}\boldsymbol{R}_Q \cdot \boldsymbol{I}_Q \,. \tag{5.21}$$

Therefore the relations

$$\boldsymbol{R}_Q^2 = \boldsymbol{I}_Q^2 \,, \quad \boldsymbol{R}_Q \cdot \boldsymbol{I}_Q = 0 \tag{5.22}$$

have to be satisfied. Then the condition (5.15) is satisfied if $\boldsymbol{R}_Q$ and $\boldsymbol{I}_Q$ have the same magnitude and are perpendicular to each other. In this case, we have

$$\boldsymbol{R}_Q^2 = \boldsymbol{I}_Q^2 = \frac{1}{4}NS^2 \,. \tag{5.23}$$

The lowest energy is then given by

$$E_{\min} = -NS^2 J(\boldsymbol{Q}) \,. \tag{5.24}$$

From (5.12), $\boldsymbol{S}_n$ has the form

$$\begin{aligned} \boldsymbol{S}_n &= 2N^{-1/2}(\boldsymbol{R}_Q \cos \boldsymbol{Q}\cdot\boldsymbol{R}_n - \boldsymbol{I}_Q \sin \boldsymbol{Q}\cdot\boldsymbol{R}_n) \\ &= S(\boldsymbol{R}\cos \boldsymbol{Q}\cdot\boldsymbol{R}_n - \boldsymbol{I}\sin \boldsymbol{Q}\cdot\boldsymbol{R}_n) \,, \end{aligned} \tag{5.25}$$

where $\boldsymbol{R}$ and $\boldsymbol{I}$ are unit vectors perpendicular to each other. Let us take the plane formed by the two unit vectors $\boldsymbol{R}$ and $\boldsymbol{I}$ as the xy plane, and the direction perpendicular to it as the z axis. With this choice we obtain

$$\begin{aligned} S_{nx} &= S\cos(\boldsymbol{Q}\cdot\boldsymbol{R}_n+\theta)\ , \\ S_{ny} &= S\sin(\boldsymbol{Q}\cdot\boldsymbol{R}_n+\theta)\ , \\ S_{nz} &= 0\ , \end{aligned} \tag{5.26}$$

where θ is the angle between $\boldsymbol{R}$ and the x axis. The spin structure described by (5.26) is called a *helical spin structure*, in which the spins on a plane perpendicular to the vector $\boldsymbol{Q}$ are all parallel and are aligned along one direction within the xy plane. The direction of the spins rotates by an angle $a|\boldsymbol{Q}|$ as one moves along $\boldsymbol{Q}$; here a is the spacing of lattice planes along the $\boldsymbol{Q}$ axis. As the above calculations show, the xy plane on which spins rotate can be selected independently of the vector $\boldsymbol{Q}$. A special case, in which the z axis and $\boldsymbol{Q}$ coincide, is called the normal helical structure. The helical spin structure was predicted by *Yoshimori* [5.1] for the spin structure of MnO_2 and also by *Kaplan* [5.2] and *Villain* [5.3]. Later it was found in other compounds and in rare-earth metals heavier than Gd. A detailed account of helical spin structures is given by *Nagamiya* [5.4].

The helical spin structure, which has been derived as the one corresponding to the lowest energy of classical spins on a Bravais lattice, is the most fundamental spin structure that contains the ferromagnetic state ($\boldsymbol{Q} = 0$) and the antiferromagnetic state (with $\boldsymbol{Q}$ at an edge of the Brillouin zone) as limiting cases. The important feature is that the period of the helix has nothing to do with the period of the underlying crystal structure and can take any value within the Brillouin zone. For a complicated crystal structure having more than two magnetic ions per unit cell, it is difficult to give a general prescription to determine the spin structure; only approximate treatments are possible. Typical examples of the latter case are provided by $CuCr_2O_4$ and $MnCr_2O_4$, which have a spinel structure. Theoretical studies of stable spin structures in such magnetic substances have been carried out by *Kaplan*, *Lyons* and others [5.5–7].

Various examples of helical structures are found in heavy rare-earth metals, Tb, Dy, Ho, Er and Tm in particular. The crystal structure of these metals is hexagonal close-packed. The $\boldsymbol{Q}$ vectors characterizing their helical spin structures are along the c axis. The plane in which the spins rotate is the c plane in the case of Tb, Dy, and Ho. In Er and Tm, however, the easy axis for spins is the c axis so that a spin-ordered state, in which the c component of spins is modulated sinusoidally, shows up in an intermediate temperature region below the transition temperature. If the nearest-neighbor exchange interaction is ferromagnetic, antiferromagnetic second and third neighbor interactions competing with the ferromagnetic one are needed to stabilize helical structures over the ferromagnetic and antiferromagnetic spin states. In the case of rare-earth metals the exchange interaction is due to an indirect exchange interaction via conduction electrons (called the RKKY interaction), as will be described later. The RKKY interaction has longer range than the superexchange interaction and changes its sign as a function of the distance. Therefore it fulfils enough conditions to cause helical spin structures.

5.3 Molecular-Field Approximation

To determine the eigenvalues of the Heisenberg Hamiltonian describing exchange - coupled spin systems is a typical example of a many-body problem. The *Hartree approximation* is the most basic approximation method to treat this problem. The spirit of this method is to replace the interaction by an average field and thereby reduce the task to a one-body problem; the average value is determined in a way consistent with the solution of the one-body problem. This idea is widely used in various fields of solid-state physics. In the case of spin systems this approximation is called the *molecular-field approximation*. The concept of the molecular field was introduced by Weiss to discuss ferromagnetism. For this reason it is sometimes called the molecular field of Weiss.

Let us first rewrite the spin operator at the nth site, $\boldsymbol{S}_n$, in terms of its average $\langle \boldsymbol{S}_n \rangle$ and the deviation from it (i.e., the fluctuation) as

$$\boldsymbol{S}_n = \langle \boldsymbol{S}_n \rangle + (\boldsymbol{S}_n - \langle \boldsymbol{S}_n \rangle) . \tag{5.27}$$

This relation is substituted into (5.1) and the square of fluctuations is ignored by assuming that fluctuations are small. The Heisenberg Hamiltonian is then approximated as

$$\mathcal{H} = 2 \sum_{\langle m,n \rangle} J(\boldsymbol{R}_m - \boldsymbol{R}_n) \langle \boldsymbol{S}_n \rangle \cdot \langle \boldsymbol{S}_m \rangle - \sum_n \boldsymbol{S}_n \cdot \Big[2 \sum_m J(\boldsymbol{R}_m - \boldsymbol{R}_n) \langle \boldsymbol{S}_m \rangle \Big] . \tag{5.28}$$

Here the first term is a constant and the second term is equivalent to the Zeeman energy of spins in an effective field (i.e., the molecular field). We assume that the spin configuration in the ground state has a helical structure. If a vector $\boldsymbol{Q}$ is introduced to characterize the helix and the xy plane is chosen as the spin rotation plane, the average value of $\boldsymbol{S}_n$ can be expressed as

$$\langle \boldsymbol{S}_n \rangle = \sigma [\cos(\boldsymbol{Q} \cdot \boldsymbol{R}_n + \alpha), \ \sin(\boldsymbol{Q} \cdot \boldsymbol{R}_n + \alpha), \ 0] , \tag{5.29}$$

where σ denotes the magnitude of the average of each spin. With this choice for $\langle \boldsymbol{S}_n \rangle$, the effective field acting on $\boldsymbol{S}_n$, which is equivalent to the second term of (5.28), is given by

$$\begin{aligned} H_{\mathrm{eff}x} &= \frac{2\sigma}{g\mu_{\mathrm{B}}} \sum_m J(\boldsymbol{R}_m - \boldsymbol{R}_n) \cos(\boldsymbol{Q} \cdot \boldsymbol{R}_m + \alpha) , \\ H_{\mathrm{eff}y} &= \frac{2\sigma}{g\mu_{\mathrm{B}}} \sum_m J(\boldsymbol{R}_m - \boldsymbol{R}_n) \sin(\boldsymbol{Q} \cdot \boldsymbol{R}_m + \alpha) . \end{aligned} \tag{5.30}$$

Using (5.14), one can write this as

$$\begin{aligned} H_{\mathrm{eff}x} &= \frac{2\sigma}{g\mu_{\mathrm{B}}} J(\boldsymbol{Q}) \cos(\boldsymbol{Q} \cdot \boldsymbol{R}_n + \alpha) , \\ H_{\mathrm{eff}y} &= \frac{2\sigma}{g\mu_{\mathrm{B}}} J(\boldsymbol{Q}) \sin(\boldsymbol{Q} \cdot \boldsymbol{R}_n + \alpha) . \end{aligned} \tag{5.31}$$

This procedure is self-consistent, in that the direction of the magnetic field $\boldsymbol{H}_{\text{eff}}$ coincides with the direction of $\langle \boldsymbol{S}_n \rangle$, which was assumed at the starting point. The magnitude of $\langle \boldsymbol{S}_n \rangle$ is given by the thermal average in the effective field (5.31). Thus we have

$$\frac{|\langle \boldsymbol{S}_n \rangle|}{S} = \frac{\sigma}{S} = \frac{\sum_{M=-S}^{S} \frac{M}{S} \exp\left(\frac{2S\sigma J(\boldsymbol{Q})}{kT}\frac{M}{S}\right)}{\sum_{M=-S}^{S} \exp\left(\frac{2S\sigma J(\boldsymbol{Q})}{kT}\frac{M}{S}\right)} = B_S\left(\frac{2J(\boldsymbol{Q})S\sigma}{kT}\right) , \tag{5.32}$$

where $B_S(x)$ is defined by

$$B_S(x) = \frac{2S+1}{2S} \coth \frac{2S+1}{2S} x - \frac{1}{2S} \coth \frac{1}{2S} x , \tag{5.33}$$

which is called the Brillouin function. For small x it can be expanded as

$$B_S(x) = \frac{S+1}{3S} x - \frac{[(S+1)^2 + S^2](S+1)}{90S^3} x^3 + \dots . \tag{5.34}$$

The magnitude σ is obtained by solving (5.32). The solution $\sigma = 0$ always exists, corresponding to the paramagnetic state. A nonzero solution for σ exists, however, if

$$T < T_Q = \frac{2J(\boldsymbol{Q})S(S+1)}{3k} . \tag{5.35}$$

In this temperature region the solution $\sigma \neq 0$ has the lower free energy. Therefore, according to the molecular-field approximation, the helical spin structure with $\boldsymbol{Q}$ corresponding to the maximum of $J(\boldsymbol{q})$ is realized at low temperatures. The average spin moment in this ordered state decreases with increasing temperature; it vanishes above $T = T_Q$ and the spin system becomes paramagnetic.

At low temperatures the value of σ is given by

$$\frac{\sigma}{S} = 1 - \frac{1}{S} \exp\left(-\frac{3}{S+1}\frac{T_Q}{T}\right) , \tag{5.36}$$

which decreases exponentially with increasing temperature. In spin-wave theory, which is a reliable approximate theory for the low-temperature region, this decrease is replaced by a power law. In the vicinity of T_Q, σ is given by

$$\left(\frac{\sigma}{S}\right)^2 = \frac{10}{3} \frac{(S+1)^2}{(S+1)^2 + S^2} \left(\frac{T_Q}{T} - 1\right) , \tag{5.37}$$

if the expansion (5.34) is used.

Further, to determine the susceptibility in the paramagnetic state, we add the Zeeman term

$$-g\mu_{\text{B}} \sum_n \boldsymbol{S}_n \cdot \boldsymbol{H} \tag{5.38}$$

to the Hamiltonian. Writing the average of $\boldsymbol{S}_n$ induced by the magnetic field as σ, we have the following form:

$$\sigma = SB_S\Big(\frac{2S\sigma J(0) + g\mu_B SH}{kT}\Big) . \tag{5.39}$$

On solving this relation for small σ, we find

$$\chi = \frac{C}{T - T_0} , \quad C = \frac{Ng^2\mu_B^2 S(S+1)}{3k} ; \tag{5.40}$$

here T_0 is defined by

$$T_0 = \frac{2J(0)S(S+1)}{3k} . \tag{5.41}$$

Thus if we plot the inverse of the susceptibility (5.40) as a function of temperature, we obtain a straight line crossing the temperature axis at T_0 as shown in Fig. 5.1. Such a relation is called the *Curie-Weiss law*; the constant C is the *Curie constant.* Except for the case of ferromagnetism ($\boldsymbol{Q} = 0$) T_0 is smaller than $T_{\boldsymbol{Q}}$. Physically, T_0 represents a lowering of energy for a parallel alignment of spins. In the case of ferromagnetism we have $T_{\boldsymbol{Q}} = T_0$ and the susceptibility diverges at the Curie temperature T_0, where the ferromagnetism appears. In a more general case the susceptibility $\chi_{\boldsymbol{Q}}$ for a rotating magnetic field with the same $\boldsymbol{Q}$ vector as the helix is given by

$$\chi_{\boldsymbol{Q}} = \frac{C}{T - T_{\boldsymbol{Q}}} \tag{5.42}$$

and it diverges at $T = T_{\boldsymbol{Q}}$, while the susceptibility χ for a uniform magnetic field does not diverge at $T_{\boldsymbol{Q}}$. For the ferromagnetic case the relation between σ and H at $T = T_0$ is given by

$$\sigma = S\Big[\frac{10}{3}\frac{(S+1)^2}{(S+1)^2 + S^2}\frac{g\mu_B}{2J(0)S}\Big]^{1/3} H^{1/3} . \tag{5.43}$$

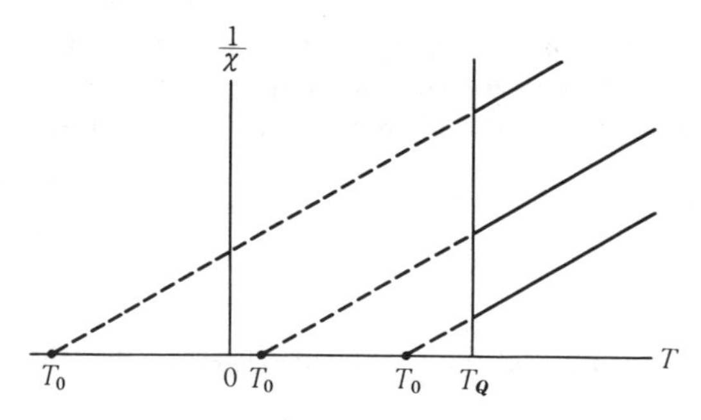

Fig. 5.1. The relation between the inverse of the susceptibility and the temperature

6. Molecular-Field Theory for Antiferromagnets

In succession to the previous chapter, we continue discussion on the basis of the molecular-field approximation to consider in more detail magnetic properties of antiferromagnets.

6.1 Susceptibility of Antiferromagnets

If the maximum of $J(\boldsymbol{q})$ is located at the edge of the Brillouin zone, the spins align either parallel or antiparallel to a particular direction. The orientation of the spin axis relative to the crystal axes is indefinite as far as the isotropic exchange interaction is concerned; it is fixed after an anisotropy energy is introduced. The spin configuration, in which all the lattice points are divided into two sublattices having spins aligned along either the + or − direction, is called the *antiferromagnetic spin configuration*; the net magnetization is zero. Many ionic crystals with iron-group elements take the antiferromagnetic spin configuration. A typical example is MnO, which has the NaCl-type structure with the Mn^{2+} ions forming a face-centered-cubic lattice; the two sublattices for + and − spins are shown in Fig. 6.1.

Let $\boldsymbol{H}_+$ and $\boldsymbol{H}_-$ be the molecular fields acting on the spins of the two sublattices. Then the Heisenberg Hamiltonian in the molecular-field approximation is

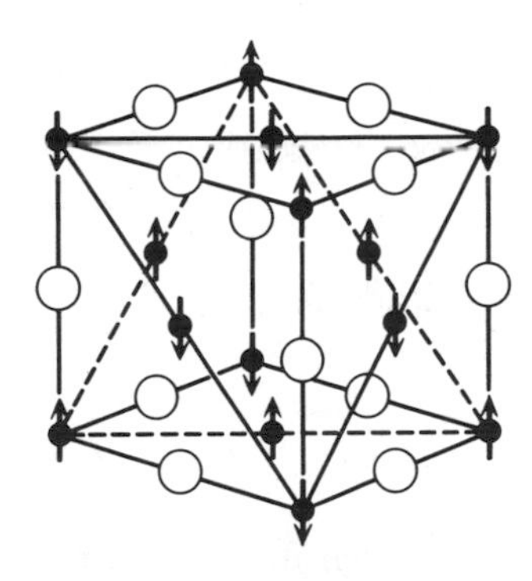

Fig. 6.1. The spin configuration of MnO. The filled and open circles represent Mn^{2+} and O^{2-}, respectively

$$\mathcal{H} = -g\mu_{\mathrm{B}} \sum_{n^+} \boldsymbol{S}_n \cdot \boldsymbol{H}_+ - g\mu_{\mathrm{B}} \sum_{n^-} \boldsymbol{S}_n \cdot \boldsymbol{H}_- \tag{6.1}$$

except for a constant term. The summation over $\boldsymbol{n}^+$ and $\boldsymbol{n}^-$ is taken for the + and − sublattices, respectively. The molecular fields $\boldsymbol{H}_\pm$ are given by

$$\begin{aligned} g\mu_{\mathrm{B}}\boldsymbol{H}_+ &= 2\sum_{m^+} J(\boldsymbol{R}_n - \boldsymbol{R}_m)\langle \boldsymbol{S}\rangle_+ + 2\sum_{m^-} J(\boldsymbol{R}_n - \boldsymbol{R}_m)\langle \boldsymbol{S}\rangle_- \,, \\ g\mu_{\mathrm{B}}\boldsymbol{H}_- &= 2\sum_{m^-} J(\boldsymbol{R}_n - \boldsymbol{R}_m)\langle \boldsymbol{S}\rangle_- + 2\sum_{m^+} J(\boldsymbol{R}_n - \boldsymbol{R}_m)\langle \boldsymbol{S}\rangle_+ \,, \end{aligned} \tag{6.2}$$

where $\langle \boldsymbol{S}\rangle_\pm$ is the average spin on the ± sublattice. In terms of the total magnetic moment

$$\boldsymbol{M}_\pm = \frac{N}{2} g\mu_{\mathrm{B}} \langle \boldsymbol{S}\rangle_\pm \tag{6.3}$$

of each sublattice, the molecular field can be expressed as

$$\begin{aligned} \boldsymbol{H}_+ &= -\Gamma \boldsymbol{M}_+ - A\boldsymbol{M}_- \,, \\ \boldsymbol{H}_- &= -\Gamma \boldsymbol{M}_- - A\boldsymbol{M}_+ \,; \end{aligned} \tag{6.4}$$

here A and Γ are given by

$$\begin{aligned} \Gamma &= -\frac{4}{g^2\mu_{\mathrm{B}}^2 N} \sum_{m^+} J(\boldsymbol{R}_n - \boldsymbol{R}_m) \,, \\ A &= -\frac{4}{g^2\mu_{\mathrm{B}}^2 N} \sum_{m^-} J(\boldsymbol{R}_n - \boldsymbol{R}_m) \,, \end{aligned} \tag{6.5}$$

where the summation over $\boldsymbol{m}^+$ or $\boldsymbol{m}^-$ is taken for each sublattice. In the first equation of (6.2), $\boldsymbol{R}_n$ is a point on the + sublattice, while in the second it is a point on the − sublattice.

The average spin on each sublattice under the molecular field $\boldsymbol{H}_\pm$, i.e., the total sublattice magnetic moment, is determined by

$$|\boldsymbol{M}_\pm| = \frac{N}{2} g\mu_{\mathrm{B}} S B_S \left[\frac{g\mu_{\mathrm{B}} S |(-\Gamma \boldsymbol{M}_\pm - A\boldsymbol{M}_\mp + \boldsymbol{H})|}{kT} \right] , \tag{6.6}$$

where $\boldsymbol{H}$ is the external magnetic field.

Let us first consider the zero-field case. Since

$$\boldsymbol{M}_+ = -\boldsymbol{M}_- = \boldsymbol{M} \tag{6.7}$$

holds in this case, this relation together with (6.6) leads to the following equation for the sublattice magnetization M:

$$M = \frac{N}{2} g\mu_{\mathrm{B}} S B_S \left[\frac{g\mu_{\mathrm{B}} S (A - \Gamma) M}{kT} \right] . \tag{6.8}$$

This relation is nothing but a special case of (5.32), which determines the magnitude of the spin in the helical spin structure; it is the same as the relation for the spontaneous magnetization in ferromagnets. The sublattice magnetization M vanishes for temperatures T greater than the Néel temperature T_N given by

$$T_N = \frac{C}{2}(A - \Gamma) \,, \tag{6.9}$$

where C is the Curie constant defined in (5.40). Above T_N a magnetic moment is induced by the external magnetic field. We can obtain the induced moment from (6.6), putting $\boldsymbol{M}_\pm = \boldsymbol{M}$, noting that $\boldsymbol{M}$ is parallel to $\boldsymbol{H}$ and expanding the Brillouin function. The result is

$$\frac{M}{H} = \chi = \frac{C}{T - T_0} \,, \tag{6.10}$$

$$T_0 = -\frac{C}{2}(A + \Gamma) < T_N \,. \tag{6.11}$$

Below T_N mutually antiparallel magnetizations appear on the two sublattices. The orientation of these magnetizations cannot be fixed by the isotropic exchange interaction. They are aligned actually along a direction for which either anisotropic interactions smaller than the isotropic one or the anisotropy energy due to an anisotropic one-ion Hamiltonian is minimal. When an external magnetic field is applied to this state, the sublattice magnetizations do not deviate from the easy axis when the field is sufficiently small. Therefore, for a weak magnetic field, we can decompose the field into two components parallel and perpendicular to the easy axis, respectively. For a magnetic field applied parallel to the spin axis, each sublattice magnetization is a sum of the magnetization in zero field and a change due to the field:

$$\boldsymbol{M}_\pm = \pm\boldsymbol{M} + \delta\boldsymbol{M}_\pm \,. \tag{6.12}$$

Substituting this relation into (6.6) and performing a Taylor expansion of the Brillouin function to first order in $\delta\boldsymbol{M}_\pm$ and the external field $\boldsymbol{H}$, we obtain

$$\delta M_\pm = -\frac{N}{2} g^2 \mu_B^2 S^2 B'_S \left[\frac{g\mu_B S(A-\Gamma)M}{kT}\right] \frac{(\Gamma\delta M_\pm + A\delta M_\mp - H)}{kT} \,, \tag{6.13}$$

where B'_S is the derivative of the Brillouin function and an even function of its argument. Solving (6.13) for δM_+, one finds the parallel susceptibility as

$$\begin{aligned}\chi_\| &= \frac{\delta M_+ + \delta M_-}{H} \\ &= \frac{2}{A+\Gamma}\left\{1 + \frac{S+1}{3S}\frac{A-\Gamma}{A+\Gamma}\frac{T}{T_N}\left[B'_S\left(\frac{3S}{S+1}\frac{T_N}{T}\frac{2M}{Ng\mu_B S}\right)\right]^{-1}\right\}^{-1} \,, \end{aligned} \tag{6.14}$$

Since the relations $M = 0$ and $B'_S(0) = (S+1)/3S$ hold at $T = T_N$, the susceptibility $\chi_\|$ at $T = T_N$ is equal to $1/A$, which is the same as the value given by (6.10) for $T = T_N$. $\chi_\|$ decreases as the temperature decreases below T_N and vanishes at $T = 0$; see Fig. 6.3.

When the external field is perpendicular to the spin axis, both sublattice magnetizations tilt in the direction of the magnetic field by the same amount. As shown in Fig. 6.2, in this state each sublattice magnetization must be parallel to the resultant vector of the molecular field and the external field. In other words,

$$(-A\boldsymbol{M}_{\mp} + \boldsymbol{H}) \parallel \boldsymbol{M}_{\pm} \tag{6.15}$$

should be satisfied, since $-\Gamma\boldsymbol{M}_{\pm}$ is parallel to $\boldsymbol{M}_{\pm}$. Therefore the tilt angle θ is determined by

$$\sin\theta = \frac{1}{2}\frac{H}{A|\boldsymbol{M}_{-}|} = \frac{H}{2AM} \,. \tag{6.16}$$

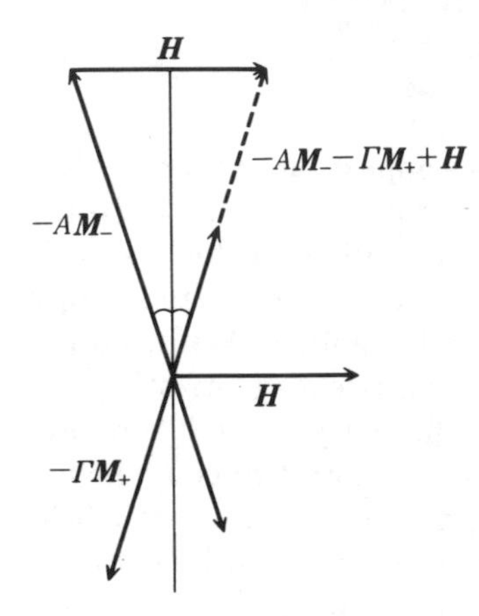

Fig. 6.2. The relation between the perpendicular magnetic field and the sublattice magnetization

In addition, the magnitude of the total magnetic field acting on $\boldsymbol{M}_{\pm}$ is given by $|A\boldsymbol{M}_{\pm} - \Gamma\boldsymbol{M}_{\pm}|$ so that the magnitude of the magnetization remains constant even when the external field is applied. From (6.16), the total magnetic moment induced by the magnetic field is

$$\boldsymbol{M}_{+} + \boldsymbol{M}_{-} = \frac{1}{A}\boldsymbol{H} \,, \tag{6.17}$$

which shows that the perpendicular susceptibility is equal to $1/A$ independent of temperature and magnetic field. When the perpendicular magnetic field is increased further, the component of the magnetization parallel to the field increases in proportion to the field until finally the magnetization becomes parallel to each other at the critical field $H_{c\perp}$ given by

$$H_{c\perp} = 2AM \,. \tag{6.18}$$

The susceptibility for a polycrystalline sample is obtained by averaging the susceptibility over the angle θ between the spin axis and the magnetic field:

$$\chi_p = \chi_{\parallel}\langle\cos^2\theta\rangle + \chi_{\perp}\langle\sin^2\theta\rangle = \frac{1}{3}(\chi_{\parallel} + 2\chi_{\perp}) \,. \tag{6.19}$$

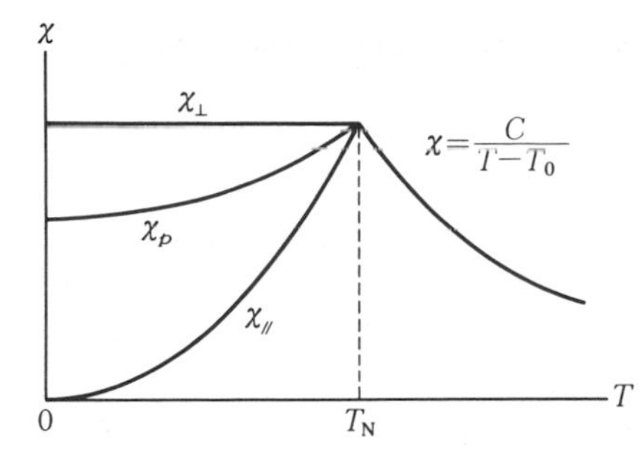

Fig. 6.3. The temperature dependence of the susceptibility of antiferromagnets

From these results the susceptibility for antiferromagnets depends on temperature as shown in Fig. 6.3. A characteristic feature of antiferromagnets is the appearance of a peak in the temperature dependence of the susceptibility. The temperature at this peak corresponds to the Néel temperature.

6.2 Rotation of Spin Axis

In the previous section we considered the magnetization of antiferromagnets in a sufficiently small magnetic field. The susceptibility thus obtained does not show any explicit dependence on the anisotropy energy. This corresponds to a limiting case where the external magnetic field is small compared with a quantity related to the anisotropy energy. For the more general case, we have to introduce explicitly the anisotropy energy. We assume, however, that the anisotropy energy is much smaller than the internal magnetic field coming from the exchange interaction. Under this assumption the magnitude of each sublattice magnetization can be regarded as determined solely by the exchange interaction.

Let us introduce the anisotropy energy for the case of orthorhombic symmetry

$$F_a = \frac{1}{2}K_1(\beta_+^2 + \beta_-^2) + \frac{1}{2}K_2(\gamma_+^2 + \gamma_-^2) , \tag{6.20}$$

where $(\alpha_\pm, \beta_\pm, \gamma_+)$ represent the direction cosines of the sublattice magnetizations relative to the crystal axes. We assume for the anisotropy constants K_1 and K_2 that

$$K_2 > K_1 > 0 , \tag{6.21}$$

taking the x axis as the most stable direction and the y axis as the second most stable. In terms of thermodynamics the anisotropy energy of (6.20) actually corresponds to the free energy. We define the direction cosines of the external field $\boldsymbol{H}$ as $(\alpha_H, \beta_H, \gamma_H)$; its magnitude H is also assumed to be smaller than the molecular field due to the exchange interaction. Then the two sublattice magnetizations are nearly antiparallel to each other; it is assumed that

$$(\alpha_+,\beta_+,\gamma_+) \simeq -(\alpha_-,\beta_-,\gamma_-) \ , \tag{6.22}$$

and the sum $\boldsymbol{M}_+ + \boldsymbol{M}_-$ is ignored compared with the difference $\boldsymbol{M}_+ - \boldsymbol{M}_-$. If the direction cosines of the difference are defined by (α,β,γ), the total free energy for the antiferromagnet is given by

$$F = -\frac{1}{2}\chi_\parallel H_\parallel^2 - \frac{1}{2}\chi_\perp H_\perp^2 + K_1\beta^2 + K_2\gamma^2 \ , \tag{6.23}$$

where $H_\parallel$ and $H_\perp$ are the parallel and perpendicular components of the external field, respectively, relative to the spin axis, which is different from the easy axis in general. The first two terms represent the change of the free energy due to the magnetic field, when the spin axis is assumed to be unchanged. Expressing (6.23) in terms of the direction cosines of $\boldsymbol{M}_+ - \boldsymbol{M}_-$ and the magnetic field $\boldsymbol{H}$, we obtain

$$F = \frac{1}{2}(\chi_\perp - \chi_\parallel)H^2\left[(\alpha\alpha_H + \beta\beta_H + \gamma\gamma_H)^2 + \kappa_1\beta^2 + \kappa_2\gamma^2 - \lambda(\alpha^2+\beta^2+\gamma^2)\right] \tag{6.24}$$

except for a constant term. Here the abbreviations

$$\kappa_1 = \frac{2K_1}{(\chi_\perp - \chi_\parallel)H^2} \ , \quad \kappa_2 = \frac{2K_2}{(\chi_\perp - \chi_\parallel)H^2} \tag{6.25}$$

are used and the last term in the square brackets has been added to impose the constraint

$$\alpha^2 + \beta^2 + \gamma^2 = 1 \ . \tag{6.26}$$

The Lagrange multiplier λ is determined so that α, β and γ, which are treated as independent variables, satisfy the condition (6.26).

The next task is to determine the orientation of the spin axis so as to minimize the free energy (6.24). Differentiating the ratio of F to $(\chi_\perp - \chi_\parallel)H^2$ with respect to α, β and γ and putting the derivatives equal to zero, we have

$$\begin{aligned}
&(\alpha\alpha_H + \beta\beta_H + \gamma\gamma_H)\alpha_H - \lambda\alpha = 0 \ , \\
&(\alpha\alpha_H + \beta\beta_H + \gamma\gamma_H)\beta_H + (\kappa_1 - \lambda)\beta = 0 \ , \\
&(\alpha\alpha_H + \beta\beta_H + \gamma\gamma_H)\gamma_H + (\kappa_2 - \lambda)\gamma = 0 \ .
\end{aligned} \tag{6.27}$$

Multiplying the three equations by α, β and γ, respectively, and summing them, we arrive at

$$(\alpha\alpha_H + \beta\beta_H + \gamma\gamma_H)^2 + \kappa_1\beta^2 + \kappa_2\gamma^2 = \lambda \ , \tag{6.28}$$

which means that λ actually gives the free energy. Therefore, among the three solutions of (6.27), the one corresponding to the minimum of λ should be chosen. The eigenvalues λ are determined by the equation

$$\lambda^3-(1+\kappa_1+\kappa_2)\lambda^2+[(\kappa_1+\kappa_2)\alpha_H^2+\kappa_2\beta_H^2+\kappa_1\gamma_H^2+\kappa_1\kappa_2]\lambda-\kappa_1\kappa_2\alpha_H^2=0\ , \tag{6.29}$$

which is obtained by setting to zero the determinant formed from the coefficients of α, β, γ in (6.27).

For simplicity we consider two cases separately, the magnetic field applied in the xy plane ($\gamma_H = 0$) and in the xz plane ($\beta_H = 0$).

6.2.1 The Case $\gamma_H = 0$

In this case (6.29) is factorized as

$$(\lambda-\kappa_2)\left[\lambda^2-(1+\kappa_1)\lambda+\kappa_1\alpha_H^2\right]=0\ , \tag{6.30}$$

which has the following three solutions:

$$\lambda_3=\kappa_2\ ,\quad \lambda_{1,2}=\frac{1}{2}\left[(1+\kappa_1)\mp\sqrt{(1+\kappa_1)^2-4\kappa_1\alpha_H^2}\right]\ . \tag{6.31}$$

Plotting these solutions as a function of α_H, we obtain Fig. 6.4. For $\gamma_H = 0$, the third equation of (6.27) gives $\gamma = 0$; α and β are determined from the first and second equations. They do not contain κ_2 and depend solely on κ_1. Multiplying the first and second equations by β and α, respectively, and taking the difference, we obtain

$$(\alpha\alpha_H+\beta\beta_H)(\alpha_H\beta-\beta_H\alpha)-\kappa_1\alpha\beta=0\ . \tag{6.32}$$

If we define the angle between the magnetic field and the x axis as θ_H, and the angle between the spin axis and the magnetic field as ψ, (6.32) leads to

$$\tan 2\psi=\frac{\sin 2\theta_H}{\cos 2\theta_H-1/\kappa_1}\ ; \tag{6.33}$$

see Fig. 6.5.

Of the two solutions of (6.33), the one corresponding to the energy eigenvalue λ_1 is the desired solution. The result (6.33) was derived in 1936 by *Néel*

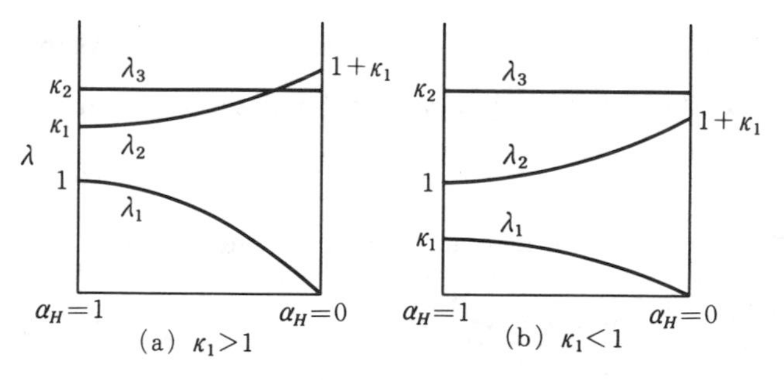

Fig. 6.4. The relation between λ and α_H ($\gamma_H = 0$)

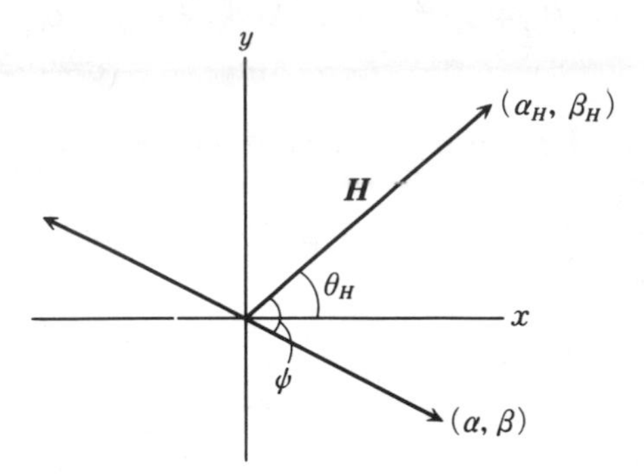

Fig. 6.5. The relation between the magnetic field and the spin axis ($\gamma_H = 0$)

[6.1]. When the external field is applied along the easy axis (the x axis), we have $\theta_H = 0$ and the solution corresponding to λ_1 is given by

$$\begin{aligned} \psi &= 0 \qquad (\kappa_1 > 1) , \\ \psi &= \frac{\pi}{2} \qquad (\kappa_1 < 1) . \end{aligned} \tag{6.34}$$

Physically this result means that the spin axis is aligned parallel to the easy axis if κ_1 is larger than 1, while it becomes perpendicular to the easy axis if κ_1 becomes smaller than 1 (which happens with the increasing external field). The critical field H_c at which the spin axis changes is

$$H_c = \sqrt{\frac{2K_1}{\chi_\perp - \chi_\parallel}} . \tag{6.35}$$

If the magnetic field is not along the x axis, the spin axis changes direction continously, from the easy axis to a direction perpendicular to it.

6.2.2 The Case $\beta_H = 0$

For the external field $\boldsymbol{H}$ lying in the xz plane, the relation $\beta_H = 0$ in the second equation of (6.27) leads to $\beta \neq 0$ and

$$\lambda_3 = \kappa_1 . \tag{6.36}$$

For the other two solutions $\beta = 0$ holds; they satisfy the quadratic equation

$$\lambda^2 - (1 + \kappa_2)\lambda + \kappa_2 \alpha_H^2 = 0 , \tag{6.37}$$

which is obtained from the first and third equations. The solutions are

$$\lambda_{1,2} = \frac{1}{2}\left[(1 + \kappa_2) \mp \sqrt{(1 + \kappa_2)^2 - 4\kappa_2 \alpha_H^2}\right] . \tag{6.38}$$

Plotting these three solutions for λ against α_H, we obtain Fig. 6.6.

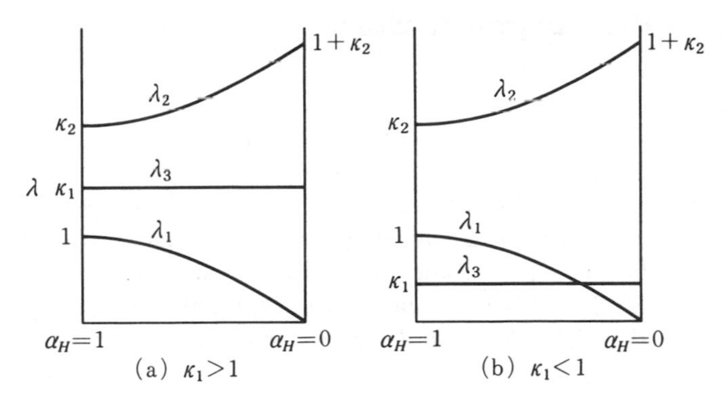

Fig. 6.6. The relation between λ and α_H ($\beta_H = 0$)

As is evident from Fig. 6.6, the eigenvalues λ_1 and λ_3 cross if $\kappa_1 < 1$ in the case $\beta_H = 0$. In the case $\kappa_1 > 1$, the spin axis stays close to the easy axis. However, when the magnetic field points close to the easy axis and becomes so large that κ_1 is smaller than 1, the λ_3 becomes the stable solution. On the other hand, if the field is tilted toward the z axis, then the stable solution swiches to λ_1 beyond a threshold and the spin axis moves back to near the easy axis. If we define the critical field for this phenomenon as (H_x, H_z), we obtain

$$\frac{H_x^2}{2K_1} - \frac{H_z^2}{2(K_2 - K_1)} = \frac{1}{\chi_\perp - \chi_\parallel} \tag{6.39}$$

by putting λ_1 in (6.38) equal to $\lambda_3 = \kappa_1$. This relation represents a hyperbola in the H_x–H_z plane, which is called the critical hyperbola (Fig. 6.7). In the hatched region on the right side of the hyperbola the spin axis is aligned along the y axis. The critical hyperbola was first derived by *Gorter* and *Haantjes* [6.2]; the above derivation is due to *Nagamiya* [6.3].

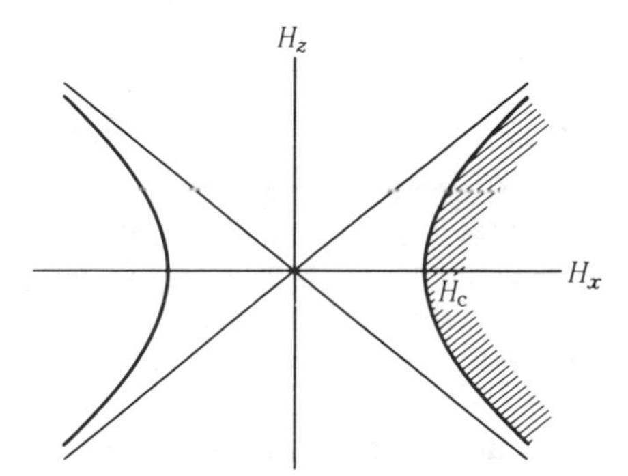

Fig. 6.7. The critical hyperbola

6.3 Transition to Paramagnetic State

For vanishing external field, the common axis of the + and − spins in the antiferromagnetic configuration points in the direction which corresponds to the minimum of the anisotropy energy (the easy axis). When the external field applied along the easy axis exceeds the critical value given by (6.35), however, the spin axis becomes perpendicular to the magnetic field. This is due to the fact that the perpendicular susceptibility $\chi_\perp$ is larger than the parallel susceptibility $\chi_\parallel$. This critical field is much larger than the anisotropy field, but is much smaller than the molecular field due to the exchange interaction. For $T = 0$, (6.35) gives

$$H_c = \sqrt{2K_1 A} = \sqrt{2\frac{K_1}{M} \cdot AM} \ , \tag{6.40}$$

since $\chi_\parallel = 0$. Therefore H_c is given by the geometrical mean of the anisotropy field K_1/M and the molecular field AM. Above this critical field, $\boldsymbol{M}_+$ and $\boldsymbol{M}_-$ tilt toward the direction of $\boldsymbol{H}$, with $\boldsymbol{M}_+ - \boldsymbol{M}_-$ remaining perpendicular to $\boldsymbol{H}$. As we showed before in the derivation of the perpendicular susceptibility, the magnitude of $\boldsymbol{M}_\pm$ stays constant when the anisotropy energy is negligible. When the external field reaches $H_{c\perp} = 2AM$ as given by (6.18), a transition from the antiferromagnetic state to the paramagnetic state occurs and $\boldsymbol{M}_+$ and $\boldsymbol{M}_-$ become parallel. At this transition there is no discontinuity in the sum $\boldsymbol{M}_+ + \boldsymbol{M}_-$.

With increasing temperature, $\chi_\parallel$ becomes nonzero, leading to a decrease in the difference between $\chi_\perp$ and $\chi_\parallel$. However, the anisotropy constant K_1 also decreases; K_1 is usually proportional to $M^{n(n+1)/2}$ at low temperatures and M^n near T_N, where n represents the order of the anisotropy energy. Thus, in the $n = 2$ case of second-order anisotropy, which is the lowest-order case, K_1 decreases in proportion to M^3 at low temperatures and to M^2 at high temperatures. The difference $\chi_\perp - \chi_\parallel$ in the susceptibility is also proportional to M^2 near T_N. Thus the critical field H_c increases with the temperature and the value at T_N (for second-order anisotropy) is given by

$$(H_c)^2_{T_N} = \frac{10}{3} \frac{(S+1)^2}{(S+1)^2 + S^2} \frac{A}{A - \Gamma} (H_c)^2_{T=0} \ . \tag{6.41}$$

If the temperature is lowered from the paramagnetic state, the antiferromagnetic state, whose spin axis is aligned along the easy axis, is realized. The transition temperature decreases proportionally to the second power of the external field, when the field is applied along the easy axis. If the magnitude of the external field exceeds $(H_c)_{T_N}$ as given by (6.41), the spins flop to a direction perpendicular to the easy axis at the temperature defined as T'_N, which also decreases proportionally to the second power of the external field. If we take into account the anisotropy energy, the two transition temperatures satisfy $T_N > T'_N$ at zero field. Because the slopes have different magnitudes ($|dT_N/dH| > |dT'_N/dH|$), the transition temperatures cross at the point given by (6.41); this situation is shown in Fig. 6.8.

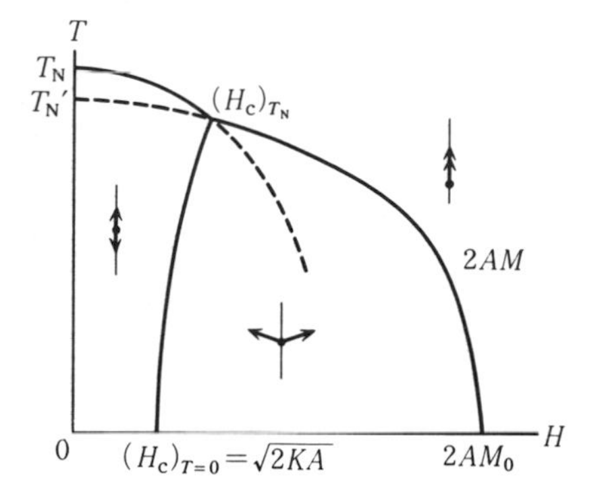

Fig. 6.8. The transition between the paramagnetic and antiferromagnetic states for a magnetic field applied along the easy axis

7. Paramagnetic Susceptibility and Critical Phenomena

In the molecular-field theory, exchange interactions are taken account of only in their averages. Therefore, exchange interactions have no effect in the paramagnetic states above the transition temeperatures in the absence of an external field. In this chapter, we consider the effect of the exchange interactions in the paramagnetic states and further discuss the critical behavior near the transition temperatures. In addition, we also derive formulae for the neutron scattering cross-sections by magnetic materials.

7.1 Susceptibility and Spin Correlation Functions

The molecular-field theory describes both ordered states and the disordered state qualitatively well and is widely used for this reason. However, in this theory the interaction is replaced by an average molecular field; deviations from the average (i.e., fluctuations of the molecular field) are ignored. Therefore this approximation is not appropriate especially in the paramagnetic region close to the transition temperature. For example, in the molecular-field approximation, the exchange interaction does not contribute to the specific heat above the transition point, if no magnetic field is present. Since the average of the exchange interaction energy is given in general by

$$-\sum_{ij} J_{ij} \langle \boldsymbol{S}_i \cdot \boldsymbol{S}_j \rangle_{\text{av}} = -N \sum_{\rho} J_{i,i+\rho} \langle \boldsymbol{S}_i \cdot \boldsymbol{S}_{i+\rho} \rangle_{\text{av}} , \tag{7.1}$$

there should be a nonvanishing contribution to the specific heat when the spin correlation function $\langle \boldsymbol{S}_i \cdot \boldsymbol{S}_{i+\rho} \rangle_{\text{av}}$ is nonvanishing. If we take into account such spin correlations, the energy of the paramagnetic state is lowered. Because of this fact, the transition temperature to the ordered state is expected to be generally lower than predicted by the molecular-field approximation.

There have been many attempts to extend the molecular-field theory to take account of the energy due to the spin correlations. Representative approaches include the method by *Weiss* [7.1], which corresponds to the Bethe approximation for the order-disorder transition in binary alloys, and the constant-coupling approximation [7.2, 3]. The moment-expansion method, in which the partition function is expanded in powers of the inverse temperature, is another method to take into account the spin correlations. Before discussing this approach, we consider first the general properties of the paramagnetic susceptibility for a spatially varying external field.

We have already derived the expression for the susceptibility $\chi(\boldsymbol{Q})$, (5.42), under the external field spatially rotating with the vector $\boldsymbol{Q}$ on the basis of the molecular-field approximation. Here we proceed, however, along the approach taken by *de Gennes* [7.4]. Suppose that a local magnetic field $\boldsymbol{H}_i$ is applied to each spin in a spin system which is described by the Heisenberg Hamiltonian:

$$\begin{aligned}\mathcal{H} &= \mathcal{H}_0 + \mathcal{H}' ,\\ \mathcal{H}_0 &= -2\sum_{\langle i,j\rangle} J_{ij}\boldsymbol{S}_i\cdot\boldsymbol{S}_j ,\\ \mathcal{H}' &= -g\mu_{\mathrm{B}}\sum_{i\alpha} S_i^\alpha H_i^\alpha ;\end{aligned} \tag{7.2}$$

here α represents the x, y or z component. If we write the eigenstate and eigenvalue of $\mathcal{H}_0$ as $|n\rangle$ and E_n, respectively, then the energy E_n' in the presence of the Zeeman energy $\mathcal{H}'$ is evaluated, to second order, as

$$\begin{aligned}E_n' = E_n &- g\mu_{\mathrm{B}}\sum_{i\alpha}\langle n|S_i^\alpha|n\rangle H_i^\alpha\\ &- g^2\mu_{\mathrm{B}}^2\sum_{ij}\sum_{\alpha\beta}\sum_{m\neq n}\frac{\langle n|S_i^\alpha|m\rangle\langle m|S_j^\beta|n\rangle}{E_m - E_n}H_i^\alpha H_j^\beta .\end{aligned} \tag{7.3}$$

Since the free energy F of the spin system is given by

$$\begin{aligned}F &= -kT\log Z ,\\ Z &= \sum_n \exp\left(-\frac{E_n'}{kT}\right) ,\end{aligned} \tag{7.4}$$

we obtain

$$F = F_0 - g\mu_{\mathrm{B}}\sum_{i\alpha}\langle S_i^\alpha\rangle H_i^\alpha - \frac{1}{2}\sum_{ij\alpha\beta}\chi_{ij}^{\alpha\beta}H_i^\alpha H_j^\beta \tag{7.5}$$

after substituting (7.3) into (7.4). In this expression, F_0 is the free energy for $\boldsymbol{H}_i = 0$; the generalized susceptibility $\chi_{ij}^{\alpha\beta}$, which represents the response of a spin S_i^α to a local field H_j^β, is given by

$$\chi_{ij}^{\alpha\beta} = \frac{(g\mu_{\mathrm{B}})^2}{kT}\left(\langle S_i^\alpha S_j^\beta\rangle - \langle S_i^\alpha\rangle\langle S_j^\beta\rangle\right) , \tag{7.6}$$

where $\langle A\rangle$ means the thermal average of a physical quantity A:

$$\langle A \rangle = \frac{\sum_n \mathrm{e}^{-E_n/kT} \langle n|A|n\rangle}{\sum_n \mathrm{e}^{-E_n/kT}} . \tag{7.7}$$

The relation $(E_m - E_n) \ll kT$ has been used to derive (7.6). The quantity $\langle (S_i^\alpha - \langle S_i^\alpha \rangle)(S_j^\beta - \langle S_j^\beta \rangle)\rangle$ on the right-hand side of (7.6) is the correlation of spins separated by $\boldsymbol{j} - \boldsymbol{i}$ and is called the *spin correlation function.*

If the local field $\boldsymbol{H}_i$ oscillates in time with an angular frequency ω, the perturbation Hamiltonian $\mathcal{H}'$ is

$$\mathcal{H}' = -\frac{g\mu_\mathrm{B}}{2} \sum_{i\alpha} S_i^\alpha (H_i^\alpha \mathrm{e}^{\mathrm{i}\omega t} + H_i^{\alpha *} \mathrm{e}^{-\mathrm{i}\omega t}) \mathrm{e}^{\eta t} , \tag{7.8}$$

where η is a positive infinitesimal. Time-dependent perturbation theory applied to this Hamiltonian gives the imaginary part of the complex susceptibilty as

$$\begin{aligned} \chi_{ij}^{\alpha\beta''} = \mathrm{Im}\, \chi_{ij}^{\alpha\beta} = \frac{1}{4} g^2 \mu_\mathrm{B}^2 \frac{1 - \mathrm{e}^{-\hbar\omega/kT}}{\hbar} \int_{-\infty}^{\infty} dt \mathrm{e}^{-\mathrm{i}\omega t} \Big[\langle S_i^\alpha(t) S_j^\beta(0)\rangle \\ + \langle S_j^\beta(t) S_i^\alpha(0)\rangle - 2\langle S_i^\alpha\rangle\langle S_j^\beta\rangle \Big] , \end{aligned} \tag{7.9}$$

which relates a time-dependent correlation function to the susceptibility; for the derivation of (7.9), see Sect. 14.4. The result in (7.9) is known as the *fluctuation-dissipation theorem* [7.5–7].

Assuming $kT \gg \hbar\omega$ so that

$$(1 - \mathrm{e}^{-\hbar\omega/kT}) \to \frac{\hbar\omega}{kT} ,$$

using the Kramers-Kronig relation

$$\chi_{ij}^{\alpha\beta'}(\omega) = -\frac{1}{\pi} \int d\omega' P\Big(\frac{1}{\omega - \omega'}\Big) \chi_{ij}^{\alpha\beta''}(\omega') , \tag{7.10}$$

we obtain again (7.6):

$$\begin{aligned} \chi_{ij}^{\alpha\beta'}(0) &= \frac{1}{\pi} \int d\omega' \frac{1}{\omega'} \chi_{ij}^{\alpha\beta''}(\omega') \\ &= \frac{(g\mu_\mathrm{B})^2}{kT} (\langle S_i^\alpha S_j^\beta \rangle - \langle S_i^\alpha\rangle\langle S_j^\beta\rangle) , \end{aligned} \tag{7.11}$$

where $\chi_{ij}^{\alpha\beta'}$ denotes the real part of the complex susceptibility. Forming the Fourier component of $\chi_{ij}^{\alpha\beta}$ by

$$\begin{aligned} \sum_{ij} \chi_{ij}^{\alpha\beta} \mathrm{e}^{\mathrm{i}\boldsymbol{q}\cdot\boldsymbol{R}_j - \mathrm{i}\boldsymbol{q}'\cdot\boldsymbol{R}_i} &= N\delta_{\boldsymbol{q}\boldsymbol{q}'} \sum_{\boldsymbol{\rho}} \chi_{i,i+\rho}^{\alpha\beta} \mathrm{e}^{\mathrm{i}\boldsymbol{q}\cdot\boldsymbol{R}_\rho} , \\ &= \chi^{\alpha\beta}(\boldsymbol{q})\delta_{\boldsymbol{q}\boldsymbol{q}'} , \end{aligned} \tag{7.12}$$

we obtain $\chi(\boldsymbol{q})$ from (7.6) or (7.11) as

$$\chi^{\alpha\beta}(\boldsymbol{q}) = \frac{N(g\mu_{\mathrm{B}})^2}{kT}\left(\langle S_{\boldsymbol{q}}^{\alpha} S_{-\boldsymbol{q}}^{\beta}\rangle - \langle S_{\boldsymbol{q}}^{\alpha}\rangle\langle S_{-\boldsymbol{q}}^{\beta}\rangle\right) ,$$
$$\boldsymbol{S}_{\boldsymbol{q}} = \frac{1}{\sqrt{N}}\sum_{i} \boldsymbol{S}_i \mathrm{e}^{-\mathrm{i}\boldsymbol{q}\cdot\boldsymbol{R}_i} . \tag{7.13}$$

$\chi(\boldsymbol{q})$ is the susceptibility $g\mu_B\langle S_{\boldsymbol{q}}\rangle/H_{\boldsymbol{q}}$ for the $\boldsymbol{q}$th Fourier component of the magnetic moment under the sinusoidally varying magnetic field $\boldsymbol{H}_i$

$$\boldsymbol{H}_i = \frac{1}{2}(\boldsymbol{H}_{\boldsymbol{q}}\mathrm{e}^{-\mathrm{i}\boldsymbol{q}\cdot\boldsymbol{R}_i} + \boldsymbol{H}_{-\boldsymbol{q}}\mathrm{e}^{\mathrm{i}\boldsymbol{q}\cdot\boldsymbol{R}_i}) , \quad \boldsymbol{H}_{-\boldsymbol{q}} = \boldsymbol{H}_{\boldsymbol{q}}^* . \tag{7.14}$$

Since $\langle \boldsymbol{S}\rangle = 0$ holds in the paramagnetic region,

$$\chi_{ii}^{\alpha\alpha} = \frac{(g\mu_{\mathrm{B}})^2}{3kT} S(S+1) \tag{7.15}$$

is obtained from (7.6). Summation of $\chi(\boldsymbol{q})$ over $\boldsymbol{q}$ in the Brillouin zone gives

$$\sum_{\boldsymbol{q},\mathrm{zone}} \chi(\boldsymbol{q}) = \frac{V}{(2\pi)^3}\int_{\mathrm{zone}} \chi(\boldsymbol{q})d\boldsymbol{q} = N^2\chi_{ii} .$$

Therefore we have from (7.15)

$$\frac{v_0}{(2\pi)^3}\int_{\mathrm{zone}} \chi(\boldsymbol{q})d\boldsymbol{q}\frac{3kT}{N(g\mu_{\mathrm{B}})^2 S(S+1)} = 1 , \tag{7.16}$$

where v_0 denotes the volume per spin.

Since the susceptibility $\chi(\boldsymbol{q})$ for free spins is given by (7.13) as $N(g\mu_B)^2 S(S+1)/3kT$, $\chi(\boldsymbol{q})$ in the molecular-field approximation is obtained as

$$\chi(\boldsymbol{q}) = \frac{C}{T - T_{\boldsymbol{q}}} ,$$
$$T_{\boldsymbol{q}} = \frac{2J(\boldsymbol{q})S(S+1)}{3k} . \tag{7.17}$$

Let the maximum value of $J(\boldsymbol{q})$ be $J(\boldsymbol{Q})$. Then, in this approximation, $\chi(\boldsymbol{Q})$ diverges at $T = T_{\boldsymbol{Q}}$ with decreasing temperature, as $(T - T_{\boldsymbol{Q}})^{-1}$. Below $T_{\boldsymbol{Q}}$ a helical spin structure having $\boldsymbol{Q}$ as its characteristic vector is established. However, a more reliable theory, the moment-expansion method, predicts $-4/3$ for the exponent.

Although the molecular-field approximation does not give the correct susceptibility, in particular near $T = T_{\boldsymbol{Q}}$, one can discuss with this theory the qualitative behavior in the vicinity of the transition temperature. For simplicity, let us restrict ourselves to the ferromagnetic and antiferromagnetic cases and assume a crystal with cubic symmetry. Expanding $J(\boldsymbol{q})$ around $\boldsymbol{Q}$ as

$$J(\boldsymbol{q}) = J(\boldsymbol{Q}) - D(\boldsymbol{q} - \boldsymbol{Q})^2 + O\left[(\boldsymbol{q} - \boldsymbol{Q})^3\right]$$

and substituting this expression into (7.17), we express $\chi(\boldsymbol{q})$ as

$$\begin{aligned} \chi(\boldsymbol{q}) &= \frac{C}{T - T_{\boldsymbol{Q}} + (\boldsymbol{q} - \boldsymbol{Q})^2 \Delta} = \frac{B}{\kappa^2 + (\boldsymbol{q} - \boldsymbol{Q})^2} , \\ B &= \frac{C}{\Delta} , \quad \kappa^2 = \frac{T - T_{\boldsymbol{Q}}}{\Delta} . \end{aligned} \tag{7.18}$$

The correlation function in the real-space representation can be calculated from (7.6) in the following way:

$$\begin{aligned} \langle S_i^\alpha S_{i+\rho}^\alpha \rangle &= \frac{kT}{(g\mu_B)^2} \chi_{i,i+\rho}^{\alpha\alpha} = \frac{kT}{N^2 (g\mu_B)^2} \sum_{\boldsymbol{q}} \chi(\boldsymbol{q}) \mathrm{e}^{-\mathrm{i}\boldsymbol{q}\cdot\boldsymbol{R}_\rho} \\ &= \frac{kT}{N^2 (g\mu_B)^2} \frac{V}{(2\pi)^3} B \int \frac{\mathrm{e}^{-\mathrm{i}\boldsymbol{q}\cdot\boldsymbol{R}_\rho}}{\kappa^2 + (\boldsymbol{q} - \boldsymbol{Q})^2} d\boldsymbol{q} \\ &\propto A \mathrm{e}^{-\mathrm{i}\boldsymbol{Q}\cdot\boldsymbol{R}_\rho} \frac{\mathrm{e}^{-\kappa R_\rho}}{R_\rho} . \end{aligned} \tag{7.19}$$

$1/\kappa$ is the effective range of the correlation, i.e., the correlation length; κ decreases as the temperature approaches $T_{\boldsymbol{Q}}$ so that the correlation length becomes infinite as $T \to T_{\boldsymbol{Q}}$. Conversely, the transition temperature can be characterized as the point where the correlation length diverges.

Let us consider the time-dependent correlation function $\langle \boldsymbol{S}_{\boldsymbol{q}}(0) \boldsymbol{S}_{-\boldsymbol{q}}(t) \rangle$; this quantity is directly related to the cross-section for diffuse scattering of neutrons. For general $\boldsymbol{q}$ the correlation function is negligible for time t greater than $\hbar/2J$, a value determined by the exchange interaction; however, the relaxation time becomes particularly long for a special value of $\boldsymbol{q}$. One of the reasons for this fact is that $S_{\boldsymbol{q}=0}$, which is the total spin, is a constant of the motion for the Heisenberg Hamiltonian and does not change with time. Thus from the continuity equation, the motion of $\boldsymbol{S}_{\boldsymbol{q}}$ is especially slow for small $\boldsymbol{q}$. A spin density $\boldsymbol{S}(\boldsymbol{r})$ which varies slowly in space can be described by the phenomenological diffusion equation

$$\frac{\partial \boldsymbol{S}(\boldsymbol{r})}{\partial t} = \Lambda \nabla^2 \boldsymbol{S}(\boldsymbol{r}) , \tag{7.20}$$

where Λ is the diffusion constant, which is inversely proportional to the susceptibility according to thermodynamic considerations [7.8]. From (7.20) the time-dependent correlation function of $\boldsymbol{S}_{\boldsymbol{q}}$ takes the form

$$\langle \boldsymbol{S}_{\boldsymbol{q}}(0) \boldsymbol{S}_{-\boldsymbol{q}}(t) \rangle = \langle \boldsymbol{S}_{\boldsymbol{q}}(0) \boldsymbol{S}_{-\boldsymbol{q}}(0) \rangle \mathrm{e}^{-\Lambda q^2 |t|} \tag{7.21}$$

for $qa \ll 1$ and $|t| \gg \hbar/2J$.

For $\boldsymbol{q} \to \boldsymbol{Q}$ and $T \to T_{\boldsymbol{Q}}$ the thermodynamic potential for $\boldsymbol{S}_{\boldsymbol{q}}$ is not much different from that for $\boldsymbol{S}_{\boldsymbol{Q}}$. Therefore the restoring force to $\boldsymbol{S}_{\boldsymbol{q}} \to 0$ is so small that $\boldsymbol{S}_{\boldsymbol{q}}$ tends to approach thermal equilibrium extremely slowly; because of this the correlation time becomes long. This phenomenon is called *critical slowing down*. In ferromagnets (which have $\boldsymbol{Q} = 0$) this phenomenon occurs together with the fact that $S_{q=0}$ is a constant of the motion for the Heisenberg Hamiltonian. The Fourier transform of (7.21) is given by

$$\frac{1}{2\pi}\int_{-\infty}^{\infty} dt e^{-i\omega t}\langle \boldsymbol{S}_q(0)\cdot \boldsymbol{S}_{-q}(t)\rangle = \frac{1}{\pi}\frac{\tau_q}{1+\omega^2\tau_q^2}\langle \boldsymbol{S}_q\cdot \boldsymbol{S}_{-q}\rangle \ . \tag{7.22}$$

where $\tau_q = 1/\Lambda q^2$; $\omega \ll 2J$ is assumed because of $t \gg \hbar/2J$. A microscopic theory of the relaxation time τ_q was presented by *Mori* and *Kawasaki* [7.9]. According to this theory, τ_q is proportional to the susceptibility χ_q, which means that $\tau_q \to \infty$ for $\boldsymbol{q} \to \boldsymbol{Q}$.

7.2 Cross-Section for Neutron Scattering

The magnetic susceptibility or spin correlation function is closely related to the scattering cross-section of neutrons by magnetic materials. For this reason neutron-scattering experiments are regarded at present as the most powerful experimental method to obtain information on spin correlation functions. In this section we explain the scattering cross-section of neutrons by electron spins, following the description due to *Van Hove* [7.10] and *de Gennes* [7.4].

A neutron and an electron interact through their magnetic moments. The interaction is the same as that between an electron and a nuclear moment, which leads to the hyperfine interaction. If we write the spin magnetic moment of the electron and the magnetic field due to the magnetic moment of the neutron by $\boldsymbol{\mu}_e$ and $\boldsymbol{H}_n$, respectively, then the interaction between a neutron and an electron spin is given by

$$\mathcal{H}_{\text{int}} = -\boldsymbol{\mu}_e \cdot \boldsymbol{H}_n \ . \tag{7.23}$$

Let us take the origin at the position of the neutron. Then the magnetic field $\boldsymbol{H}_n$ due to the magnetic moment $\boldsymbol{\mu}_n$ of the neutron has the form

$$\boldsymbol{H}_n = \text{curl}\left(\boldsymbol{\mu}_n \times \frac{\boldsymbol{r}}{r^3}\right) \ . \tag{7.24}$$

If one takes for $\boldsymbol{r}$ the ionic radius, the magnitude of the interaction (7.23) is of the order of 10^{-3}K, which is much smaller than the energy 10^3K of thermal neutrons. Therefore the scattering of neutrons by electron spins can be treated within the Born approximation. The interaction of a neutron with an electron orbital moment has a form different from (7.23). We will not go into this problem, but in the limit of small scattering vector $\boldsymbol{q}$, the orbital moment has the same effect as the spin moment.

Let the initial and final states of the total system consisting of a neutron and the electron spins be n and m, respectively. According to the golden rule, the transition probability from n to m caused by (7.23) is given by

$$W_{n\to m} = \frac{2\pi}{\hbar}|\langle n|\mathcal{H}_{\text{int}}|m\rangle|^2 \rho(E_m)\delta(E_n - E_m) \ , \tag{7.25}$$

where $\rho(E_m)$ is the density of final states of energy E_m.

Let us assume that the neutron has wavevectors $\boldsymbol{k}_0$ in the initial state and $\boldsymbol{k}$ in the final state. The energy change $\hbar\omega$ of the neutron satisfies

$$\frac{\hbar^2 k^2}{2M} - \frac{\hbar^2 k_0^2}{2M} = \hbar\omega \ . \tag{7.26}$$

Further, we express the neutron spin in the initial and final states as σ and σ', respectively, and the corresponding states of the electron spin system as n and n'. From (7.25), the transition probability from $(n\boldsymbol{k}_0\sigma)$ to $(n'\boldsymbol{k}\sigma')$ takes the form

$$W_{n\to m} = \frac{2\pi}{\hbar^2}|\langle n\boldsymbol{k}_0\sigma|\mathcal{H}_{\text{int}}|n'\boldsymbol{k}\sigma'\rangle|^2 dk_x dk_y dk_z \delta\Big(\frac{E_{n'}-E_n}{\hbar}+\omega\Big) \ . \tag{7.27}$$

Using the relation

$$dk_x dk_y dk_z = k^2 dk\, d\Omega = \frac{M}{\hbar} k\, d\Omega\, d\omega \ , \tag{7.28}$$

we obtain for the differential cross-section of scattering of neutrons (per unit solid angle and unit energy) the expression

$$\frac{d^2\sigma}{d\Omega\, d\omega} = \frac{2\pi M k}{\hbar^3}\sum_{\sigma\sigma'}\sum_{nn'} p_\sigma p_n \Big|\Big\langle n\boldsymbol{k}_0\sigma\Big|\mathcal{H}_{\text{int}}\Big|n'\boldsymbol{k}\sigma'\Big\rangle\Big|^2 \delta\Big(\frac{E_{n'}-E_n}{\hbar}+\omega\Big) \ . \tag{7.29}$$

Since the density of states of neutrons in the final state has been expressed with (7.28), we take for the plane-wave state with wave vector $\boldsymbol{k}$ the one normalized in the whole space

$$\psi_{\boldsymbol{k}} = \frac{1}{(2\pi)^{3/2}} \mathrm{e}^{\mathrm{i}\boldsymbol{k}\cdot\boldsymbol{r}_\mathrm{n}} \tag{7.30}$$

and for the incident wave the plane wave normalized with unit flux density

$$\psi_{\boldsymbol{k}_0} = \sqrt{\frac{M}{\hbar k_0}} \mathrm{e}^{\mathrm{i}\boldsymbol{k}_0\cdot\boldsymbol{r}_\mathrm{n}} \ . \tag{7.31}$$

On substituting (7.30, 31) into the matrix element of (7.29), one finds that (7.29) can be rewritten as

$$\frac{d^2\sigma}{d\Omega\, d\omega} = \Big(\frac{M}{2\pi\hbar^2}\Big)^2\Big(\frac{k}{k_0}\Big)\sum_{\sigma\sigma'}\sum_{nn'} p_\sigma p_n \Big|\int \langle n\sigma|\mathrm{e}^{-\mathrm{i}\boldsymbol{q}\cdot\boldsymbol{r}_\mathrm{n}}\mathcal{H}_{\text{int}}|n'\sigma'\rangle d\boldsymbol{r}_\mathrm{n}\Big|^2$$
$$\times\, \delta\Big(\frac{E_{n'}-E_n}{\hbar}+\omega\Big) \ . \tag{7.32}$$

Here $\boldsymbol{q}$ is the scattering vector defined by $\boldsymbol{q} = \boldsymbol{k}_0 - \boldsymbol{k}$ and p_σ and p_n are the probability of the neutron spin and the electron spin system in the initial state, respectively. In particular, p_n is given by the Boltzmann factor

$$p_n = \exp\Big(-\frac{E_n}{kT}\Big)\Big/\sum_n \exp\Big(-\frac{E_n}{kT}\Big) . \tag{7.33}$$

The integration over the coordinate $\boldsymbol{r}_{\mathrm{n}}$ of the neutron is replaced by an integration over $\boldsymbol{r}_{\mathrm{n}} - \boldsymbol{r}_{\mathrm{e}} = \boldsymbol{r}$, the coordinate relative to the electron coordinate $\boldsymbol{r}_{\mathrm{e}}$. Then the matrix element has the form

$$\begin{aligned}\int \langle n\sigma|\mathrm{e}^{-\mathrm{i}\boldsymbol{q}\cdot\boldsymbol{r}_{\mathrm{n}}}\mathcal{H}_{\mathrm{int}}|n'\sigma'\rangle d\boldsymbol{r}_{\mathrm{n}} \\ = \Big\langle n\sigma\Big|\sum_{\mathrm{e}}(-\mathrm{e}^{-\mathrm{i}\boldsymbol{q}\cdot\boldsymbol{r}_{\mathrm{e}}})\int d\boldsymbol{r}\mathrm{e}^{-\mathrm{i}\boldsymbol{q}\cdot\boldsymbol{r}}\boldsymbol{\mu}_{\mathrm{e}}\cdot\boldsymbol{H}_{\mathrm{n}}(\boldsymbol{r})\Big|n'\sigma'\Big\rangle ,\end{aligned} \tag{7.34}$$

where the summation over e means summation over all the electrons. Using (7.24) for $\boldsymbol{H}_{\mathrm{n}}(\boldsymbol{r})$, we can calculate the integral on the right-hand side as

$$\begin{aligned}\int d\boldsymbol{r}\mathrm{e}^{-\mathrm{i}\boldsymbol{q}\cdot\boldsymbol{r}}\boldsymbol{\mu}_{\mathrm{e}}\cdot\boldsymbol{H}_{\mathrm{n}}(\boldsymbol{r}) &= 4\pi\Big(\boldsymbol{\mu}_{\mathrm{e}}\cdot\boldsymbol{\mu}_{\mathrm{n}} - \frac{(\boldsymbol{\mu}_{\mathrm{n}}\cdot\boldsymbol{q})(\boldsymbol{\mu}_{\mathrm{e}}\cdot\boldsymbol{q})}{q^2}\Big) \\ &= 4\pi\boldsymbol{\mu}_{\mathrm{e}\perp}\cdot\boldsymbol{\mu}_{\mathrm{n}} ,\end{aligned} \tag{7.35}$$

where the symbol $\perp$ denotes the component perpendicular to the scattering vector $\boldsymbol{q}$.

Let us define the $\boldsymbol{q}$ component of the spin magnetic moment of the electrons by

$$\boldsymbol{M}(\boldsymbol{q}) = \frac{1}{\sqrt{N}}\sum_{\mathrm{e}}\boldsymbol{\mu}_{\mathrm{e}}\mathrm{e}^{-\mathrm{i}\boldsymbol{q}\cdot\boldsymbol{r}_{\mathrm{e}}} . \tag{7.36}$$

In the localized spin model, the total spin magnetic moment of each ion $\boldsymbol{M}_n$ is given from its spin density $\rho_{\mathrm{s}}(\boldsymbol{r})$ by

$$\boldsymbol{M}_n = \boldsymbol{\mu}_{\mathrm{e}}\int \rho_{\mathrm{s}}(\boldsymbol{r})d\boldsymbol{r} , \tag{7.37}$$

if lattice vibrations are neglected. By using $\rho_{\mathrm{s}}(\boldsymbol{r})$, $\boldsymbol{M}(\boldsymbol{q})$ takes the following form:

$$\boldsymbol{M}(\boldsymbol{q}) = \frac{F_{\mathrm{s}}(\boldsymbol{q})}{\sqrt{N}}\sum_n \boldsymbol{M}_n\mathrm{e}^{-\mathrm{i}\boldsymbol{q}\cdot\boldsymbol{R}_{\mathrm{n}}} , \tag{7.38}$$

where $F_{\mathrm{s}}(\boldsymbol{q})$ is the atomic form factor normalized to 1 at $q = 0$:

$$F_{\mathrm{s}}(\boldsymbol{q}) = \frac{\int \rho_{\mathrm{s}}(\boldsymbol{r})\mathrm{e}^{-\mathrm{i}\boldsymbol{q}\cdot\boldsymbol{r}}d\boldsymbol{r}}{\int \rho_{\mathrm{s}}(\boldsymbol{r})d\boldsymbol{r}} . \tag{7.39}$$

Using (7.35) and $\boldsymbol{M}(\boldsymbol{q})$ in (7.36), we can rewrite (7.34) as

$$-4\pi\sqrt{N}\langle n\sigma|\boldsymbol{M}_{\perp}(\boldsymbol{q})\cdot\boldsymbol{\mu}_{\mathrm{n}}|n'\sigma'\rangle . \tag{7.40}$$

Next, we substitute the representation

$$\delta\Big(\frac{E_{n'}-E_n}{\hbar}+\omega\Big)=\frac{1}{2\pi}\int_{-\infty}^{\infty}\exp\Big[-\mathrm{i}\Big(\omega+\frac{E_{n'}-E_n}{\hbar}\Big)t\Big]dt \tag{7.41}$$

of the δ function into (7.29). Then, by noting that

$$\Big\langle n\Big|\exp\Big(-\mathrm{i}\frac{E_{n'}-E_n}{\hbar}t\Big)\boldsymbol{M}(\boldsymbol{q})\Big|n'\Big\rangle=\langle n|\boldsymbol{M}(\boldsymbol{q},t)|n'\rangle\ ,$$

the differential cross section can be expressed as

$$\frac{d^2\sigma}{d\Omega\,d\omega}=\Big(\frac{2M}{\hbar^2}\Big)^2\frac{k}{k_0}\frac{N}{2\pi}\int_{-\infty}^{\infty}dt e^{-\mathrm{i}\omega t}\langle[\boldsymbol{\mu}_\mathrm{n}\cdot\boldsymbol{M}_\perp(\boldsymbol{q},0)]\,[\boldsymbol{\mu}_\mathrm{n}\cdot\boldsymbol{M}_\perp(-\boldsymbol{q},t)]\rangle\ , \tag{7.42}$$

where $\langle\ldots\rangle$ means the average over the spin states of the neutrons and the thermal average over the electron spins.

In the case of unpolarized neutron spins, we average also over the polarization. Equation (7.42) becomes

$$\frac{d^2\sigma}{d\Omega\,d\omega}=\frac{k}{k_0}\Big(1.91\frac{e}{\hbar c}\Big)^2\frac{N}{2\pi}\int_{-\infty}^{\infty}dt e^{-\mathrm{i}\omega t}\langle\boldsymbol{M}_\perp(\boldsymbol{q},0)\cdot\boldsymbol{M}_\perp(-\boldsymbol{q},t)\rangle\ , \tag{7.43}$$

where $\mu_\mathrm{n}=-1.91(e\hbar/Mc)\boldsymbol{s}_\mathrm{n}$ (magnitude of $\boldsymbol{s}_\mathrm{n}=1/2$) has been used.

The results of (7.42, 43) describe the scattering of neutrons by electron spins. In addition to this process, neutrons interact also with nuclei and are scattered by them. The interaction between a neutron and a nucleus can be divided into a part independent of the nuclear spin and another part depending on the nuclear spin and the neutron spin. If the nuclear spins are unpolarized, the contribution depending on the neutron spin is not important; therefore we can ignore this part. In such a situation the scattering cross-section consists of three parts, the part due purely to nuclear scattering, the part (7.42) due to the electron spins, and an interference term between the two. For a polarized neutron beam the interference term gives an effect which cannot be neglected; however, it vanishes after averaging over the neutron spin in the case of unpolarized neutrons. Therefore, for the unpolarized case, the cross-sections due to nuclear scattering and to electron spins can be treated independently.

The result of (7.43) shows that the scattering cross-section of a neutron beam is given by the spectrum of the $\boldsymbol{q}$-component of the spin density. Since the correlation between $\boldsymbol{M}(\boldsymbol{q},t)$ and $\boldsymbol{M}(\boldsymbol{q}',0)$ vanishes for $t\to\infty$, the relation

$$\lim_{t\to\infty}\langle\boldsymbol{M}_\perp(\boldsymbol{q},0)\cdot\boldsymbol{M}_\perp(-\boldsymbol{q},t)\rangle=\langle\boldsymbol{M}_\perp(\boldsymbol{q})\rangle\langle\boldsymbol{M}_\perp(-\boldsymbol{q})\rangle \tag{7.44}$$

holds. Thus the cross-section of (7.43) can be separated into two parts:

$$\frac{d^2\sigma}{d\Omega\, d\omega} = \left(\frac{d^2\sigma}{d\Omega\, d\omega}\right)_{\text{Bragg}} + \left(\frac{d^2\sigma}{d\Omega\, d\omega}\right)_{\text{diffuse}} , \tag{7.45}$$

$$\left(\frac{d^2\sigma}{d\Omega\, d\omega}\right)_{\text{Bragg}} = \frac{k}{k_0}\left(1.91\frac{e}{\hbar c}\right)^2 N\delta(\omega)\langle \boldsymbol{M}_\perp(\boldsymbol{q})\rangle\langle \boldsymbol{M}_\perp(-\boldsymbol{q})\rangle , \tag{7.46}$$

$$\left(\frac{d^2\sigma}{d\Omega\, d\omega}\right)_{\text{diffuse}} = \frac{k}{k_0}\left(1.91\frac{e}{\hbar c}\right)^2 \frac{N}{2\pi}\int_{-\infty}^{\infty} \mathrm{e}^{-\mathrm{i}\omega t} dt$$
$$\times \left[\langle \boldsymbol{M}_\perp(\boldsymbol{q},0)\cdot \boldsymbol{M}_\perp(-\boldsymbol{q},t)\rangle - \langle \boldsymbol{M}_\perp(\boldsymbol{q})\rangle\langle \boldsymbol{M}_\perp(-\boldsymbol{q})\rangle\right] . \tag{7.47}$$

Equation (7.46) corresponds to the elastic scattering because of the factor $\delta(\omega)$. If the spin system is in an ordered state whose spin structure is helical with characteristic vector $\boldsymbol{Q}$, the average of $\boldsymbol{S}_{\pm \boldsymbol{Q}}$ is nonzero. In this case, the average of $\boldsymbol{M}(\boldsymbol{q})$ given by (7.38) is nonvanishing only for $\boldsymbol{q}$ satisfying

$$\boldsymbol{q} = \pm\boldsymbol{Q} + \boldsymbol{K}_n , \tag{7.48}$$

where $\boldsymbol{K}_n$ is a reciprocal lattice vector. From this we find that the selection rule for magnetic Bragg scattering in (7.46) is given by (7.48) and that the spin structure and the spin orientation can be determined from the measurement of magnetic Bragg reflection. In contrast, (7.47) has no selection rule either for ω or for $\boldsymbol{q}$. Therefore it gives diffuse scattering in both energy and angular distribution. Equation (7.47) integrated over ω can be expressed from (7.13, 38) as

$$\frac{d\sigma}{d\Omega} = \left(\frac{k}{k_0}\right)\left|1.91\frac{e}{\hbar c}F(\boldsymbol{q})\right|^2 kT\sum_{\alpha\beta}\chi^{\alpha\beta}(\boldsymbol{q})\left(\delta_{\alpha\beta} - \frac{q_\alpha q_\beta}{q^2}\right) , \tag{7.49}$$

which is proportional to $\chi(\boldsymbol{q})$. This means that the scattering cross-section for small $\boldsymbol{q} - \boldsymbol{Q}$ increases according to (7.18), when the temperature T approaches the Néel temperature in the paramagnetic region ($\langle M(\boldsymbol{q})\rangle = 0$). This phenomenon is known as *critical scattering.*

The inelastic scattering in (7.47) is proportional to the Fourier transform of the time correlation function of $\boldsymbol{S}_{\boldsymbol{q}}$, i.e., (7.22). Thus we can observe with inelastic neutron scattering the phenomenon of critical slowing down directly. Moreover, a direct measurement of the excitation energy of spin waves is possible with this method in the ferromagnetic or antiferromagnetic state.

7.3 Moment-Expansion Method

In the previous two sections we discussed the interrelations among the magnetic susceptibility, spin correlation functions and neutron scattering cross-sections, and also explained qualitative features of critical phenomena near the critical point on the basis of the molecular-field approximation. In this section we discuss the *moment-expansion method* for calculating the magnetic susceptibility and the specific heat above the critical point.

Let us consider the Heisenberg Hamiltonian under a magnetic field applied along the z-axis

$$\mathcal{H} = -\sum_{ij} J_{ij} \boldsymbol{S}_i \cdot \boldsymbol{S}_j - g\mu_{\mathrm{B}} H \sum_i S_{iz} \,. \tag{7.50}$$

The partition function Z for this Hamiltonian is given by

$$Z = \mathrm{Trace}\Big\{\exp\Big(-\frac{\mathcal{H}}{kT}\Big)\Big\} \,. \tag{7.51}$$

Expanding formally the exponential function on the right-hand side in terms of the inverse of kT, we have

$$Z = \mathrm{Trace}\{1\}\Big[\sum_{n=0}^{\infty} \frac{1}{n!}\Big(\frac{-1}{kT}\Big)^n \langle\mathcal{H}^n\rangle\Big] \,, \tag{7.52}$$

where $\langle\mathcal{H}^n\rangle$ is defined by

$$\langle\mathcal{H}^n\rangle = \frac{\mathrm{Trace}\{\mathcal{H}^n\}}{\mathrm{Trace}\{1\}} \,. \tag{7.53}$$

By taking the logarithm of (7.52) one finds

$$\begin{aligned}\log Z = N\log(2S+1) + \frac{1}{2(kT)^2}\Big[\langle\mathcal{H}^2\rangle - \frac{1}{3kT}\langle\mathcal{H}^3\rangle \\ + \frac{1}{12(kT)^2}\left(\langle\mathcal{H}^4\rangle - 3\langle\mathcal{H}^2\rangle^2\right) + \dots\Big] \,.\end{aligned} \tag{7.54}$$

Here $\mathrm{Trace}\{1\} = (2S+1)^N$ and $\mathrm{Trace}\{\mathcal{H}\} = 0$ have been used; S is the magnitude of each spin.

Since the free energy F is given by

$$F = -kT\log Z \,, \tag{7.55}$$

the magnetic moment M induced by the magnetic field, and the specific heat C_V are

$$\begin{aligned}M = kT\frac{\partial}{\partial H}\log Z = \frac{1}{2kT}\frac{\partial}{\partial H}\Big[\langle\mathcal{H}^2\rangle - \frac{1}{3kT}\langle\mathcal{H}^3\rangle \\ + \frac{1}{12k^2T^2}\left(\langle\mathcal{H}^4\rangle - 3\langle\mathcal{H}^2\rangle^2\right) + \dots\Big] \,,\end{aligned} \tag{7.56}$$

$$\begin{aligned}C_V &= 2kT\frac{\partial}{\partial T}\log Z + kT^2\frac{\partial^2}{\partial T^2}\log Z \\ &= \frac{1}{kT^2}\langle\mathcal{H}^2\rangle - \frac{1}{k^2T^3}\langle\mathcal{H}^3\rangle + \frac{1}{2k^3T^4}\left(\langle\mathcal{H}^4\rangle - 3\langle\mathcal{H}^2\rangle^2\right) + \dots \,.\end{aligned} \tag{7.57}$$

The expansion of $\log Z$ is in terms of cumulants (or semi-invariants). The average $\langle\mathcal{H}^n\rangle$ contains, in addition to terms proportional to N, terms proportional to $N^2, N^3 \dots N^n$; however terms higher than N cancel with each other in the cumulants appearing in (7.54), leaving only terms proportional to N. This is an expected result since the free energy should be proportional to N. In this way the moment expansion of (7.52) gives a high-temperature expansion in powers of J/kT. If we note that $\langle \mathcal{H}^2 \rangle = (2/3)NS^2(S+1)^2 \sum_j J_{ij}^2$, then the first term in (7.57) for the specific heat reduces to

$$C_V^{(2)} = \frac{2N}{3kT^2} S^2(S+1)^2 \sum_j J_{ij}^2 + \frac{N}{3kT^2} g^2 \mu_{\mathrm{B}}^2 H^2 S(S+1) , \tag{7.58}$$

including the Zeeman term. This is the leading part of the specific heat above the transition point.

For the magnetic susceptibility we pick up the coefficient of the H^2 term from (7.56), ignoring the saturation effect. Thus we obtain

$$\chi = \frac{Ng^2\mu_{\mathrm{B}}^2 S(S+1)}{3kT} \left\{ 1 + \frac{2JzS(S+1)}{3kT} + \frac{2J^2 zS(S+1)}{3k^2T^2} \left[\frac{2}{3}(z-1)S(S+1) - \frac{1}{2} \right] + \ldots \right\} ,$$

where the exchange interaction connects only nearest-neighbor spin pairs and z denotes the number of nearest-neighbor spins. The inverse of this quantity has the form

$$\frac{1}{\chi} = \frac{3kT}{Ng^2\mu_{\mathrm{B}}^2 S(S+1)} \left\{ 1 - \frac{2JzS(S+1)}{3kT} + \frac{J^2 zS(S+1)}{3k^2T^2} \left[1 + \frac{4}{3}S(S+1) \right] - \ldots \right\} . \tag{7.59}$$

To this order, the lattice structure appears only through z. In higher orders, details of the lattice structure naturally appear. Keeping the first and second terms in (7.59) and determining the temperature at which the susceptibility diverges for $J > 0$, we obtain

$$T_c = \frac{2JzS(S+1)}{3k} ,$$

which recovers the value in the molecular-field approximation. If we keep the terms to J^2, the equation to determine T_c is quadratic; the physical solution, which corresponds to the higher T_c, is

$$\frac{Jz}{kT_c} = \frac{z}{1+\frac{4}{3}S(S+1)} \left\{ 1 - \sqrt{1 - 3\left[1 + \frac{4}{3}S(S+1)\right]\left[zS(S+1)\right]^{-1}} \right\} . \tag{7.60}$$

For $S = 1/2$ the right-hand side of (7.60) is

$$\frac{z}{2}\left(1 - \sqrt{1 - \frac{8}{z}}\right) ,$$

equalling 2.54 for $z = 12$. The corresponding value in the molecular-field approximation is 2. This result is given in the original paper of Heisenberg. Generally speaking, the Curie temperature approaches the true value in oscillatory fashion with increasing order of the approximation.

The calculation of higher-order terms in the high temperature expansion of the susceptibility and the specific heat has been carried out by many researchers, in particular *Domb* and *Sykes* [7.11, 12] and *Rushbrook* and *Wood* [7.13]. We express the susceptibility as

$$\frac{3kT\chi}{Ng^2\mu_B^2 S(S+1)} = \sum_{n=0}^{\infty} \frac{h_n(S)}{S^{2n}} K^n ; \tag{7.61}$$

for nearest-neighbor interactions ($J > 0$) on the fcc lattice, the coefficients $h_n(S)$ are

$$\begin{aligned}
h_0(S) &= 1 , \\
h_1(S) &= 4X , \\
h_2(S) &= \frac{X}{3}(44X - 3) , \\
h_3(S) &= \frac{X}{45}(2328X^2 - 382X + 12) , \\
h_4(S) &= \frac{X}{540}(96368X^3 - 26432X^2 + 2352X - 45) , \\
h_5(S) &= \frac{X}{14175}(8600616X^4 - 3377996X^3 + 526440X^2 - 30060X + 432) , \\
h_6(S) &= \frac{X}{340200}(694722560X^5 - 358715504X^4 \\
&\quad + 81267018X^3 - 8691207X^2 + 367290X - 4347) ,
\end{aligned} \tag{7.62}$$

where

$$K = \frac{2S^2 J}{kT} , \quad X = S(S+1) .$$

The coefficients $h_n(S)/S^{2n}$ of K^n in (7.61) are positive and remain finite for $S \to \infty$. The singular point z_c of a function of z, which is given by power series with positive coefficients a_n as

$$f(z) = \sum_{n=0}^{\infty} a_n z^n , \tag{7.63}$$

is located on the positive real axis. z_c determines the convergence radius, which is given by

$$\frac{1}{z_c} = \lim_{n\to\infty} |a_n|^{1/n} . \tag{7.64}$$

However, since the convergence of (7.64) for large n is slow, we use instead the relation

$$\frac{1}{z_c} = \lim_{n\to\infty} \frac{a_n}{a_{n-1}} \tag{7.65}$$

to determine z_c; this procedure is called the *ratio method.* The ratios of the expansion coefficients in (7.61) can be expressed as

$$\frac{a_n}{a_{n-1}} = \frac{1}{K_c}\left[1 + \frac{\gamma-1}{n} + O\left(\frac{1}{n^2}\right)\right] , \tag{7.66}$$

where K_c and $\gamma - 1$ are constants. The a_n satisfying (7.66) are the expansion coefficients of the function

$$f(K) = \frac{A}{(1-K/K_c)^\gamma} . \tag{7.67}$$

Therefore, for $T \to T_c$, χ has a singularity of the form

$$\chi \propto \frac{1}{(1-K/K_c)^\gamma} \propto \frac{1}{(T-T_c)^\gamma} , \tag{7.68}$$

where the transition temperature T_c is given by

$$T_c = \frac{2S^2 J}{kK_c} . \tag{7.69}$$

If we plot a_n/a_{n-1} against $1/n$, it starts from $4X/S^2$ at $1/n = 1$ and depends on n mildly. As shown in Fig. 7.1, this dependence is almost linear for a large n. Extrapolating this to $1/n \to 0$, we find $1/K_c$ from the intercept with the ordinate and $(\gamma-1)/K_c$ from the slope near $1/n = 0$. In this way we obtain for $S \to \infty$ the values

$$\frac{1}{K_c} = 3.1902 , \qquad \gamma - 1 = 0.33 \sim \frac{1}{3} . \tag{7.70}$$

The value of γ turns out to be almost independent of S for general S. Further, the value of γ is close to 4/3 also for the bcc lattice. Thus, for the

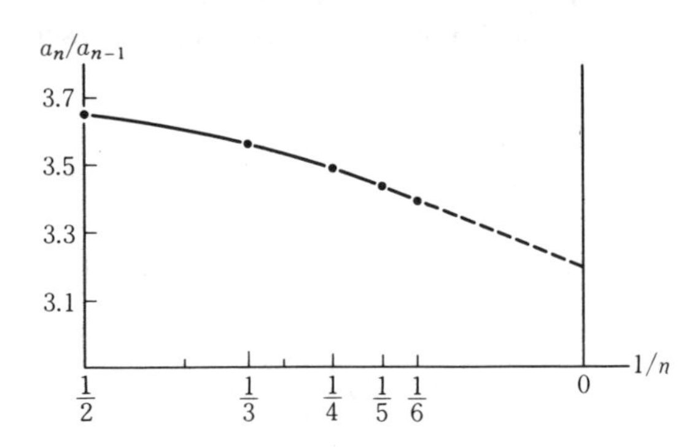

Fig. 7.1. a_n/a_{n-1} vs. $1/n$ for the $S = \infty$ Heisenberg model on the fcc lattice

three-dimensional spin system with nearest-neighbor ferromagnetic exchange interactions, the susceptibility near T_c can be described by (7.68) together with $\gamma = 4/3$ independent of the lattice structure and the magnitude of the spin. This singularity at $T = T_c$ is significantly different from the result of the molecular-field theory ($\gamma = 1$). Similarly, we can obtain an expansion for the specific heat. It appears that the nature of the singularity does not change with S in this case also. However, the singularity seems weaker than a power law in $1 - T_c/T$; it is speculated to be logarithmic.

A similar calculation can be carried out for the Ising model (or its generalization, in which the z-component can take $2S + 1$ values) with nearest-neighbor interactions. For the three-dimensional Ising model, the susceptibility and specific heat have been obtained as

$$\chi \propto \left(1 - \frac{T_c}{T}\right)^{-5/4}, \quad C_V \propto \left(1 - \frac{T_c}{T}\right)^{-1/8} \tag{7.71}$$

with much higher accuracy than those for the Heisenberg model [7.12].

For the two-dimensional lattice, the celebrated exact solution of *Onsager* [7.14] is available for the Ising model with nearest-neighbor interactions. For the two-dimensional Heisenberg model with $J > 0$, the ferromagnetic state does not occur at finite temperatures; the ordered phase becomes possible only after introducing the anisotropy energy.

7.4 Scaling Hypothesis and Critical Exponents

Magnetic materials generally undergo a phase transition at a critical temperature T_c, which is accompanied by singular behavior in physical quantities such as the susceptibility and the specific heat. The singularities are characterized by critical exponents which are universal quantities. In this section we consider this problem from a general viewpoint, taking ferromagnets as an example.

The most basic physical quantity showing a singularity at the Curie point T_c is the spin correlation function. Here we assume that spins are distributed continuously in space instead of being localized at lattice points. The reason is that the structure on scales shorter than the spin correlation length ξ is considered to be unimportant for critical phenomena. In this case the spin correlation function is defined as

$$G(\boldsymbol{x}) = \langle [\sigma(\boldsymbol{x}) - \langle\sigma\rangle][\sigma(0) - \langle\sigma\rangle] \rangle \, , \tag{7.72}$$

where $\sigma(\boldsymbol{x})$ is the spin density at $\boldsymbol{x}$ and $\langle \ldots \rangle$ represents the thermal average. The average value $\langle\sigma\rangle$ is nonzero for $T < T_c$ or under a finite magnetic field. In (7.72) σ is the component parallel to $\langle\sigma\rangle$. On Fourier transforming $\sigma(\boldsymbol{x})$ as

$$\sigma(\boldsymbol{x}) = L^{-d/2}\sum_{\boldsymbol{k}} \mathrm{e}^{\mathrm{i}\boldsymbol{k}\cdot\boldsymbol{x}}\sigma_{\boldsymbol{k}} \ , \quad \sigma_{\boldsymbol{k}} = \sigma^*_{-\boldsymbol{k}} \ , \tag{7.73}$$

we obtain the Fourier $\boldsymbol{k}$-component as

$$\sigma_{\boldsymbol{k}} = L^{-d/2}\int d\boldsymbol{x}\ \sigma(\boldsymbol{x})\mathrm{e}^{-\mathrm{i}\boldsymbol{k}\cdot\boldsymbol{x}} \ . \tag{7.74}$$

Here L denotes the linear size of the magnet and d is the dimension of the system. We define $G(\boldsymbol{k})$ as the Fourier transform of the correlation function in (7.72)

$$G(\boldsymbol{k}) = \int d\boldsymbol{x}\big\langle\big[\sigma(\boldsymbol{x}) - \langle\sigma\rangle\big]\big[\sigma(0) - \langle\sigma\rangle\big]\big\rangle\mathrm{e}^{-\mathrm{i}\boldsymbol{k}\cdot\boldsymbol{x}} \ . \tag{7.75}$$

Using (7.74) to evaluate $\langle|\sigma_{\boldsymbol{k}}|^2\rangle$, we have

$$\begin{aligned}\langle|\sigma_{\boldsymbol{k}}|^2\rangle &= L^{-d}\int d\boldsymbol{x}_1 d\boldsymbol{x}_2\langle\sigma(\boldsymbol{x}_1)\sigma(\boldsymbol{x}_2)\rangle\mathrm{e}^{-\mathrm{i}\boldsymbol{k}\cdot(\boldsymbol{x}_1-\boldsymbol{x}_2)} \\ &= G(\boldsymbol{k})\end{aligned} \tag{7.76}$$

for $\boldsymbol{k} \neq 0$. Since one can assume for isotropic media that $\langle|\sigma_{\boldsymbol{k}}|^2\rangle$ is independent of the direction of $\boldsymbol{k}$, the correlation function $G(\boldsymbol{k})$ can be regarded as a function of the magnitude of $\boldsymbol{k}$ only. $G(k)$ should diverge as $k \to 0$ and $T \to T_c$. In the limit $k \to 0$ we have

$$G(0) = \int d\boldsymbol{x}\big\langle\big[\sigma(\boldsymbol{x}) - \langle\sigma\rangle\big]\big[\sigma(0) - \langle\sigma\rangle\big]\big\rangle \ , \tag{7.77}$$

which is related to the susceptibility χ as

$$\chi = \frac{G(0)}{T} \ . \tag{7.78}$$

Let us define by ξ the correlation length over which the spin correlation is maintained. The quantity ξ increases with decreasing temperature and diverges at $T = T_c$. The phase transition can be characterized by the divergence of the correlation length. In the mean-field approximation for the exchange interaction in (7.1), the correlation length ξ is given by the inverse of κ in (7.18). Since $G(k)$, the Fourier transform of the correlation function, has a sharp peak at $k = 0$, ξ can be defined generally, independent of the details of the model, from the width of the peak:

$$\xi^2 = -\frac{1}{2}G^{-1}(0)\Big(\frac{d^2G(k)}{dk^2}\Big)_{k=0} \ . \tag{7.79}$$

According to experiments, ξ diverges at T_c as

$$\xi \propto |T - T_c|^{-\nu} \ . \tag{7.80}$$

Since one can prove that the critical exponent ν has the same value for both $T > T_c$ and $T < T_c$, this relation is assumed henceforth. Note in this connection that the two proportionality coefficients in (7.80) may be different, however, and that for $T < T_c$ one should take as σ_k in $G(k) = \langle|\sigma_k|^2\rangle$ the component parallel to the spontaneous magnetization.

We now introduce the hypothesis that the singularity of all physical quantities at T_c comes from the singularity of the correlation length ξ of spin fluctuations. This hypothesis means that all physical quantities are scaled with one parameter ξ. For this reason it is called the *scaling hypothesis.* Let us consider $G(k)$, the Fourier component of the spin correlation function. $G(k)$ depends on k and also on microscopic lengths $b_1, b_2, \ldots$. According to the scaling hypothesis these are scaled solely by ξ. Therefore $G(k)$ should be a function of $k\xi, b_1/\xi, b_2/\xi, \ldots$ as

$$G(k) = f(k\xi, b_1/\xi, b_2/\xi, \ldots) \ . \tag{7.81}$$

When $|T - T_c|$ is sufficiently small, $b_1, b_2, \ldots$ are much smaller than ξ. Thus we can expand $G(k)$ in terms of $b_1/\xi, b_2/\xi \ldots$ and approximate it with the leading term. Namely, we have

$$\begin{aligned} G(k) &\simeq \xi^y \left[g(k\xi) + \text{higher order terms in } b_1/\xi, b_2/\xi, \ldots\right] \\ &\simeq \xi^y g(k\xi) \ , \end{aligned} \tag{7.82}$$

$$g(k\xi) = \tilde{g}(k\xi) b_1^{x_1} b_2^{x_2} \ldots \ , \tag{7.83}$$

where $\tilde{g}(k\xi)$ is the coefficient of the leading term

$$f(k\xi, b_1/\xi, b_2/\xi, \ldots) \to \tilde{g}(k\xi)(b_1/\xi)^{x_1}(b_2/\xi)^{x_2} \ldots$$

in the limit $b_1/\xi, b_2/\xi, \ldots \to 0$. Therefore we obtain $-y = x_1 + x_2 + \ldots$; here y must be larger than 0. The singular temperature dependence in all physical quantities comes from ξ. The function $g(k\xi)$ also depends on temperature, but the dependence is smooth, and g can be considered constant in a narrow temperature region near T_c. On the other hand, k^{-1} is arbitrary, so that $k\xi$ can take any value.

The scaling hypothesis does not determine the value of y, nor does it tell us the precise functional form of g. However, this hypothesis does give relations between y and critical exponents.

Let us define the critical exponents of the susceptibility and the correlation function $G(k)$ at $T = T_c$ by

$$\chi(T) \propto |T - T_c|^{-\gamma} \ , \tag{7.84}$$

$$G(k) \propto k^{-2+\eta} \ , \qquad T = T_c \ . \tag{7.85}$$

If we let $k \to 0$, we obtain

$$G(0) = \xi^y g(0) \propto |T - T_c|^{-\nu y}$$

on using (7.80) for ξ. Since $G(0)$ is proportional to the susceptibility, we obtain the following relation from (7.84):

$$\nu y = \gamma \ . \tag{7.86}$$

Another important relation can be obtained from (7.82) by taking $T \to T_c, \xi \to \infty$. Since we have

$$\lim_{k\xi\to\infty} G(k) = \lim_{k\xi\to\infty} \xi^y g(k\xi) \propto k^{-2+\eta} ,$$

$\lim_{k\xi\to\infty} g(k\xi) = (k\xi)^{-y}$ should hold. Thus we obtain

$$y = 2 - \eta = \frac{\gamma}{\nu} . \tag{7.87}$$

This relation involving η, γ and ν is one of the scaling relations for critical exponents.

Next we apply a scale transformation, keeping in mind the scaling hypothesis which states that ξ is the only important length. If we multiply the length scale by s, the length x and the wavevector k are transformed as $x \to x' = s^{-1}x$ and $k \to k' = sk$, respectively. Let us define in this scale transformation the dimension of x as -1 and that of k as 1. In general, we define the dimension of A as λ, when A behaves under this transformation as

$$A \to A' = s^\lambda A . \tag{7.88}$$

Clearly the dimension of the correlation length ξ is -1; the dimension of $G(k)$ is the same as that of ξ^y, i.e., $-y$. Since $G(k)$ has the dimension of (spin density)$^2\times$volume, the relation $2d_\sigma - d = -y$ holds on using the dimension d_σ of spin density. Therefore, noting (7.87), we obtain for d_σ

$$d_\sigma = \frac{1}{2}(d - y) = \frac{1}{2}(d - 2 + \eta) . \tag{7.89}$$

If we know the dimension of one quantity, we can derive in this way the dependence on ξ and consequently the dependence on $|T - T_c|$.

Since the total free energy is invariant under the scale transformation, its dimension is zero. Therefore the free energy F per unit volume has the dimension of d. Thus we obtain $F \propto \xi^{-d} \propto |T - T_c|^{\nu d}$ for $T \to T_c$. The specific heat behaves as

$$C = -T\frac{\partial^2 F}{\partial T^2} \propto |T - T_c|^{\nu d - 2} . \tag{7.90}$$

If we define the critical exponent α of the specific heat by $C \propto |T - T_c|^{-\alpha}$, we find

$$\alpha = 2 - \nu d . \tag{7.91}$$

Let us consider next the magnetization m for $T < T_c$. Since m is the average of the spin density, we obtain from (7.89)

$$m \propto \xi^{-(1/2)(d-2+\eta)} \propto |T - T_c|^{(1/2)\nu(d-2+\eta)} .$$

Introducing the critical exponent β by $m \propto (T_c - T)^\beta$, we have

$$\beta = \frac{1}{2}\nu(d - 2 + \eta) . \tag{7.92}$$

When an external field h is present, m is given by $-\partial F/\partial h$; therefore d_h, the dimension of h, is given by

$$d_h = d - d_\sigma = \frac{1}{2}(d + 2 - \eta) . \tag{7.93}$$

For $T = T_c$ and $h \neq 0$, m is finite. Since $\xi \to \infty$ as $T \to T_c$, the critical exponent δ defined by $m \propto (h^{1/d_h})^{d_\sigma} = h^{1/\delta}$ is given as

$$\delta = \frac{d_h}{d_\sigma} = \frac{d+2-\eta}{d-2+\eta} . \tag{7.94}$$

We note also that the dimension of $h\xi^{d_h}$ is zero. Therefore, for $T \neq T_c$ and $h \neq 0$, m has the form

$$m = \xi^{-d_\sigma} W_\pm(h\xi^{d_h}) . \tag{7.95}$$

Similarly, we obtain for the free energy

$$F = \xi^{-d} f_\pm(h\xi^{d_h}) . \tag{7.96}$$

Here $\pm$ signifies the function for $T > T_c$ and that for $T < T_c$.

So far we have derived relations among various critical exponents, i.e., scaling laws, on the basis of the scaling hypothesis. The results are summarized as follows:

$$\begin{aligned}
\nu &= \nu' = \frac{\gamma}{2-\eta} , \\
\alpha &= \alpha' = 2 - \nu d , \\
\beta &= \frac{1}{2}\nu(d-2+\eta) , \\
\delta &= \frac{d+2-\eta}{d-2+\eta} .
\end{aligned} \tag{7.97}$$

These relations are satisfied within experimental accuracy in various ferromagnets. For the two-dimensional Ising model on the square lattice, the exact solution by *Onsager* [7.14] is available. The exact solution tells us that $\alpha = 0$ (logarithmic), $\beta = 1/8$, $\gamma = 7/4$, $\delta = 15$, $\eta = 1/4$ and $\nu = 1$. These values satisfy the above relations exactly. This fact strongly supports the validity of the scaling hypothesis. Incidentally, we can derive from (7.97) the following relations:

$$\alpha + 2\beta + \gamma = \alpha + \beta(1 + \delta) = 2 , \tag{7.98}$$

which can be obtained from thermodynamic arguments.

We note however that the scaling hypothesis is merely an assumption, whose validity has to be clarified. The renormalization group theory, in fact, has been developed to justify the assumption. Our description in this section is based on Ma's textbook [7.15], in which theoretical developments of the renormalization group on critical phenomena are described in detail.

The scaling hypothesis has been used in various fields of physics and has achieved many successes; its application to critical phenomena is a successful example.

8. Spin-Wave Theory of Ferromagnets

In order to study various properties of spin systems described by the Heisenberg Hamiltonian, the moment-expansion method presented previously is useful at high temperatures. On the other hand, spin-wave theory is a good approximation at low temperatures. In the latter theory one starts from the ground state in which all spins are aligned and then describes excited states as a collection of spin waves. Let us consider an excited state in a $S = 1/2$ system in which the z component of a single spin is flipped from the complete ferromagnetic state where all spins are aligned along one direction. This excited state is not an eigenstate of the Heisenberg Hamiltonian. Due to the transverse component of the exchange interaction, the flipped spin moves around in the lattice; the eigenstate is a state in which a wave of spins is excited. At low temperatures the amplitude of the spin wave is small and consequently the interaction among waves is negligible. Therefore states excited above the ground state can be well approximated as a collection of independent spin waves. Such a spin wave theory was proposed for ferromagnets in 1930 by Bloch [8.1].[1] The $T^{3/2}$ law for the decrease of the spontaneous magnetization of ferromagnets was derived with this theory. The spin-wave theory, which is similar to the Debye approximation for lattice vibrations, is a basic method to treat excited states in the vicinity of the ground state. It is an effective theoretical method to study the low temperature region for antiferromagnets, ferrimagnets and helical spin structures.

8.1 Bloch Spin-Wave Theory

Following the theory of lattice vibrations, we consider the equation of motion of the spin $\boldsymbol{S}_j$ on the jth lattice point:

$$\frac{d\boldsymbol{S}_j}{dt} = \frac{\mathrm{i}}{\hbar}[\mathcal{H}, \boldsymbol{S}_j] = -\frac{1}{\hbar}(\boldsymbol{H}_j \times \boldsymbol{S}_j) \ . \tag{8.1}$$

Here $\mathcal{H}$ represents the Heisenberg Hamiltonian including the Zeeman energy $-g\mu_{\mathrm{B}}\boldsymbol{H} \cdot \boldsymbol{S}_j$. The second relation of (8.1) can be obtained from the commu-

[1] In the same year Slater discussed the ground state of the Heisenberg model for $J < 0$ [8.2].

tation relations for spin operators. The magnetic field $\boldsymbol{H}_j$ acting on the jth spin is

$$\boldsymbol{H}_j = 2\sum_i J_{ij}\boldsymbol{S}_i + g\mu_{\rm B}\boldsymbol{H} \; ; \tag{8.2}$$

it includes the effective field due to the exchange interaction. Equation (8.1) represents the classical equation of motion of an angular momentum, relating the time dependence of the spin to the torque due to the magnetic field. To take into account the anisotropy energy, the anisotropy field $\boldsymbol{H}_{\rm an}$ coming from the anisotropy energy should be added to $\boldsymbol{H}_j$.

Substituting (8.2) into (8.1) and writing down the equation of motion for the x, y, and z components, we obtain the following relations:

$$\begin{aligned}
\hbar\frac{dS_{jx}}{dt} &= -2\sum_i J_{ij}(S_{iy}S_{jz} - S_{iz}S_{jy}) + g\mu_{\rm B}HS_{jy} \; , \\
\hbar\frac{dS_{jy}}{dt} &= -2\sum_i J_{ij}(S_{iz}S_{jx} - S_{ix}S_{jz}) - g\mu_{\rm B}HS_{jx} \; , \\
\hbar\frac{dS_{jz}}{dt} &= -2\sum_i J_{ij}(S_{ix}S_{jy} - S_{iy}S_{jx}) \; .
\end{aligned} \tag{8.3}$$

Here the direction of the external field $\boldsymbol{H}$ is chosen as the z axis. In ferromagnets ($J > 0$) under such an external field each spin is aligned along the z axis; that is, $S_{jz} = S$ holds. If we restrict ourselves to low-energy excited states, the deviations S_{jx} and S_{jy} of the spin from the z direction may be regarded as small quantities of first order. From the third relation of (8.3), the change of the z component of spin in time is then a small quantity of second order. Therefore one can approximate S_{jz} as

$$S_{jz} \sim S \; . \tag{8.4}$$

Substituting (8.4) into the first and second relations of (8.3), we obtain the linear equations

$$\begin{aligned}
\hbar\frac{dS_{jx}}{dt} &= -2S\sum_i J_{ij}(S_{iy} - S_{jy}) + g\mu_{\rm B}HS_{jy} \; , \\
\hbar\frac{dS_{jy}}{dt} &= -2S\sum_i J_{ij}(S_{jx} - S_{ix}) - g\mu_{\rm B}HS_{jx} \; .
\end{aligned} \tag{8.5}$$

Putting

$$S_{jx} \pm {\rm i}S_{jy} = S_{j\pm} \; , \tag{8.6}$$

we have from (8.5)

$$\hbar\frac{dS_{j\pm}}{dt} = \pm{\rm i}\big[-2S\sum_i J_{ij}(S_{j\pm} - S_{i\pm}) - g\mu_{\rm B}HS_{j\pm}\big] \; . \tag{8.7}$$

Assuming a Bravais lattice, we apply the Fourier transform

$$S_{\mu-} = \frac{1}{\sqrt{N}} \sum_j \mathrm{e}^{-\mathrm{i}\boldsymbol{\mu}\cdot\boldsymbol{R}_j} S_{j-} \qquad (S_{\mu-} = S^*_{-\mu+}) \,, \tag{8.8}$$

which expresses $S_{\mu\pm}$ ($\boldsymbol{\mu}$: wave vector in the first Brillouin zone) in terms of the spin operator $S_{j\pm}$ at each lattice point. Then we have

$$\begin{aligned} \frac{\hbar}{\mathrm{i}} \frac{dS_{\mu-}}{dt} &= \Big[2S \sum_i J_{ij}(1 - \mathrm{e}^{-\mathrm{i}\boldsymbol{\mu}\cdot(\boldsymbol{R}_i - \boldsymbol{R}_j)}) + g\mu_\mathrm{B} H\Big] S_{\mu-} \\ &= \Big[2S(J(0) - J(\boldsymbol{\mu})) + g\mu_\mathrm{B} H\Big] S_{\mu-} \,, \end{aligned} \tag{8.9}$$

where $J(\boldsymbol{\mu})$ is defined in (5.14). If we put

$$S_{\mu-} \propto \delta S_\mu \mathrm{e}^{\mathrm{i}\omega_\mu t + \mathrm{i}\alpha} \,, \tag{8.10}$$

then the eigenfrequency of the spin waves is obtained as

$$\hbar\omega_\mu = 2S[J(0) - J(\boldsymbol{\mu})] + g\mu_\mathrm{B} H \,. \tag{8.11}$$

The first term, which depends on the exchange interaction, goes to zero as μ^2 in the limit $\boldsymbol{\mu} \to 0$.

In the state where the wave with wavenumber $\boldsymbol{\mu}$ is excited, $\boldsymbol{S}_j$ and $\boldsymbol{S}_\mu$ are related by

$$S_{j-} = \frac{1}{\sqrt{N}} \mathrm{e}^{\mathrm{i}\boldsymbol{\mu}\cdot\boldsymbol{R}_j} S_{\mu-} = \frac{1}{\sqrt{N}} \delta S_\mu \mathrm{e}^{\mathrm{i}(\boldsymbol{\mu}\cdot\boldsymbol{R}_j + \omega_\mu t + \alpha)} \,. \tag{8.12}$$

Thus the x and y components of $\boldsymbol{S}_j$ are given by

$$\begin{aligned} S_{jx} &= \frac{1}{\sqrt{N}} \delta S_\mu \cos(\boldsymbol{\mu}\cdot\boldsymbol{R}_j + \omega_\mu t + \alpha) \,, \\ S_{jy} &= -\frac{1}{\sqrt{N}} \delta S_\mu \sin(\boldsymbol{\mu}\cdot\boldsymbol{R}_j + \omega_\mu t + \alpha) \,; \end{aligned} \tag{8.13}$$

the spins $\boldsymbol{S}_j$ viewed from the z direction are shown in Fig. 8.1.

On the other hand, from (8.1, 9) the commutation relation between the Hamiltonian $\mathcal{H}$ and $S_{\mu-}$ is

$$[\mathcal{H}, S_{\mu-}] = \hbar\omega_\mu S_{\mu-} \tag{8.14}$$

in the linear approximation. Applying this operator equation to the ground-state wave function ψ_g (in which the spin on each lattice point is aligned along the z axis), we obtain

$$\mathcal{H} S_{\mu-}\psi_\mathrm{g} - S_{\mu-}\mathcal{H}\psi_\mathrm{g} = \hbar\omega_\mu S_{\mu-}\psi_\mathrm{g} \,. \tag{8.15}$$

Since $S_{\mu-}\psi_\mathrm{g}$ is the eigenfunction with a spin wave excited, we have the energy of this excited state E_μ after operating $\mathcal{H}$ onto this state. Therefore the relation (8.15) reduces to

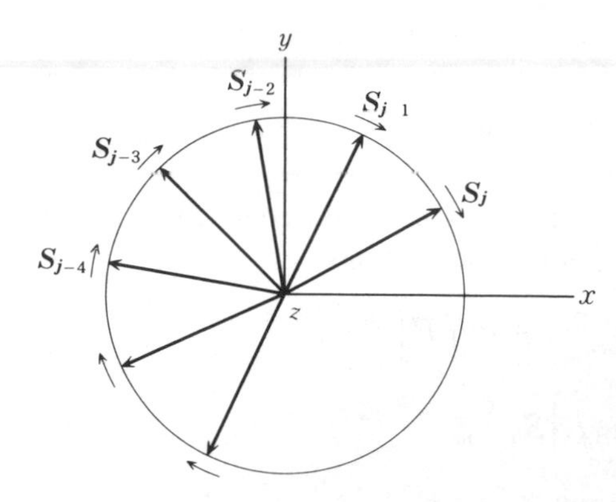

Fig. 8.1. The relation among spins $\mathbf{S}_j$ in the spin-wave state

$$E_\mu - E_\mathrm{g} = \hbar\omega_\mu \tag{8.16}$$

if we define the energy eigenvalue of the ground state to be E_g. Equation (8.16) means that the eigenfrequency ω_μ is equal to the excitation energy of the spin wave with the wavevector $\boldsymbol{\mu}$. In particular the mode for $\boldsymbol{\mu} = 0$ corresponds to the in-phase precession of all the spins; its energy is independent of the exchange interaction. The mode which can be excited by microwaves under standard conditions is this uniform one, which is often called the Kittel mode [8.3]. The excitation energy of this mode is given by the Zeeman energy. In practice one has to take into account the anisotropy energy and the effect of the demagnetizing field due to magnetic charges appearing on the surface of ferromagnets. If we assume a ferromagnet in the shape of an ellipsoid having demagnetizing factors N_x, N_y, N_z along its principal axes, the demagnetizing field is given by

$$(-N_x M_x, -N_y M_y, -N_z M_z) , \tag{8.17}$$

where M_x, M_y, M_z are the components of the magnetic moment (per unit volume) of the ferromagnet. Adding (8.17) to the external field parallel to the z axis, we can obtain the frequency ω_K of the Kittel mode from the equation of motion of M_x, M_y, M_z directly as

$$\hbar\omega_\mathrm{K} = g\mu_\mathrm{B}\sqrt{[H + (N_x - N_z)M][H + (N_y - N_z)M]} . \tag{8.18}$$

When a uniaxial anisotropy field H_A is present, H_A has to be added to H in the above expression. The relation (8.18) was first derived by Kittel.

The demagnetizing field originates from the magnetic dipole interaction. The eigenfrequency of spin waves with finite wavelength is also affected by the magnetic dipole interaction which is of long range. Let us consider two cases in which the wavevector $\boldsymbol{\mu}$ is either parallel or perpendicular to the z axis. As shown in Fig. 8.2(a), in the former case the demagnetizing field along the transverse (x, y) direction is cancelled and negligible, if the wavelength

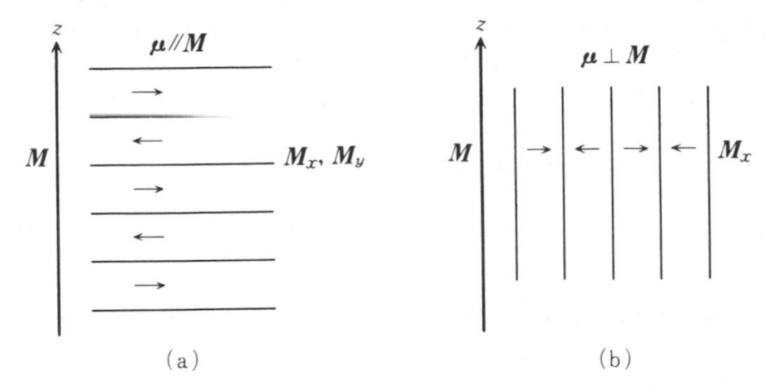

Fig. 8.2. Spin-wave excitations with wavevector parallel and perpendicular to $\boldsymbol{M}$

is much shorter than the size of the sample. Thus the spin-wave frequency is given by

$$\hbar\omega_\mu = \hbar\omega_\mu^0 + g\mu_B H - g\mu_B N_z M \ , \tag{8.19a}$$

where $\hbar\omega_\mu^0$ is the part due to the exchange interaction only.

On the other hand, in case (b) where $\boldsymbol{\mu}$ is perpendicular to the z axis, the demagnetizing field is absent for the direction perpendicular to $\boldsymbol{\mu}$ and the z axis, while we have $-4\pi M_\mu$ in the direction of $\boldsymbol{\mu}$. Therefore the frequency for this case is given by

$$\hbar\omega_\mu = \sqrt{[\hbar\omega_\mu^0 + g\mu_B H + g\mu_B(4\pi - N_z)M][\hbar\omega_\mu^0 + g\mu_B H - g\mu_B N_z M]} \ . \tag{8.19b}$$

When $\boldsymbol{\mu}$ is tilted from the z axis by an angle θ_μ, the term 4π in the first parenthesis of (8.19b) is replaced by $4\pi \sin^2 \theta_\mu$.

If the wavelength of the spin wave is comparable to the sample size, the transverse demagnetizing field has to be taken into account. In other words, the surface effect of the sample is so important that the eigenmodes are then described as standing waves. In this case the spatial variation of the magnetization is extremely gradual. Thus we can ignore the exchange interaction compared with the dipole-dipole interaction, and the problem can be treated classically. Such a characteristic oscillation mode, which is called the *magnetostatic mode*, was first observed in a ferrite by *White* and *Solt* [8.4]. Theoretical treatments of this mode were made by *Mercereau* and *Feynman* [8.5] and by *Walker* [8.6]; the mode is often called the Walker mode. If the dipole-dipole interaction is taken into account, the eigenfrequency of spin waves has a finite width for a given magnitude of $\boldsymbol{\mu}$. The upper and lower limits are given by (8.19b, a), respectively. The frequency of the Kittel mode lies within this width for $\boldsymbol{\mu} \to 0$.

When the dipole-dipole interaction and the anisotropy energy are neglected, the excitation energy (8.11) of spin wave and the eigenfunction $S_{\mu-}\psi_g$

of the excited state are exact for a single spin flip. In the more general case, where we have several flipped spins, interactions among spin waves become important. If the number of excited spin waves is not large as at sufficiently low temperatures, one is allowed to treat these spin waves as independent. In this case, the energy of the excited state is given by

$$E = \sum_{\mu} \langle n_\mu \rangle \hbar\omega_\mu , \tag{8.20}$$

where n_μ represents the number of excited spin waves. According to Boltzmann statistics the thermal average of the number of spin waves with wavevector $\boldsymbol{\mu}$ has the following form:

$$\langle n_\mu \rangle = \frac{\sum_{n_\mu} n_\mu \exp(-n_\mu \hbar\omega_\mu / kT)}{\sum_{n_\mu} \exp(-n_\mu \hbar\omega_\mu / kT)} = \frac{1}{\mathrm{e}^{\hbar\omega_\mu/kT} - 1} . \tag{8.21}$$

The number of flipped spins is obtained by summing (8.21) over $\boldsymbol{\mu}$ as

$$\sum_{\mu} \langle n_\mu \rangle = \sum_{\mu} \frac{1}{\mathrm{e}^{\hbar\omega_\mu/kT} - 1} = \frac{V}{8\pi^3} \cdot 4\pi \int_0 \frac{\mu^2 d\mu}{\mathrm{e}^{C\mu^2/kT} - 1} , \tag{8.22}$$

where $\hbar\omega_\mu$ has been approximated for $H = 0$ and small μ as

$$\hbar\omega_\mu \simeq C\mu^2 . \tag{8.23}$$

The value of C is given by

$$C = 2SJa^2 \tag{8.24}$$

(for the simple cubic, body-centered cubic, and face-centered cubic lattices), if the exchange integral between nearest-neighbor spins is J and the exchange integrals for farther spins are vanishing. Here a is the length of the unit cell for each case, i.e., the lattice constant.

The summation over $\boldsymbol{\mu}$ in (8.22) is taken within the first Brillouin zone. If this is replaced by an integral, the integrand becomes small for large $\boldsymbol{\mu}$ at low temperatures. Therefore we are allowed to take the upper bound of the integral as infinity:

$$\sum_{\mu} \langle n_\mu \rangle = \frac{V}{2\pi^2} \left(\frac{kT}{C} \right)^{3/2} \int_0^\infty \frac{x^2 dx}{\mathrm{e}^{x^2} - 1} = \frac{V}{a^3} \left(\frac{kT}{8\pi JS} \right)^{3/2} \zeta(\tfrac{3}{2}) , \tag{8.25}$$

where $\zeta(a) = \sum_{n=1}^{\infty} n^{-a}$ is Riemann's zeta function and

$$\zeta(\tfrac{3}{2}) = \sum_{n=1}^{\infty} \frac{1}{n^{3/2}} = 2.612 . \tag{8.26}$$

Thus the spontaneous magnetization of ferromagnets decreases with increasing temperature, from the completely saturated value NS at zero temperature; the decrease is proportional to $T^{3/2}$. A similar calculation shows that the corresponding specific heat is proportional to $T^{3/2}$. Because of the μ^4 and μ^6 terms, which can be obtained by expanding $\hbar\omega_\mu$ in terms of $\boldsymbol{\mu}$, there are also contributions proportional to $T^{5/2}$ and $T^{7/2}$. Spin-wave interactions, which have been ignored here, lead to contributions proportional to T^4 and higher order, as will be discussed later.

The integral in (8.25) is convergent for three-dimensional lattices. However, in the case of two-dimensional and one-dimensional lattices the integral is divergent at the lower bound, since the power of x in the numerator is smaller than 2. This means that the number of flipped spins diverges at finite temperatures, suggesting an instability of the ferromagnetic state. The absence of phase transition in one dimension is a general property of cooperative phenomena, while the absence of ferromagnetism in two dimensions is specific to the Heisenberg Hamiltonian. If we consider anisotropy energies, ferromagnetism becomes possible just as in the Ising model. In any dimension the ground state corresponding to the lowest energy is the ferromagnetic state with all spins aligned.

8.2 Interactions Among Spin Waves I – Bethe Theory

In the previous section we described low-energy excited states of ferromagnets as a collection of independent spin waves. To study the effect of interactions among spin waves, we consider the one-dimensional spin system with nearest-neighbor exchange interactions and explain Bethe's method [8.7] for determining the excited states.

For simplicity, we assume that the spin at each lattice point is 1/2. Let $\psi(m_1, m_2, \ldots, m_r)$ be the wave function for an N-spin system in which r spins are reversed. Here $m_1, m_2, \ldots, m_r$ represent the lattice points at which a reversed spin is located. Then the eigenstate of the Heisenberg Hamiltonian for r reversed spins is given by a linear combination of $\psi(m_1, m_2, \ldots, m_r)$:

$$\psi = \sum_{m_1 m_2 \ldots m_r} a(m_1, m_2, \ldots, m_r)\psi(m_1, m_2, \ldots, m_r) , \tag{8.27}$$

where $m_1, m_2, \ldots, m_r$ take values from 1 through N under the constraint

$$m_1 < m_2 < m_3 < \ldots < m_r . \tag{8.28}$$

The number of independent functions $\psi(m_1, m_2, \ldots, m_r)$ is the number of ways to choose r lattice points from N lattice points, i.e., $\binom{N}{r}$.

First we determine the diagonal element of the Hamiltonian for the state $\psi(m_1, m_2, \ldots, m_r)$. If we assume a one-dimensional ring and define the first and Nth lattice points to be nearest neighbors of each other, the diagonal element is given by

$$\mathcal{H}_{m_1 m_2 \ldots m_r; m_1 m_2 \ldots m_r} = \frac{JN}{2} - JN_{\parallel} , \tag{8.29}$$

where $N_{\parallel}$ denotes the number of nearest-neighbor parallel-spin pairs.

Off-diagonal matrix elements $-J$ exist between two states in which an antiparallel spin pair change their spins via exchange. Let the two states be $(m_1, m_2, \ldots, m_i, \ldots, m_r)$ and $(m_1, m_2, \ldots, m_i+1, \ldots, m_r)$. Then the relation $m_{i+1} \neq m_i + 1$ must be satisfied. Using these off-diagonal elements and the diagonal elements in (8.29), we obtain from Schrödinger equation for ψ the following relation involving the coefficient a:

$$2\varepsilon a(m_1, m_2, \ldots, m_r) + \sum_{m'_1, m'_2, \ldots, m'_r} [a(m'_1, m'_2, \ldots, m'_r) \\ - a(m_1, m_2, \ldots, m_r)] = 0 , \tag{8.30}$$

where the summation over $m'_1, m'_2, \ldots, m'_r$ is taken for states which are connected by exchanging opposite spin nearest-neighbor pairs from $(m_1, m_2, \ldots, m_r)$. The parameter ε is related to the energy eigenvalue E by

$$2\varepsilon J = E + \frac{NJ}{2} . \tag{8.31}$$

The coefficients a must satisfy, in addition to (8.30), the periodic boundary condition

$$a(m_1, m_2, \ldots, m_i, \ldots, m_r) = a(m_1, m_2, \ldots, m_{i-1}, m_{i+1}, \ldots, m_r, m_i+N) . \tag{8.32}$$

If there is just one reversed spin, we put $a(m) = \mathrm{e}^{ikm}$ and obtain from (8.30)

$$\varepsilon = 1 - \cos k . \tag{8.33}$$

The boundary condition (8.32) leads to

$$k = \frac{2\pi}{N}\lambda \quad (\lambda = 0, 1, 2, \ldots, N-1) . \tag{8.34}$$

The excitation energy $\hbar\omega_k$ of spin waves can be obtained as $\hbar\omega_k = 2\varepsilon J$ by subtracting from E the ground-state energy $-NJ/2$. The state for $k = 0$ corresponds to the total spin NS and its z component $NS - 1$, while the states for $k \neq 0$ belong to the total spin $NS - 1$ and its z component $NS - 1$.

If there are two reversed spins, the plane-wave solution is modified by the inteaction between the two spin waves. For $r = 2$ we have

$$-2\varepsilon a(m_1, m_2) = a(m_1 + 1, m_2) + a(m_1 - 1, m_2) + a(m_1, m_2 + 1) \\ + a(m_1, m_2 - 1) - 4a(m_1, m_2) , \tag{8.35a}$$

if the reversed spins are separated from each other. However,

$$-2\varepsilon a(m_1, m_1+1) = a(m_1-1, m_1+1) + a(m_1, m_1+2) - 2a(m_1, m_1+1) \tag{8.35b}$$

holds, if they are located on nearest-neighbor sites. If we impose on the coefficients a the condition

$$a(m_1, m_1) + a(m_1 + 1, m_1 + 1) - 2a(m_1, m_1 + 1) = 0 \,, \tag{8.36}$$

then (8.35b) can be reduced to (8.35a). The solution of (8.35a) is given by

$$a(m_1, m_2) = c_1 \mathrm{e}^{\mathrm{i}(f_1 m_1 + f_2 m_2)} + c_2 \mathrm{e}^{\mathrm{i}(f_2 m_1 + f_1 m_2)} \,, \tag{8.37}$$

$$\varepsilon = 1 - \cos f_1 + 1 - \cos f_2 \,. \tag{8.38}$$

The coefficients c_1 and c_2 are determined to satisfy (8.36) and the parameters f_1 and f_2 are fixed by the boundary condition (8.32). Substituting (8.37) into (8.36), we obtain

$$\frac{c_1}{c_2} = \frac{\sin\left(\frac{f_1 - f_2}{2}\right) + \mathrm{i}\left(\cos\frac{f_1+f_2}{2} - \cos\frac{f_1-f_2}{2}\right)}{\sin\left(\frac{f_1 - f_2}{2}\right) - \mathrm{i}\left(\cos\frac{f_1+f_2}{2} - \cos\frac{f_1-f_2}{2}\right)} \,. \tag{8.39}$$

Therefore, putting

$$c_1 = \mathrm{e}^{\mathrm{i}\varphi/2}, \qquad c_2 = \mathrm{e}^{-\mathrm{i}\varphi/2} \,, \tag{8.40}$$

one finds

$$2\cot\frac{\varphi}{2} = \cot\frac{f_1}{2} - \cot\frac{f_2}{2} \,, \tag{8.41}$$

where $\varphi/2$ represents the phase shift due to the interaction. In terms of φ, a has the form

$$a(m_1, m_2) = \mathrm{e}^{\mathrm{i}(f_1 m_1 + f_2 m_2 + \varphi/2)} + \mathrm{e}^{\mathrm{i}(f_2 m_1 + f_1 m_2 - \varphi/2)} \,. \tag{8.42}$$

Taking m_1 and m_2 between 1 and N and substituting this relation into the boundary condition

$$a(m_1, m_2) = a(m_2, m_1 + N) \,, \tag{8.43}$$

we obtain

$$\begin{aligned} &\mathrm{e}^{\mathrm{i}(f_1 m_1 + f_2 m_2 + \varphi/2)} + \mathrm{e}^{\mathrm{i}(f_2 m_1 + f_1 m_2 - \varphi/2)} \\ &\quad = \mathrm{e}^{\mathrm{i}[f_1 m_2 + f_2(m_1+N) + \varphi/2]} + \mathrm{e}^{\mathrm{i}[f_2 m_2 + f_1(m_1+N) - \varphi/2]} \,. \end{aligned} \tag{8.44}$$

In order that this relation holds for all m_1 and m_2, the first term on the left-hand side must equal the second term on the right-hand side and the second term on the left-hand side must equal the first term on the right-hand side. Thus we have

$$\begin{aligned} N f_1 - \varphi &= 2\pi\lambda_1 \,, \\ N f_2 + \varphi &= 2\pi\lambda_2 \,. \end{aligned} \qquad (\lambda_1, \lambda_2 = 0, 1, 2, \ldots, N-1) \tag{8.45}$$

Therefore the sum k of independent wavenumbers is given by

$$k + 2n\pi = f_1 + f_2 = \frac{2\pi}{N}(\lambda_1 + \lambda_2) \,, \quad (0 \le k \le 2\pi) \tag{8.46}$$

where a suitable integer is chosen for n. If both the reversed spins move to their neighboring sites respectively, the coefficient $a(m_1, m_2)$ is multiplied by $e^{\pm ik}$.

Let us restrict the region for φ to

$$-\pi \leq \varphi \leq \pi \ . \tag{8.47}$$

φ changes sign if f_1 and f_2 are interchanged. When f_1 is increased from 0 with f_2 fixed, $\cot(\varphi/2)$ decreases from ∞ to a small positive value and becomes 0 at $f_1 = f_2$. Therefore φ changes from 0 to π in this case; when f_1 passes through f_2, φ jumps from $+\pi$ to $-\pi$ and increases further to 0 for $f_1 = 2\pi$. Thus φ is equal to $\pm\pi$ for $f_1 = f_2$, where we have from (8.45)

$$\begin{aligned} \varphi &= \pi \ , & \lambda_1 &= \lambda_2 - 1 = \frac{Nf_1}{2\pi} - \frac{1}{2} \ , \\ \varphi &= -\pi \ , & \lambda_1 &= \lambda_2 + 1 = \frac{Nf_1}{2\pi} + \frac{1}{2} \ , \end{aligned} \tag{8.48}$$

respectively. From (8.42) we then find

$$a(m_1, m_2) = e^{if_1(m_1+m_2)}(e^{\pm i\pi/2} + e^{\mp i\pi/2}) \equiv 0 \tag{8.49}$$

for $f_1 = f_2$, showing that there is no solution in this case. For this reason, the allowed values of λ_1, for fixed f_2, are given by

$$\lambda_1 = 0, 1, 2, \ldots, \lambda_2 - 2, \lambda_2 + 2, \ldots, N - 1 \ .$$

Since the state does not change under the interchange of f_1 and f_2, we can assume $f_1 < f_2$. In this case the number of allowed values for λ_1 is $\lambda_2 - 1$ and λ_2 can change from 2 to $N - 1$. Then the total number of independent solutions is given by

$$\sum_{\lambda_2=2}^{N-1} (\lambda_2 - 1) = \binom{N-1}{2} \ .$$

In fact, the total number of independent solutions is the number of values which m_1 and m_2 can take, namely $\binom{N}{2}$. Therefore there should be $N - 1$ different solutions in addition to the above solutions.

These additional $N - 1$ solutions can be found by assuming complex conjugate values for f_1 and f_2:

$$f_1 = u + iv \ , \quad f_2 = u - iv \ , \tag{8.50}$$

$$\cot\frac{f_1}{2} = \frac{\cos\frac{u}{2}\cosh\frac{v}{2} - i\sin\frac{u}{2}\sinh\frac{v}{2}}{\sin\frac{u}{2}\cosh\frac{v}{2} + i\cos\frac{u}{2}\sinh\frac{v}{2}} = \frac{\sin u - i\sinh v}{\cosh v - \cos u} \ . \tag{8.51}$$

Substituting (8.50) into (8.45), we obtain

$$N(f_1 - f_2) = 2iNv = 2\pi(\lambda_1 - \lambda_2) + 2\varphi \ .$$

Thus, putting

$$\varphi = \psi + \mathrm{i}\chi \;, \tag{8.52}$$

we have

$$\psi = \pi(\lambda_2 - \lambda_1) \;, \quad \chi = Nv \;. \tag{8.53}$$

Since for finite v, χ becomes a large number of the order of N, we obtain

$$\cot\frac{\varphi}{2} = \frac{\sin\psi - \mathrm{i}\sinh\chi}{\cosh\chi - \cos\psi} \simeq -\mathrm{i}(1 + 2\mathrm{e}^{-\chi+\mathrm{i}\psi}) \;. \tag{8.54}$$

If we put, as the first approximation, $-\mathrm{i}$ for the right hand side, the following relation is obtained from (8.41, 51):

$$2\cot\frac{\varphi}{2} = -2\mathrm{i} = \frac{\sin u - \mathrm{i}\sinh v}{\cosh v - \cos u} - \frac{\sin u + \mathrm{i}\sinh v}{\cosh v - \cos u} \;, \tag{8.55}$$

which results in

$$\sinh v = \cosh v - \cos u \;, \quad \mathrm{e}^{-v} = \cos u \;. \tag{8.56}$$

From (8.38) the energy eigenvalue ε is

$$\varepsilon = \sin^2 u = \frac{1}{2}(1 - \cos 2u) \;. \tag{8.57}$$

As evident from (8.56), $\cos u$ is positive. Therefore, if we restrict u to take only positive values, u must lie in the ranges

$$0 \le u \le \frac{\pi}{2} \;, \quad \frac{3}{2}\pi \le u < 2\pi \;.$$

Then, defining $k = 2u + 2n\pi$, we find

$$u = \frac{k}{2} \qquad \text{for} \quad 0 \le k < \pi \;,$$
$$u = \frac{k}{2} + \pi \quad \text{for} \quad \pi \le k \le 2\pi \;.$$

Let us put

$$v = v_0 + \Delta v \tag{8.58}$$

in the second approximation; here v_0 is the value in the first approximation. In this case, (8.55) can be written, to first order in Δv, as

$$2\cot\frac{\varphi}{2} = -2\mathrm{i} - 4\mathrm{i}\mathrm{e}^{-\chi+\mathrm{i}\psi} = -2\mathrm{i}\frac{\sinh v}{\cosh v - \cos u}$$
$$= -2\mathrm{i}(1 + 2\Delta v\cot^2 u) \;,$$

from which we obtain

$$\Delta v = \tan^2 u\mathrm{e}^{-\chi+\mathrm{i}\psi} \;. \tag{8.59}$$

Here χ may be approximated as Nv_0.

Next, we put

$$k = \frac{2\pi}{N}\lambda \quad (\lambda = 0, 1, 2, \dots, N-1) \ .$$

If λ is smaller than $N/2$, the region for k is $0 \le k < \pi$ so that $u = (2\pi/N)(\lambda/2)$ holds. Therefore, for even λ, $\lambda_1 = \lambda_2 = \lambda/2$ and $\psi = 0$ are obtained. On the other hand, for odd λ, $\lambda_1 = (\lambda-1)/2$, $\lambda_2 = (\lambda+1)/2$ and $\psi = \pi$ are found. If λ is larger than $N/2$, the region for k is $\pi \le k \le 2\pi$, from which we have $u = (2\pi/N)[(N+\lambda)/2]$. Therefore, if $N+\lambda$ is even, we have $\lambda_1 = \lambda_2 = (N+\lambda)/2$ and $\psi = 0$, while for odd $N+\lambda$, $\lambda_1 = (N+\lambda-1)/2$, $\lambda_2 = (N+\lambda+1)/2$ and $\psi = \pi$ are obtained.

Using in (8.59) the value of ψ thus determined, we find

$$\Delta v = \pm \tan^2 u \mathrm{e}^{-Nv_0} \ . \tag{8.60}$$

In (8.60), $v > v_0$ holds for the + sign (corresponding to even λ and $N+\lambda$). Therefore, if v_0 is replaced in the next approximation by the value for the second approximation v, Δv becomes smaller and converges rapidly to the exact value.

In the case of the $-$ sign (corresponding to odd λ and $N+\lambda$) Δv becomes larger in absolute magnitude as the approximation proceeds. However, as far as v_0 is finite, this correction remains small. If u is either very small or close to 2π so that $\cos u \sim 1$ holds, v_0 is given by

$$v_0 = -\log\cos u \simeq 1 - \cos u \simeq \frac{1}{2}u^2 \quad \text{or} \quad \frac{1}{2}(2\pi - u)^2$$

and is sufficiently small. For u of the order of $1/\sqrt{N}$, Nv_0 is finite and $\Delta v = -u^2\mathrm{e}^{-Nv_0}$ becomes larger in magnitude than v_0 when $Nv_0 < \log 2$, i.e., $u^2 < 2\log 2/N$, holds. In this case $v_1 = v_0 + \Delta v$ is negative so that the solution with a pair of complex conjugate wavevectors does not exist. In such a case, or to be more exact, when λ is an odd number and $u < 2\sin^{-1}(1/\sqrt{N}) \simeq 2/\sqrt{N}$ holds (or when $N+\lambda$ is an odd number and $2\pi - u < 2\sin^{-1}(1/\sqrt{N})$ holds), a solution with a real wave vector appears instead of the bound-state solution.

If u is close to zero and $(N/2\pi)k = \lambda$ is odd, then

$$N\frac{k}{2} - \pi = 2\pi\frac{1}{2}(\lambda - 1)$$

holds. Then, on putting $\lambda_1 = (1/2)(\lambda-1)$ and $\lambda_2 = (1/2)(\lambda+1)$, we find solutions of (8.41, 45), which satisfy $f_1 < f_2$ and $\varphi \ne \pi$, besides the ones satisfying $f_1 = f_2$ and $\varphi = \pi$. The amplitude $a(m_1, m_2)$ is nonvanishing for this solution. Let us put $f_1 = f - 2\delta/N$ and $f_2 = f + 2\delta/N$. Assuming f to be small and approximating $\tan f_1$ as f_1, and $\cos f_1$ as 1, one can rewrite (8.41) as

$$2\cot\frac{\varphi}{2} = \frac{2}{f - 2\frac{\delta}{N}} - \frac{2}{f + 2\frac{\delta}{N}} = \frac{8\delta}{Nf^2} \ . \tag{8.61}$$

Moreover, one obtains from (8.45)

$$2\varphi = 2\pi(\lambda_2 - \lambda_1) - N(f_2 - f_1) = 2\pi - 4\delta \ . \tag{8.62}$$

Substituting this relation into (8.61), we have

$$\frac{\tan\delta}{\delta} = \frac{4}{Nf^2} \ . \tag{8.63}$$

From (8.63), δ satisfying $\delta < \pi/2$ is determined for $f < 2/\sqrt{N}$. Thus we obtain scattering-wave solutions satisfying $\varphi > 0$ and $Nf_1 > 2\pi\lambda_1$.

From the above results we conclude that for each λ an additional solution with a real or complex wave number exists. The maximum value of λ is $N-2$, for which $\lambda_1 = \lambda_2 = N-1$ holds. If λ is equal to $N-1$, we have $\lambda_1 = N-1$ and $\lambda_2 = N$; then λ_2 is outside the allowed region. In this way we find that there are $N-1$ additional solutions for each $\lambda = 0, 1, 2, \ldots, N-2$.

The amplitude of the complex-wavenumber solution is given by

$$a(m_1, m_2) = \mathrm{e}^{\mathrm{i}u(m_1+m_2)} \begin{cases} \cosh v\left[\frac{1}{2}N - (m_2 - m_1)\right] \\ \sinh v\left[\frac{1}{2}N - (m_2 - m_1)\right] \ ; \end{cases} \tag{8.64}$$

The cosh or sinh function is chosen, according to whether λ (for $\lambda < N/2$) and $N + \lambda$ (for $\lambda > N/2$) are even or odd. As is evident from (8.64), the probability of finding two reversed spins, $|a(m_1, m_2)|^2$, is the largest when two spins are the closest and decays exponentially with $m_2 - m_1$. Thus this solution represents a bound state of two reversed spins. The energy eigenvalue of this solution is given by

$$\varepsilon_\mathrm{b} = \sin^2 u \ . \tag{8.57}$$

On the other hand, we have

$$\varepsilon_\mathrm{s} = 2 - \cos f_1 - \cos(k - f_1) \ . \tag{8.38}$$

for real wavenumbers. The minimum value of (8.38) is

$$f_1 = \frac{1}{2}k \ , \qquad (0 < k < \pi)$$
$$f_1 = \frac{1}{2}k + \pi \ . \quad (\pi \le k \le 2\pi)$$

Both of them give

$$\varepsilon_\mathrm{min} = 2(1 - \cos u) \ ,$$

if we put $f_1 = f_2 = u$. Therefore the energy of the bound state is always lower than that of the scattering-wave states. For small u, however, the difference is negligible.

The interactions between two spin waves in the three-dimensional lattice, scattering wave states in particular, were discussed in detail by *Dyson* [8.8]. According to the study of the simple-cubic lattice by *Wortis* [8.9], the domain in which bound states do not exist extends over a finite region centered at the origin or $(2\pi, 2\pi, 2\pi)$ in k-space. For the one-dimensional case, Bethe discussed a more general case with r reversed spins. This study will be described later in connection with the ground state of antiferromagnets.

8.3 Interactions Among Spin Waves II – Holstein-Primakoff Method

In order to take into account the interaction among spin waves automatically, it is convenient to use the *Holstein-Primakoff method* [8.10]. We now introduce operators $S_{l\pm}$ and n_l by

$$\begin{aligned} S_{l\pm} &= S_{lx} \pm \mathrm{i}S_{ly} \ , \\ n_l &= S - S_{lz} \ . \end{aligned} \tag{8.65}$$

The eigenvalues of n_l are integers from 0 to $2S$ according to the eigenvalue of S_{lz}, $M = -S, -S+1, \dots, S$. Let $\psi(M)$ be the eigenfunction for the eigenvalue M of S_{lz}. Then we obtain

$$\begin{aligned} S_{l-}\psi(M) &= \sqrt{(S+M)(S-M+1)}\psi(M-1) \\ &= (2S)^{1/2}\Big[(n_l+1)\big(1-\frac{n_l}{2S}\big)\Big]^{1/2}\psi(M-1) \ . \end{aligned}$$

Therefore, by using n_l for M, we obtain

$$S_{l-}\psi(n_l) = (2S)^{1/2}(n_l+1)^{1/2}\Big(1-\frac{n_l}{2S}\Big)^{1/2}\psi(n_l+1) \ . \tag{8.66a}$$

Similarly, we have

$$S_{l+}\psi(n_l) = (2S)^{1/2}\Big(1-\frac{n_l-1}{2S}\Big)^{1/2} n_l^{1/2}\psi(n_l-1) \ . \tag{8.66b}$$

Introducing new Bose operators $a_l^\dagger$ and a_l by $a_l^\dagger a_l = n_l$, we find the following relations:

$$\begin{aligned} a_l^\dagger a_l \psi(n_l) &= n_l\psi(n_l) \ , \\ a_l^\dagger\psi(n_l) &= \sqrt{n_l+1}\psi(n_l+1) \ , \\ a_l\psi(n_l) &= \sqrt{n_l}\psi(n_l-1) \ . \end{aligned} \tag{8.67}$$

Therefore, on comparing (8.66, 67), we obtain

$$
\begin{aligned}
S_{l-} &= (2S)^{1/2} a_l^\dagger \Big(1 - \frac{a_l^\dagger a_l}{2S}\Big)^{1/2} , \\
S_{l+} &= (2S)^{1/2} \Big(1 - \frac{a_l^\dagger a_l}{2S}\Big)^{1/2} a_l , \\
S_{lz} &= S - a_l^\dagger a_l .
\end{aligned}
\tag{8.68}
$$

The eigenvalue of $a_l^\dagger a_l$ is any integer $(0, 1, 2, \ldots)$, while that of n_l is restricted to the values $0, 1, 2, \ldots, 2S$. However, as evident from (8.68), the matrix element of S_{l-} and S_{l+} between $2S$ and $2S + 1$ vanishes automatically, assuring that there is no mixing of states with $n_l > 2S$ into those with $n_l \leq 2S$.

Writing the exchange interaction as

$$
\begin{aligned}
\mathcal{H} = -2J \sum_{\langle i,j \rangle} \Big[&(S - a_i^\dagger a_i)(S - a_j^\dagger a_j) + S a_i^\dagger \Big(1 - \frac{a_i^\dagger a_i}{2S}\Big)^{1/2} \Big(1 - \frac{a_j^\dagger a_j}{2S}\Big)^{1/2} a_j \\
&+ S \Big(1 - \frac{a_i^\dagger a_i}{2S}\Big)^{1/2} a_i a_j^\dagger \Big(1 - \frac{a_j^\dagger a_j}{2S}\Big)^{1/2} \Big]
\end{aligned}
$$

and expanding formally the square roots, we have

$$
\begin{aligned}
\mathcal{H} = -NzJS^2 &+ 2JS \sum_{\langle i,j \rangle} (a_i^\dagger a_i + a_j^\dagger a_j - a_i^\dagger a_j - a_i a_j^\dagger) \\
&-2J \sum_{\langle i,j \rangle} \Big[a_i^\dagger a_i a_j^\dagger a_j - \frac{1}{4} (a_i^\dagger a_i^\dagger a_i a_j + a_i^\dagger a_j^\dagger a_j a_j \\
&\qquad + a_i^\dagger a_i a_i a_j^\dagger + a_i a_j^\dagger a_j^\dagger a_j) \Big] + \ldots .
\end{aligned}
\tag{8.69}
$$

This is clearly an expansion in terms of $1/S$. The first, second and third terms are proportional to the second, first and zeroth powers of S, respectively. Since the second term is quadratic in a_j and $a_j^\dagger$, it can be easily diagonalized by the transformation

$$
\begin{aligned}
a_\mu^\dagger &= \frac{1}{\sqrt{N}} \sum_l \mathrm{e}^{+\mathrm{i}\mu \cdot R_l} a_l^\dagger , \\
a_\mu &= \frac{1}{\sqrt{N}} \sum_l \mathrm{e}^{-\mathrm{i}\mu \cdot R_l} a_l .
\end{aligned}
\tag{8.70}
$$

The energy eigenvalue is then given by

$$
E = -NzJS^2 + \sum_\mu \hbar\omega_\mu a_\mu^\dagger a_\mu , \tag{8.71}
$$

where $\hbar\omega_\mu$ is the spin-wave energy defined in (8.11). The third term in (8.69) describes the interaction among spin waves; its effects can be, in principle, taken into account by perturbation theory. In this type of calculation one should sum up all the contributions of the same order in $1/S$.

In the above expansion, the transverse component is expressed in terms of n_l by taking the z-component of S as a basis. Conversely, one may take the transverse component as a basis and express the z-component in terms of the transverse one, as was shown by *Anderson* [8.11]. In this case an iterative approximation leads to the following expression for $a_l^\dagger$ and a_l in terms of $S_{l\pm}$

$$a_l^\dagger = \frac{1}{\sqrt{2S}} S_{l-} + \frac{1}{4S} \frac{1}{(2S)^{3/2}} S_{l-} S_{l-} S_{l+} + \dots ,$$

$$a_l = \frac{1}{\sqrt{2S}} S_{l+} + \frac{1}{4S} \frac{1}{(2S)^{3/2}} S_{l-} S_{l+} S_{l+} + \dots .$$

Substituting this relation into the third equation of (8.68), we can expand S_{lz} in terms of the transverse component as

$$S_{lz} = S - \frac{1}{2S} S_{l-} S_{l+} + \dots . \tag{8.72}$$

If we add the second term of this equation to (8.5), the term

$$\hbar \frac{dS_{j\pm}}{dt} \to \pm \frac{\mathrm{i}}{S} \sum_i J_{ij} (S_{i-} S_{i+} S_{j\pm} - S_{i\pm} S_{j-} S_{j+})$$

is added to (8.7). Therefore

$$-\frac{1}{S} \frac{1}{\sqrt{N}} \sum_i \sum_j J_{ij} (\mathrm{e}^{-\mathrm{i}\boldsymbol{\mu}\cdot\boldsymbol{R}_j} S_{i-} S_{i+} S_{j-} - \mathrm{e}^{-\mathrm{i}\boldsymbol{\mu}\cdot\boldsymbol{R}_i} S_{j-} S_{i-} S_{i+})$$

should be added to (8.9). Rewriting the added term with the Fourier transform

$$S_{i\pm} = \frac{1}{\sqrt{N}} \sum_\rho \mathrm{e}^{\mp \mathrm{i}\boldsymbol{\rho}\cdot\boldsymbol{R}_i} S_{\rho\pm} ,$$

we have

$$-\frac{1}{S} \frac{1}{N^2} \sum_{\lambda\nu\rho} \sum_{ij} J_{ij} (\mathrm{e}^{\mathrm{i}(\boldsymbol{\nu}-\boldsymbol{\rho})\cdot\boldsymbol{R}_{ij}} - \mathrm{e}^{\mathrm{i}(\boldsymbol{\nu}-\boldsymbol{\rho}-\boldsymbol{\mu})\cdot\boldsymbol{R}_{ij}})$$
$$\times \mathrm{e}^{\mathrm{i}(\boldsymbol{\lambda}-\boldsymbol{\mu}+\boldsymbol{\nu}-\boldsymbol{\rho})\cdot\boldsymbol{R}_j} S_{\lambda-} S_{\nu-} S_{\rho+} , \tag{8.73}$$

where $\boldsymbol{R}_{ij} = \boldsymbol{R}_i - \boldsymbol{R}_j$. Since J_{ij} is a function of $\boldsymbol{R}_{ij}$, one can carry out the summation over $\boldsymbol{R}_j$, with which one obtains from (8.73)

$$\text{the additional term} = \frac{1}{SN} \sum_i J_{ij} \sum_{\lambda\rho} \mathrm{e}^{-\mathrm{i}\boldsymbol{\lambda}\cdot\boldsymbol{R}_{ij}} (1 - \mathrm{e}^{\mathrm{i}\boldsymbol{\mu}\cdot\boldsymbol{R}_{ij}})$$
$$\times S_{\lambda-} S_{\rho+\mu-\lambda-} S_{\rho+} . \tag{8.74}$$

We keep only the terms, in which the wave vectors of $S_{\lambda-}$ and $S_{\rho+}$ in the product of three $S_{\lambda\pm}$'s coincide, and ignore other terms having different wavevectors. The reason is that the latter terms oscillate with different phases so that their contributions are expected to be small. This is a widely used approximation, which is called the *Random Phase Approximation* (RPA). There are two cases with the same phase, i.e., $\boldsymbol{\lambda} = \boldsymbol{\mu}$ and $\boldsymbol{\lambda} = \boldsymbol{\rho}$. Therefore, in the RPA, (8.74) reduces to

$$\text{the additional term} = \frac{1}{SN}\sum_i J_{ij}\sum_{\rho}\left(\mathrm{e}^{-\mathrm{i}\boldsymbol{\mu}\cdot\boldsymbol{R}_{ij}} - 1 + \mathrm{e}^{-\mathrm{i}\boldsymbol{\rho}\cdot\boldsymbol{R}_{ij}} - \mathrm{e}^{\mathrm{i}(\boldsymbol{\mu}-\boldsymbol{\rho})\cdot\boldsymbol{R}_{ij}}\right) \times S_{\mu-}S_{\rho-}S_{\rho+} \,. \tag{8.75}$$

We now replace $S_{\rho-}S_{\rho+}$ by its average for the noninteracting case:

$$\langle S_{\rho-}S_{\rho+}\rangle = 2S\langle n_\rho\rangle \,, \tag{8.76}$$

where $\langle n_\rho\rangle$ is the average number of spin waves with wavevector $\boldsymbol{\rho}$. Rewriting (8.75) as

$$\text{Eq.(8.75)} = -\frac{2S}{SN}\sum_{\rho}\sum_i J_{ij}(1 - \mathrm{e}^{-\mathrm{i}\boldsymbol{\mu}\cdot\boldsymbol{R}_{ij}})(1 - \mathrm{e}^{\mathrm{i}\boldsymbol{\rho}\cdot\boldsymbol{R}_{ij}})\langle n_\rho\rangle S_{\mu-} \,,$$

and adding this to the right-hand side of (8.9) (for $H = 0$), we obtain the following eigenfrequency:

$$\hbar\omega_\mu = 2S\sum_i J_{ij}\left[1 - \frac{1}{SN}\sum_{\rho}(1 - \mathrm{e}^{\mathrm{i}\boldsymbol{\rho}\cdot\boldsymbol{R}_{ij}})\langle n_\rho\rangle\right](1 - \mathrm{e}^{-\mathrm{i}\boldsymbol{\mu}\cdot\boldsymbol{R}_{ij}}) \,. \tag{8.77}$$

Equation (8.77) contains the lowest-order effect of interaction between spin waves. Here one should note that $\langle n_\rho\rangle$ is accompanied by the factor $1 - \mathrm{e}^{\mathrm{i}\boldsymbol{\rho}\cdot\boldsymbol{R}_{ij}}$. If the latter factor was absent, the eigenfrequency would be the same as the one for the noninteracting case, in which S is replaced by its thermal average

$$\langle S\rangle = S\big(1 - \frac{1}{SN}\sum_{\rho}\langle n_\rho\rangle\big) \,.$$

However, because of the factor $1 - \mathrm{e}^{\mathrm{i}\boldsymbol{\rho}\cdot\boldsymbol{R}_{ij}}$, this interpretation is not correct. The correction term in (8.77) is rather a sum of the spin-wave energy over the wave vector $\boldsymbol{\rho}$, giving a contribution which is proportional to $T^{5/2}$. Therefore the effect of the interaction, which shows up in the decrease of the spontaneous magnetization ΔM due to the temperature, starts from $T^{3/2+5/2} = T^4$. This power is much higher than the Bloch value 3/2 so that Bloch's theory is expected to be a fairly good approximation in a wide range of the low-temperature region. Studies on the interaction among spin waves have also been carried out by *Keffer* and *Loudon* [8.12], *Oguchi* [8.13], *Kanamori* and *Tachiki* [8.14] and others.

these contributions are expected to be small. This is a widely used approximation, which is called the [illegible] approximation [illegible]

[illegible] returns to [illegible]

The [illegible] term [illegible]

[illegible]

We now replace [illegible] by its average for the [illegible]

$$\langle S_k S_{-k} \rangle = 2S \langle n_k \rangle [illegible]$$

where $\langle n_k \rangle$ is the average number of spin waves with wavevector k [illegible]

(8.3.[illegible])

[illegible]

and [illegible] we obtain the [illegible]

[illegible] (8.[illegible])

In addition [illegible] interaction between [illegible] waves. Here [illegible] by the factor [illegible]

[illegible]

[illegible]

However [illegible]

[illegible]

9. Spin-Wave Theory of Antiferromagnets

In the preceeding chapter, we developed the spin-wave theory of ferromagnetism. In this chapter, we present the spin-wave theory of antiferromagnetism. In the antiferromagnets, the ordered states described by the molecular-field theory are not the ground state of the Heisenberg Hamiltonian for quantum spins. The antiferromagnetic spin-wave theory is also a powerful method to investigate the ground states of quantum antiferromagnets.

9.1 Ground State of Antiferromagnets

In the case of ferromagnetic spin systems ($J > 0$), the state having the maximal degeneracy, i.e., the state in which all spins are aligned along one direction, becomes the ground state of the Heisenberg Hamiltonian. Spin-wave theory has shown that low-energy excitations from this ground state can be described as a collection of spin waves.

In the case of antiferromagnetic spin systems ($J < 0$) on a bipartite lattice, the ground state in the molecular-field approximation consists of two sublattices, in which all spins on one of the sublattices are aligned along one direction, while all spins on the other sublattice point in the opposite direction. However, this state is not an eigenstate of the Heisenberg Hamiltonian. If we operate on this state with the transverse part of the (nearest-neighbor) exchange interaction, a neighboring spin pair (aligned antiparallel) is changed to a state in which the + and − spins are exchanged. In this way the off-diagonal terms in the Hamiltonian modify the ground state, lowering at the same time the energy relative to the value obtained in the molecular-field approximation. The spin-wave theory for antiferromagnets does not only give the excitation energy spectrum from the ground state as in the case of ferromagnets, but is also a powerful method to study the ground state and its energy.

In the molecular-field approximation for antiferromagnetic spin systems, the ground-state energy is determined solely by the longitudinal part of the exchange interaction. Therefore the molecular-field value

$$E_{\mathrm{M}} = NzS^2J \tag{9.1}$$

is an upper bound for the ground-state energy of antiferromagnetic spin systems. Here z represents the number of nearest neighbors and the exchange interaction is present only for nearest-neighbor spins.

Let us assume that, like the simple-cubic lattice and the body centered cubic lattice, our lattice is bipartite and has the following property: the nearest neighbors of the lattice points of one sublattice belong to the other sublattice. We consider a cluster consisting of a spin $\boldsymbol{S}_j$ on one sublattice and its z neighboring spins $\boldsymbol{S}_{j+\delta}$ on the other sublattice, which are interacting with the former. The Hamiltonian for this cluster is given by

$$\mathcal{H}_{\mathrm{c}} = -2J\boldsymbol{S}_j \cdot \sum_{\delta} \boldsymbol{S}_{j+\delta} \; . \tag{9.2}$$

If we put $\sum_\delta \boldsymbol{S}_{j+\delta} = \boldsymbol{S}_{\mathrm{T}}$, the energy of this spin cluster is given by

$$\begin{aligned} -2J(\boldsymbol{S}_j \cdot \boldsymbol{S}_{\mathrm{T}}) &= -J\left[(\boldsymbol{S}_j + \boldsymbol{S}_{\mathrm{T}})^2 - \boldsymbol{S}_j^2 - \boldsymbol{S}_{\mathrm{T}}^2\right] \\ &= -J\left[(\boldsymbol{S}_j + \boldsymbol{S}_{\mathrm{T}})^2 - S(S+1) - S_{\mathrm{T}}(S_{\mathrm{T}}+1)\right] \; . \end{aligned}$$

For $J < 0$ this energy is the lowest when the magnitude of $\boldsymbol{S}_j + \boldsymbol{S}_{\mathrm{T}}$ is $S_{\mathrm{T}} - S$. The energy for this case, $-J[(S_{\mathrm{T}} - S)(S_{\mathrm{T}} - S + 1) - S(S+1) - S_{\mathrm{T}}(S_{\mathrm{T}}+1)] = 2JS(S_{\mathrm{T}} + 1)$, is lowest when S_{T} has its maximum value zS. Thus the lowest energy is given by

$$2JS(zS+1) \; . \tag{9.3}$$

Since the total Hamiltonian is the sum of (9.2) over $\boldsymbol{j}$, the ground-state energy is clearly larger than $N/2$ times (9.3), so that this value gives a lower bound to the ground-state energy E_{g}. From (9.1, 3), we find the following inequalities:

$$NzJS^2 > E_{\mathrm{g}} > NzJS^2\left(1 + \frac{1}{zS}\right) \; . \tag{9.4}$$

The right-hand inequality, which was derived by *Anderson* [9.1], suggests that the energy lowering from the molecular-field approximation, i.e., the quantum effect, increases as zS becomes smaller. The result suggests, in particular, that in the one-dimensional lattice with spin $1/2$ (giving the smallest zS), the ground-state energy may become as low as about twice the value in the molecular-field approximation.

9.2 Spin-Wave Theory of Antiferromagnets

Anderson [9.2] extended spin-wave theory to antiferromagnets and discussed the ground state. Dividing the crystal lattice into two sublattices, we represent points on one sublattice by $\boldsymbol{j}$ and points on the other sublattice by $\boldsymbol{l}$. The Heisenberg Hamiltonian is then expressed, together with the Zeeman energy and the anisotropy energy, as

$$\mathcal{H} = -2\sum_{jl} J_{jl}\boldsymbol{S}_j \cdot \boldsymbol{S}_l - g\mu_{\mathrm{B}}\boldsymbol{H} \cdot \Big(\sum_j \boldsymbol{S}_j + \sum_l \boldsymbol{S}_l\Big)$$
$$- \frac{1}{2}D\Big(\sum_j S_{jz}^2 + \sum_l S_{lz}^2\Big) , \tag{9.5}$$

where the easy axis of spins is taken as the z axis and the anisotropy energy is assumed, for simplicity, as the form of a one-ion anisotropy. We assume that $S \geq 1$ and that the exchange interaction is present only between spins on different sublattices.

As for ferromagnets, the equations of motion for the spins $\boldsymbol{S}_j$ and $\boldsymbol{S}_l$ on the two sublattices are

$$\frac{\hbar}{\mathrm{i}}\frac{d\boldsymbol{S}_j}{dt} = [\mathcal{H}, \boldsymbol{S}_j] = \mathrm{i}(\boldsymbol{H}_j \times \boldsymbol{S}_j) , \tag{9.6a}$$

$$\frac{\hbar}{\mathrm{i}}\frac{d\boldsymbol{S}_l}{dt} = [\mathcal{H}, \boldsymbol{S}_l] = \mathrm{i}(\boldsymbol{H}_l \times \boldsymbol{S}_l) , \tag{9.6b}$$

where the effective fields $\boldsymbol{H}_j$ and $\boldsymbol{H}_l$ acting on $\boldsymbol{S}_j$ and $\boldsymbol{S}_l$ are given by

$$\boldsymbol{H}_j = 2\sum_l J_{jl}\boldsymbol{S}_l + g\mu_{\mathrm{B}}\boldsymbol{H} + DS_{jz}\boldsymbol{z} , \tag{9.7a}$$

$$\boldsymbol{H}_l = 2\sum_j J_{jl}\boldsymbol{S}_j + g\mu_{\mathrm{B}}\boldsymbol{H} + DS_{lz}\boldsymbol{z} , \tag{9.7b}$$

respectively; $\boldsymbol{z}$ is the unit vector along the z direction.

Now we decompose $\boldsymbol{S}_j$ and $\boldsymbol{S}_l$ into the ground-state value in the molecular-field theory ($\pm\boldsymbol{S}$) and small deviations from it ($\delta\boldsymbol{S}_j$ and $\delta\boldsymbol{S}_l$):

$$\boldsymbol{S}_j = \boldsymbol{S} + \delta\boldsymbol{S}_j , \quad \boldsymbol{S}_l = -\boldsymbol{S} + \delta\boldsymbol{S}_l . \tag{9.8}$$

We assume that $\pm\boldsymbol{S}$ is aligned along the z direction. In this case $\delta\boldsymbol{S}_j$ and $\delta\boldsymbol{S}_l$ lie necessarily in the xy plane. Using (9.8) in (9.7a, b), we obtain from (9.6)

$$\hbar\delta\dot{\boldsymbol{S}}_j = -\Big[\Big(-2\sum_l J_{jl}\boldsymbol{S} + g\mu_{\mathrm{B}}\boldsymbol{H} + D\boldsymbol{S}\Big) \times \delta\boldsymbol{S}_j + 2\sum_l J_{jl}\delta\boldsymbol{S}_l \times \boldsymbol{S}\Big] , \tag{9.9a}$$

$$\hbar\delta\dot{\boldsymbol{S}}_l = -\Big[\Big(2\sum_j J_{jl}\boldsymbol{S} + g\mu_{\mathrm{B}}\boldsymbol{H} - D\boldsymbol{S}\Big) \times \delta\boldsymbol{S}_l - 2\sum_j J_{jl}\delta\boldsymbol{S}_j \times \boldsymbol{S}\Big] . \tag{9.9b}$$

Here we have neglected small terms of second order in $\delta\boldsymbol{S}_j$ and $\delta\boldsymbol{S}_l$ on the right-hand sides. The magnetic field is assumed to be applied along the z axis.

Let us introduce $\boldsymbol{A}_\mu$ and $\boldsymbol{B}_\mu$ by the Fourier transformation

$$\boldsymbol{A}_\mu = \Big(\frac{2}{N}\Big)^{1/2}\sum_j \mathrm{e}^{-\mathrm{i}\boldsymbol{\mu}\cdot\boldsymbol{R}_j}\delta\boldsymbol{S}_j , \tag{9.10a}$$

$$\boldsymbol{B}_\mu = \Big(\frac{2}{N}\Big)^{1/2}\sum_l \mathrm{e}^{-\mathrm{i}\boldsymbol{\mu}\cdot\boldsymbol{R}_l}\delta\boldsymbol{S}_l . \tag{9.10b}$$

The equations of motion for $\boldsymbol{A}_\mu$ and $\boldsymbol{B}_\mu$ are given by

$$\hbar\dot{\boldsymbol{A}}_\mu = -\Big[\big(-2\sum_l J_{jl}\boldsymbol{S} + g\mu_B\boldsymbol{H} + D\boldsymbol{S}\big)\times\boldsymbol{A}_\mu + 2\sum_j J_{jl}\mathrm{e}^{-\mathrm{i}\boldsymbol{\mu}\cdot(\boldsymbol{R}_j-\boldsymbol{R}_l)}\boldsymbol{B}_\mu\times\boldsymbol{S}\Big], \tag{9.11a}$$

$$\hbar\dot{\boldsymbol{B}}_\mu = -\Big[\big(2\sum_j J_{jl}\boldsymbol{S} + g\mu_B\boldsymbol{H} - D\boldsymbol{S}\big)\times\boldsymbol{B}_\mu - 2\sum_l J_{jl}\mathrm{e}^{-\mathrm{i}\boldsymbol{\mu}\cdot(\boldsymbol{R}_l-\boldsymbol{R}_j)}\boldsymbol{A}_\mu\times\boldsymbol{S}\Big], \tag{9.11b}$$

where $\boldsymbol{\mu}$ takes $N/2$ values in the first Brillouin zone for the sublattice. For simplicity we assume that the interaction is present only between nearest neighbor sites. We define a quantity γ_μ by

$$\sum_j J_{jl}\mathrm{e}^{-\mathrm{i}\boldsymbol{\mu}\cdot(\boldsymbol{R}_j-\boldsymbol{R}_l)} = zJ\gamma_\mu , \tag{9.12}$$

where z is the number of nearest neighbors. From (9.11) the equations for the quantities $A_{\mu\pm}$ and $B_{\mu\pm}$ defined by

$$A_{\mu\pm} = A_{\mu x} \pm \mathrm{i}A_{\mu y} ,$$
$$B_{\mu\pm} = B_{\mu x} \pm \mathrm{i}B_{\mu y}$$

become

$$\hbar\dot{A}_{\mu\pm} = \mp\mathrm{i}\Big[(-2JzS + g\mu_B H + DS)A_{\mu\pm} - 2JzS\gamma_\mu B_{\mu\pm}\Big] , \tag{9.13a}$$

$$\hbar\dot{B}_{\mu\pm} = \mp\mathrm{i}\Big[(2JzS + g\mu_B H - DS)B_{\mu\pm} + 2JzS\gamma_\mu A_{\mu\pm}\Big] . \tag{9.13b}$$

In this way $A_{\mu+}$ is coupled to $B_{\mu+}$, while $A_{\mu-}$ is coupled to $B_{\mu-}$. If we take the time dependence of A_μ and B_μ as $\exp(\mathrm{i}\omega_\mu t)$, the relation determining the eigenfrequency ω_μ is given by

$$\begin{vmatrix} \hbar\omega_\mu - (2JzS - g\mu_B H - DS) & -2JzS\gamma_\mu \\ 2JzS\gamma_\mu & \hbar\omega_\mu + (2JzS + g\mu_B H - DS) \end{vmatrix} = 0 . \tag{9.14}$$

The solutions of this equation are

$$\hbar|\omega_\mu^\pm| = \sqrt{(2JzS)^2(1-\gamma_\mu^2) + 4|J|zS^2D + S^2D^2} \pm g\mu_B H . \tag{9.15}$$

Since $1-\gamma_\mu^2 \propto \boldsymbol{\mu}^2$ holds in the limit $\boldsymbol{\mu}\to 0$, we obtain

$$\hbar|\omega_{\mu=0}^\pm| = \sqrt{4|J|zS^2D + S^2D^2} \pm g\mu_B H \tag{9.16}$$

or

$$\hbar|\omega_{\mu=0}^\pm| = g\mu_B\big[\sqrt{(2H_E + H_A)H_A} \pm H\big] \tag{9.17}$$

in terms of the molecular field $H_{\rm E} = 2|J|zS/g\mu_{\rm B}$ and the anisotropy field $H_{\rm A} = DS/g\mu_{\rm B}$. This spatially uniform mode of the spin wave with $\mu = 0$ couples with the electromagnetic field so that it can be observed in microwave resonance absorption experiments. The result of (9.17) is exactly the relation derived by *Nagamiya* [9.3], *Kittel* [9.4] and *Keffer* and *Kittel* [9.5] before the spin-wave theory for antiferromagnets was established. The frequency of the mode corresponding to the minus sign in (9.17) vanishes at

$$H_{\rm c} = \sqrt{(2H_{\rm E} + H_{\rm A})H_{\rm A}} \; . \tag{9.18}$$

The critical field $H_{\rm c}$ corresponds to the magnetic field at which the common axis of the spins on the two sublattices rotates from the easy axis to a direction perpendicular to it.

In the case $H = 0$, the frequencies for the clockwise and anti-clockwise precessions of the spins with respect to the z axis coincide so that the two modes are degenerate. If neither the external field nor the anisotropy field is present, $\hbar\omega_\mu^\pm$ vanishes proportionally to $|\mu|$. To excite spin waves with energy $\hbar\omega_{\mu=0}^\pm$ is equivalent classically to rotating the common axis of the spins on the + and − sublattices from the z axis to an arbitrary direction. Therefore $\hbar\omega_{\mu=0} = 0$ means that this rotation does not cost any energy. The exchange-interaction part of the Hamiltonian commutes with the total spin operator; consequently, it is invariant under the spin rotation. In this case, if each sublattice has a finite expectation value of spin, all the states generated by rotation of the spin axis are degenerate with each other. This situation is similar to that in ferromagnets.

In the Holstein-Primakoff method, the spin operator on the + sublattice is expressed, similarly to the case for ferromagnets, by (8.68) with Bose operators $a_j^\dagger$ and a_j. On the other hand, for the − sublattice we introduce Bose operators $b_l^\dagger$ and b_l by

$$\begin{aligned} S_{lz} &= -S + b_l^\dagger b_l \; , \\ S_{l+} &= \sqrt{2S}\, b_l^\dagger \left[1 - \frac{b_l^\dagger b_l}{2S}\right]^{1/2} , \\ S_{l-} &= \sqrt{2S}\left[1 - \frac{b_l^\dagger b_l}{2S}\right]^{1/2} b_l \; . \end{aligned} \tag{9.19}$$

Using (8.68) and (9.19) in the Heisenberg Hamiltonian (9.5) and neglecting higher-order terms describing the interaction between spin waves, we obtain

$$\begin{aligned} \mathcal{H} = JzNS^2 - \frac{D}{2}NS^2 - 2JS\sum_{\langle j,l\rangle}(a_j^\dagger a_j + b_l^\dagger b_l + a_j b_l + a_j^\dagger b_l^\dagger) \\ + g\mu_{\rm B}H\Big(\sum_j a_j^\dagger a_j - \sum_l b_l^\dagger b_l\Big) + DS\Big(\sum_j a_j^\dagger a_j + \sum_l b_l^\dagger b_l\Big) \; , \end{aligned} \tag{9.20}$$

where the summation over $\langle \boldsymbol{j}, \boldsymbol{l} \rangle$ means that we take the summation of $\boldsymbol{l}$ over the nearest-neighbor lattice points of $\boldsymbol{j}$ and then carry out the summation over $\boldsymbol{j}$. As before, Fourier transforms are introduced as

$$\begin{aligned} a_{\mu}^{\dagger} &= \sqrt{\frac{2}{N}} \sum_{j} \mathrm{e}^{\mathrm{i}\mu\cdot R_j} a_j^{\dagger} , \qquad a_{\mu} = \sqrt{\frac{2}{N}} \sum_{j} \mathrm{e}^{-\mathrm{i}\mu\cdot R_j} a_j , \\ b_{\mu}^{\dagger} &= \sqrt{\frac{2}{N}} \sum_{l} \mathrm{e}^{-\mathrm{i}\mu\cdot R_l} b_l^{\dagger} , \qquad b_{\mu} = \sqrt{\frac{2}{N}} \sum_{l} \mathrm{e}^{\mathrm{i}\mu\cdot R_l} b_l . \end{aligned} \tag{9.21}$$

Expressing (9.20) in terms of $a_{\mu}^{\dagger}$, $b_{\mu}^{\dagger}$ etc., we obtain

$$\begin{aligned} \mathcal{H} = JzNS^2 - \frac{D}{2}NS^2 - 2JzS \sum_{\mu} \big[&(1+h+d)a_{\mu}^{\dagger}a_{\mu} + (1-h+d)b_{\mu}^{\dagger}b_{\mu} \\ &+ \gamma_{\mu}(a_{\mu}^{\dagger}b_{\mu}^{\dagger} + a_{\mu}b_{\mu}) \big] ; \end{aligned} \tag{9.22}$$

here we have defined h and d by

$$h = \frac{g\mu_{\mathrm{B}}H}{2|J|zS} , \qquad d = \frac{D}{2|J|z} . \tag{9.23}$$

Equation (9.22) can be diagonalized by the Bogoliubov transformation

$$\begin{aligned} a_{\mu} &= \cosh\theta_{\mu}\alpha_{\mu} - \sinh\theta_{\mu}\beta_{\mu}^{\dagger} , \qquad & a_{\mu}^{\dagger} &= \cosh\theta_{\mu}\alpha_{\mu}^{\dagger} - \sinh\theta_{\mu}\beta_{\mu} , \\ b_{\mu} &= -\sinh\theta_{\mu}\alpha_{\mu}^{\dagger} + \cosh\theta_{\mu}\beta_{\mu} , \qquad & b_{\mu}^{\dagger} &= -\sinh\theta_{\mu}\alpha_{\mu} + \cosh\theta_{\mu}\beta_{\mu}^{\dagger} , \\ \tanh 2\theta_{\mu} &= \frac{\gamma_{\mu}}{1+d} . \end{aligned} \tag{9.24}$$

The resulting Hamiltonian is

$$\begin{aligned} \mathcal{H} = &JzNS(S+1) - \frac{D}{2}NS(S+1) \\ &+ \sum_{\mu} \Big[\hbar\omega_{\mu}^{+}(\alpha_{\mu}^{+}\alpha_{\mu} + \frac{1}{2}) + \hbar\omega_{\mu}^{-}(\beta_{\mu}^{+}\beta_{\mu} + \frac{1}{2}) \Big] , \end{aligned} \tag{9.25}$$

where $\hbar\omega_{\mu}^{\pm}$ are the eigenfrequencies of the spin system, which are given by (9.15).

The ground-state energy in spin-wave theory is obtained from the zero-point energy of the third term of (9.25) as follows:

$$\begin{aligned} E_{\mathrm{g}} &= JzNS(S+1) - 2JzS \sum_{\mu} (1-\gamma_{\mu}^2)^{1/2} \\ &= JzNS^2 \left[1 + \frac{1}{S} \Big(1 - \frac{2}{N} \sum_{\mu} \sqrt{1-\gamma_{\mu}^2} \Big) \right] , \end{aligned} \tag{9.26}$$

where the external field and the anisotropy field have been put to zero. For the one-dimensional chain, the two-dimensional square lattice, and the three-dimensional simple-cubic lattice, γ_{μ} is given by

$$\gamma_\mu = \frac{1}{D}\sum_{i=1}^{D} \cos \mu_i a \qquad (D = 1, 2, 3) \tag{9.27}$$

where a is the lattice constant. The summation over μ in (9.26) is taken over all the $N/2$ points in the first Brillouin zone of the sublattice. Therefore, we can write the sum as

$$\begin{aligned} I_D &= \frac{2}{N}\sum_{\mu}\sqrt{1-\gamma_\mu^2} \\ &= (2\pi)^{-D}\int_{-\pi}^{\pi}\cdots\int_{-\pi}^{\pi} d\lambda_1 \ldots d\lambda_D\left[1-\left(\frac{1}{D}\sum_{i=1}^{D}\cos\lambda_i\right)^2\right]^{1/2} . \end{aligned} \tag{9.28}$$

Here $(2\pi)^D$ is the volume of the domain of the λ_i $(-\pi < \lambda_i < \pi)$. The numerical values of (9.28) are

$$I_1 = \frac{2}{\pi}\,, \qquad I_2 = 1 - 0.158\,, \qquad I_3 = 1 - 0.097\,. \tag{9.29}$$

Thus the ground-state energy is obtained as [9.2]

$$\begin{aligned} (E_{\rm g})_{D=1} &= 2NJS^2\left(1+\frac{0.363}{S}\right), \\ (E_{\rm g})_{D=2} &= 4NJS^2\left(1+\frac{0.158}{S}\right), \\ (E_{\rm g})_{D=3} &= 6NJS^2\left(1+\frac{0.097}{S}\right). \end{aligned} \tag{9.30}$$

These numerical values satisfy the inequality (9.4), as they should. In the case of one dimension, in particular, the value is very close to the exact value for $S = 1/2$ obtained by Bethe and Hulthén $(E_{\rm g})_{D=1} = (1/2)NJ(4\log 2 - 1) = (1/2)NJ \times 1.7726$, which we shall discuss in the next section. Even for the one-dimensional system (the most unfavorable) the theory gives the ground-state energy extremely close to the exact value. It is expected from this fact that spin-wave theory is a very good approximation also for antiferromagnets at low temperatures.

Next, let us examine the expectation value of the spin on each sublattice:

$$\begin{aligned} \langle S_{z\rm tot}\rangle &= \sum_j \langle S_{jz}\rangle = \frac{1}{2}NS - \sum_\mu \langle a_\mu^\dagger a_\mu\rangle \\ &= \frac{1}{2}NS - \sum_\mu \Big[\cosh^2\theta_\mu\langle\alpha_\mu^\dagger\alpha_\mu\rangle + \sinh^2\theta_\mu\langle\beta_\mu\beta_\mu^\dagger\rangle \\ &\qquad - \sinh\theta_\mu\cosh\theta_\mu\big(\langle\alpha_\mu^\dagger\beta_\mu^\dagger\rangle + \langle\alpha_\mu\beta_\mu\rangle\big)\Big] \\ &= \frac{1}{2}NS - \sum_\mu \sinh^2\theta_\mu\,. \end{aligned} \tag{9.31}$$

Since, from (9.24), $\sinh^2\theta_\mu$ is given by

$$\sinh^2\theta_\mu = \frac{1}{2}\left[\frac{1}{2}\left(\frac{1-\gamma_\mu}{1+\gamma_\mu}\right)^{1/2} + \frac{1}{2}\left(\frac{1+\gamma_\mu}{1-\gamma_\mu}\right)^{1/2} - 1\right], \tag{9.32}$$

$\langle S_{z\mathrm{tot}}\rangle$ is then expressed as

$$\langle S_{z\mathrm{tot}}\rangle = \frac{NS}{2}\left[1 - \frac{1}{2S}\left(\frac{2}{N}\sum_\mu \frac{1}{\sqrt{1-\gamma_\mu^2}} - 1\right)\right]. \tag{9.33}$$

If we define the sum over μ as J_D, then J_D can be calculated as

$$\begin{aligned} J_D &= \frac{2}{N}\sum_\mu \frac{1}{\sqrt{1-\gamma_\mu^2}} \\ &= \frac{1}{(2\pi)^D}\int_{-\pi}^{\pi}\cdots\int_{-\pi}^{\pi} d\lambda_1\ldots d\lambda_D\left[1-\left(\frac{1}{D}\sum_{i=1}^{D}\cos\lambda_i\right)^2\right]^{-1/2}. \end{aligned} \tag{9.34}$$

J_1 diverges logarithmically. This means that the sublattice magnetization has zero expectation value in one dimension; in other words the ground state, in which the molecular-field theory predicts that two sublattices have antiparallel spins, is not stable. This result agrees also with the exact solution for one dimension. If the anisotropy energy is present, J_1 is calculated as

$$J_1 = \frac{1}{2\pi}\int_{-\pi}^{\pi} d\lambda\left[1 - \frac{1}{(1+d)^2}\cos^2\lambda\right]^{-1/2} = \frac{2}{\pi}K\left(\frac{1}{1+d}\right), \tag{9.35}$$

where $K(k)$ is the complete elliptic integral of the first kind. In the limit $d \to 0^+$, J_1 behaves as

$$J_1 \to \frac{1}{\pi}\log\frac{1+d}{d}. \tag{9.36}$$

Thus, in the presence of the anisotropy energy, the sublattice moment is finite, but J_1 diverges at $d = 0$. This means that $d \to 0$ is a singular point in one dimension. On the other hand, the integral is convergent in two and three dimensions. The values are obtained as

$$J_2 = 1.393\,, \qquad J_3 = 1.156\,,$$

from which the average spin on each sublattice is given by

$$\begin{aligned} \langle S_{z\mathrm{tot}}\rangle &= \frac{NS}{2}\left(1 - \frac{0.197}{S}\right) \quad (D=2)\,, \\ &= \frac{NS}{2}\left(1 - \frac{0.078}{S}\right) \quad (D=3)\,. \end{aligned} \tag{9.37}$$

The second term represents the reduction of the average spin due to quantum effects.

Let us consider next the transverse components of the spins. From (8.68), (9.19, 21, 24) we obtain

$$\sum_j S_{jx} + \sum_l S_{lx} = \frac{1}{2}(NS)^{1/2}(\cosh\theta_0 - \sinh\theta_0)(\alpha_0 + \beta_0^\dagger + \alpha_0^\dagger + \beta_0)\,, \tag{9.38a}$$

$$\sum_j S_{jy} + \sum_l S_{ly} = -\frac{\mathrm{i}}{2}(NS)^{1/2}(\cosh\theta_0 - \sinh\theta_0)(\alpha_0 + \beta_0^\dagger - \alpha_0^\dagger - \beta_0)\,, \tag{9.38b}$$

$$\sum_j S_{jx} - \sum_l S_{lx} = \frac{1}{2}(NS)^{1/2}(\cosh\theta_0 + \sinh\theta_0)(\alpha_0 - \beta_0^\dagger + \alpha_0^\dagger - \beta_0)\,, \tag{9.38c}$$

$$\sum_j S_{jy} - \sum_l S_{ly} = -\frac{\mathrm{i}}{2}(NS)^{1/2}(\cosh\theta_0 + \sinh\theta_0)(\alpha_0 - \beta_0^\dagger - \alpha_0^\dagger + \beta_0)\,. \tag{9.38d}$$

From these relations we obtain the average of the square of (9.38a, b) as

$$\left\langle\left(\sum_j S_{jx}+\sum_l S_{lx}\right)^2\right\rangle = \frac{NS}{2}(\cosh 2\theta_0 - \sinh 2\theta_0) = \frac{NS}{2}\left(\sqrt{\frac{1-\gamma_\mu}{1+\gamma_\mu}}\right)_{\mu=0}. \tag{9.39}$$

Therefore the average of the square of $(S_{x,y})_{\text{tot}}$ depends on N with a power smaller than N. On the other hand, the average of the square of (9.38c, d) is given by

$$\left\langle\left(\sum_j S_{jx}-\sum_l S_{lx}\right)^2\right\rangle = \frac{NS}{2}(\cosh 2\theta_0 + \sinh 2\theta_0) = \frac{NS}{2}\left(\sqrt{\frac{1+\gamma_\mu}{1-\gamma_\mu}}\right)_{\mu=0}, \tag{9.40}$$

which diverges at $\mu = 0$ as $1/\mu$. This divergence is related to the fact that the spin axis is free to rotate in the ground state; the expression can be easily made convergent by introducing the anisotropy energy. When the anisotropy energy is introduced, γ_μ is replaced by $\gamma_\mu/(1+d)$ and

$$\left\langle\left(\sum_j S_{jx} - \sum_l S_{lx}\right)^2\right\rangle \simeq \frac{NS}{(2d)^{1/2}} \tag{9.41}$$

is obtained. Therefore, if d is larger than N^{-2}, the square root of the average of the square of $\sum_j S_{jx}$ or $\sum_l S_{lx}$ is smaller than N so that it can be ignored compared with $(1/2)NS$. Thus, the divergence in (9.40) can be eliminated by an infinitesimal anisotropy energy.

In this way spin-wave theory predicts that except for one dimension the state in which spins on the two sublattices are aligned antiparallel is the ground state. However, the magnitude of the spin on each sublattice is reduced due to the quantum effect. The magnitude of the total spin in this state is much smaller than N so that one can regard it as a quantity almost equal to zero in the $N \to \infty$ limit. Strictly speaking, however, S_{tot} is not zero. On the other hand, the Heisenberg Hamiltonian commutes with S_{tot} and the ground state should be a spin-singlet state with $S_{\text{tot}} = 0$. Such a spin singlet-state in a strict sense can be constructed from a linear combination of all the states that can be generated by rotating the spin axis of the sublattice. In this case the energy lowering due to the linear combination is of order of $1/N$ and can be ignored in the large N limit. In other words, the frequency $J/\hbar N$ of spin axis rotation in the state with a sublattice moment becomes infinitesimally small as $N \to \infty$ and this state becomes practically the ground state. In real antiferromagnets the presence of sublattice moment is confirmed by neutron scattering and susceptibility measurements on single crystals. This fact is in accord with the conclusion of spin-wave theory. However, since the anisotropy energy is present in real antiferromagnets, it is not certain whether the anisotropy energy is truly required to produce a finite sublattice moment. Judging from the behavior of the exact solution for one dimension, we think that a finite anisotropy energy is not necessary to have a sublattice magnetization and that the conclusion of spin-wave theory is correct.

The temperature dependence of the sublattice magnetic moment, free energy and susceptibility was calculated by *Kubo* [9.6] on the basis of spin-wave theory. If the anisotropy energy is ignored, the decrease of the sublattice magnetic moment, the specific heat and the parallel susceptibility are proportional to T^2, T^3 and T^2, respectively. The perpendicular susceptibility is the same as predicted by the molecular-field theory; however, due to the interaction among spin waves, the value becomes a little smaller and at the same time a negative term proportional to T^2 is added [9.6, 7]. These correction terms are both proportional to $1/S$. In real antiferromagnets the anisotropy energy cannot be ignored. The reason is that, as (9.15, 17) show, the product of the anisotropy field and the molecular field due to the exchange interaction appears and this term is much larger than the anisotropy field. In particular, the spin-wave spectrum has an energy gap of $g\mu_{\text{B}}\sqrt{2H_{\text{E}}H_{\text{A}}}$. The calculation of the sublattice magnetization, specific heat and susceptibility, which takes into account the anisotropy energy, was made by *Eisele* and *Keffer* [9.8]. The interaction among spin waves in antiferromagnets was studied by *Kanamori* and *Tachiki* [9.9].

9.3 One-Dimensional Heisenberg Model – Bethe-Hulthén Theory

In the case of the one-dimensional lattice we can determine exactly the ground-state energy by Bethe's method described in Sect. 8.2, without starting from the ground state in the molecular-field theory. Using this method, which is generally called the Bethe Ansatz, *Hulthén* [9.10] obtained the exact solution of the ground state of the one-dimensional antiferromagnet. Let us describe this *Bethe-Hulthén theory* below.

We assume that the interaction is present only between nearest-neighbor spins and that the magnitude of the spin $\boldsymbol{S}$ is $1/2$. In antiferromagnets the exchange coefficient J is negative and the spin state for the lowest energy corresponds to a state with $N/2$ reversed spins. Therefore the problem reduces to determining the lowest energy state for $r = N/2$ in (8.30). Equation (8.30) can be written as follows:

$$\begin{aligned}-2\varepsilon a(m_1,\dots,m_i,\dots,m_r) &= \sum_{i=1}^{r}\big[a(m_1,\dots,m_i+1,\dots,m_r)\\ &+a(m_1,\dots,m_i-1,\dots,m_r)-2a(m_1,\dots,m_i,\dots,m_r)\big]\ .\end{aligned} \tag{9.42}$$

Here it is assumed that $m_1, m_2 \dots, m_r$ are not nearest neighbors of each other. If two of the m are nearest neighbors of each other, for instance in the case of $m_{k+1} = m_k + 1$, (9.42) is replaced by

$$\begin{aligned}&-2\varepsilon a(m_1,\dots,m_i,\dots,m_k,m_k+1,\dots,m_r) = a(\dots,m_k-1,m_k,\dots)\\ &+a(\dots,m_k,m_k+2,\dots)-2a(\dots,m_k,m_k+1,\dots)\\ &+\sum_{i\neq k,k+1}\big[a(\dots,m_i+1,\dots)+a(\dots,m_i-1,\dots)\\ &-2a(m_1,\dots,m_i,\dots,m_r)\big]\ .\end{aligned} \tag{9.43}$$

Therefore, as in the case of two reversed spins, we formally impose on the coefficients $a(m)$ the relation

$$\begin{aligned}&2a(m_1,\dots,m_k,m_k+1,\dots,m_r)\\ &\quad= a(\dots,m_k,m_k,\dots)+a(\dots,m_k+1,m_k+1,\dots)\ .\end{aligned} \tag{9.44}$$

Then the relation (9.42) is always satisfied.

Extending (8.42), we write the solution of (9.42) as

$$a(m_1,\dots,m_r) = \sum_{\mathrm{P}=1}^{r!}\exp\mathrm{i}\Big(\sum_{k=1}^{r} f_{\mathrm{P}k}m_k + \frac{1}{2}\sum_{k<l}\varphi_{\mathrm{P}k,\mathrm{P}l}\Big)\ . \tag{9.45}$$

P represents the permutation operator for $1, 2, \dots, r$ and Pk is the kth number of the sequence obtained after P operates on $(1, 2, \dots, r)$. Equation (9.45) is called the *Bethe Ansatz.* On substituting (9.45) into (9.42), we immediately obtain

$$\varepsilon = \sum_{k=1}^{r}(1 - \cos f_k) \ . \tag{9.46}$$

We now substitute (9.45) into (9.44) and rewrite the summation over the permutation P as that over a pair of a permutation P and another permutaion P$'$, which satisfies $\mathrm{P}'l = \mathrm{P}l, l \neq (k, k+1)$, $\mathrm{P}'k = \mathrm{P}(k+1)$, $\mathrm{P}'(k+1) = \mathrm{P}k$. Since (9.44) must be satisfied for each term of this pair, the following relation is obtained:

$$\begin{aligned} &\mathrm{e}^{(\mathrm{i}/2)\varphi_{\mathrm{P}k,\mathrm{P}(k+1)}}\left[2\mathrm{e}^{\mathrm{i}f_{\mathrm{P}(k+1)}} - 1 - \mathrm{e}^{\mathrm{i}(f_{\mathrm{P}k}+f_{\mathrm{P}(k+1)})}\right] \\ &+ \mathrm{e}^{(-\mathrm{i}/2)\varphi_{\mathrm{P}k,\mathrm{P}(k+1)}}\left[2\mathrm{e}^{\mathrm{i}f_{\mathrm{P}k}} - 1 - \mathrm{e}^{\mathrm{i}(f_{\mathrm{P}k}+f_{\mathrm{P}(k+1)})}\right] = 0 \ . \end{aligned} \tag{9.47}$$

If we rewrite this relation as

$$\begin{aligned} &\frac{\cos\frac{1}{2}\varphi_{\mathrm{P}k,\mathrm{P}(k+1)} + \mathrm{i}\sin\frac{1}{2}\varphi_{\mathrm{P}k,\mathrm{P}(k+1)}}{\cos\frac{1}{2}\varphi_{\mathrm{P}k,\mathrm{P}(k+1)} - \mathrm{i}\sin\frac{1}{2}\varphi_{\mathrm{P}k,\mathrm{P}(k+1)}} \\ &= \frac{\sin\frac{1}{2}(f_{\mathrm{P}k} - f_{\mathrm{P}(k+1)}) + \mathrm{i}\left[\cos\frac{1}{2}(f_{\mathrm{P}k} + f_{\mathrm{P}(k+1)}) - \cos\frac{1}{2}(f_{\mathrm{P}k} - f_{\mathrm{P}(k+1)})\right]}{\sin\frac{1}{2}(f_{\mathrm{P}k} - f_{\mathrm{P}(k+1)}) - \mathrm{i}\left[\cos\frac{1}{2}(f_{\mathrm{P}k} + f_{\mathrm{P}(k+1)}) - \cos\frac{1}{2}(f_{\mathrm{P}k} - f_{\mathrm{P}(k+1)})\right]} \ , \end{aligned}$$

we immediately obtain the relation

$$\cot\frac{1}{2}\varphi_{\mathrm{P}k,\mathrm{P}(k+1)} = \frac{1}{2}\left(\cot\frac{1}{2}f_{\mathrm{P}k} - \cot\frac{1}{2}f_{\mathrm{P}(k+1)}\right)$$

or, corresponding to (8.41),

$$\cot\frac{1}{2}\varphi_{k,l} = \frac{1}{2}\left(\cot\frac{1}{2}f_k - \cot\frac{1}{2}f_l\right) \quad (-\pi \leq \varphi_{k,l} \leq \pi) \ . \tag{9.48}$$

The coefficients $a(m)$ must satisfy also the periodic boundary condition (8.32). Substituting (9.45) into (8.32), we obtain

$$\begin{aligned} &\sum_{\mathrm{P}}\exp\mathrm{i}\Big(\sum_{k=1}^{r} f_{\mathrm{P}k}m_k + \frac{1}{2}\sum_{k<l}\varphi_{\mathrm{P}k,\mathrm{P}l}\Big) \\ &\quad = \sum_{\mathrm{P}'}\exp\mathrm{i}\Big[\sum_{k=2}^{r} f_{\mathrm{P}'(k-1)}m_k + f_{\mathrm{P}'r}(m_1 + N) + \frac{1}{2}\sum_{k<l}\varphi_{\mathrm{P}'k,\mathrm{P}'l}\Big] \ . \end{aligned}$$

In order that this relation holds for all $(m_1, m_2, \dots, m_r)$, the two terms on the left-hand and right-hand sides having the same dependence on m_k must be equal to each other. The term on the left-hand side corresponding to one P has its counterpart on the right-hand side in the term of P$''$ satisfying $\mathrm{P}''(k-1) = \mathrm{P}k \ \ (k = 2, \dots, r)$ and $\mathrm{P}''r = \mathrm{P}1$. The condition that these two terms be the same is given by

$$N f_{\mathrm{P}''r} + \frac{1}{2}\sum_{k<l}\varphi_{\mathrm{P}''k,\mathrm{P}''l} - \frac{1}{2}\sum_{k<l}\varphi_{\mathrm{P}k,\mathrm{P}l} = 2\pi\lambda \ ,$$

where λ is an integer. After expressing P$''$ in terms of P, one finds

$$
\begin{aligned}
&Nf_{\mathrm{P}1} + \frac{1}{2}\sum_{k<l<r-1} \varphi_{\mathrm{P}(k+1),\mathrm{P}(l+1)} + \frac{1}{2}\sum_{k=1}^{r-1} \varphi_{\mathrm{P}(k+1),\mathrm{P}1} \\
&\quad - \frac{1}{2}\sum_{2\le k<l} \varphi_{\mathrm{P}k,\mathrm{P}l} - \frac{1}{2}\sum_{l=2}^{r} \varphi_{\mathrm{P}1,\mathrm{P}l} \\
&= Nf_{\mathrm{P}1} - \sum_{k=2}^{r} \varphi_{\mathrm{P}1,\mathrm{P}k} \ ;
\end{aligned}
$$

here the relation $\varphi_{k,l} = -\varphi_{l,k}$ has been used. Therefore

$$
Nf_k = 2\pi\lambda_k + \sum_l \varphi_{k,l} \tag{9.49}
$$

must hold for all k; λ_k in the above takes the values $0, 1, 2, \ldots, N-1$.

The previous discussions for the case of two reversed spins show that in the solution for real wavenumber f, the difference of two λ's has to be 2 or larger than 2 (i.e., $\lambda_2 - \lambda_1 \ge 2$). If the difference is 1 or 0, we have a bound-state solution with 2 complex conjugate f's. However, in the case $J < 0$, the energy of this solution is higher than the scattering wave solutions for real f, and we need not consider such a case. In this way $N/2$ values for λ_k, which correspond to the lowest-energy state, are determined uniquely as follows:

$$
\{1, 3, 5, 7, \ldots, N-1\} \ , \quad \text{i.e.,} \ \{\lambda_1, \lambda_2, \ldots, \lambda_{N/2}\} \ \ (\lambda_j = 2j-1) \ .
$$

If all the lattice points are shifted by 1, the total wave function for this set $\{\lambda_j\}$ acquires an extra factor $\exp(ik_0)$, where k_0 is given by

$$
k_0 = \frac{2\pi}{N}\sum_j \lambda_j = \frac{N}{2}\pi \ .
$$

Thus the factor is either $+1$ or -1, respectively, depending on whether $N/2$ is even or odd. The state for $\{0, 2, 4, \ldots, N-2\}$ belongs to the spin triplet $S_{\mathrm{tot}} = 1$ with its z component $S_{z\mathrm{tot}} = 0$, since $\lambda_1 = 0$.

Putting $\lambda_j/N = x$ and treating x as a continuous variable, we can rewrite (9.49) as

$$
f(x) = 2\pi x + \frac{1}{2}\int_0^1 \varphi(x, y)dy \ . \tag{9.50}
$$

At the same time (9.48, 46) are expressed as

$$
\cot\frac{1}{2}\varphi(x, y) = \frac{1}{2}\left[\cot\frac{1}{2}f(x) - \cot\frac{1}{2}f(y)\right] , \tag{9.51}
$$

$$
2\varepsilon = N\int_0^1 [1 - \cos f(x)]\, dx \ , \tag{9.52}
$$

where $-\pi \le \varphi(x, y) < \pi$ is assumed. Let us introduce

$$\cot \frac{1}{2} f(x) = \xi \tag{9.53}$$

and define $g(\xi)$ by

$$\frac{d\xi}{dx} = -\frac{1}{g(\xi)} . \tag{9.54}$$

Then, using (9.51), we have from (9.50)

$$f(x) = 2\pi x + \int_{-\infty}^{\infty} \cot^{-1} \frac{\xi - \eta}{2} \, g(\eta) \, d\eta . \tag{9.55}$$

Since $\varphi(x, y)$ is defined to lie between $-\pi$ and π, $\cot^{-1}(\xi - \eta)/2$ has a discontinuity of $\mp\pi/2$ at $\xi - \eta = 0\mp$. Noting this fact and differentiating both sides with respect to x, we obtain

$$\frac{df(x)}{dx} = \pi + \int_{-\infty}^{\infty} \frac{2}{4 + (\xi - \eta)^2} \frac{g(\eta)}{g(\xi)} d\eta . \tag{9.56}$$

By differentiating (9.53) with respect to x and using (9.54), one obtains the relation

$$\frac{df}{dx} = \frac{2}{1 + \xi^2} \frac{1}{g(\xi)} .$$

With this relation, (9.56) is rewritten as

$$\frac{2}{1 + \xi^2} = \pi g(\xi) + 2 \int_{-\infty}^{\infty} \frac{1}{4 + (\xi - \eta)^2} g(\eta) d\eta . \tag{9.57}$$

The solution of this integral equation for $g(\xi)$ is found easily by Fourier transformation; the transform G of g is

$$G(u) = \int_{-\infty}^{\infty} g(\xi) \mathrm{e}^{\mathrm{i}u\xi} d\xi = \frac{1}{\cosh u} . \tag{9.58}$$

The energy eigenvalue ε is obtained from (9.52) as

$$\varepsilon = N \int_0^1 \sin^2 \frac{f(x)}{2} dx = N \int_{-\infty}^{\infty} \frac{g(\xi)}{1 + \xi^2} d\xi . \tag{9.59}$$

After substituting the inverse transform of (9.58) into $g(\xi)$, one obtains ε explicitly as

$$\begin{aligned} \varepsilon &= \frac{N}{2\pi} \int_{-\infty}^{\infty} G(u) du \int_{-\infty}^{\infty} \frac{d\xi}{1 + \xi^2} \mathrm{e}^{-\mathrm{i}u\xi} \\ &= \frac{N}{2} \int_{-\infty}^{\infty} \frac{\mathrm{e}^{-|u|}}{\cosh u} du = 2N \int_0^{\infty} \frac{\mathrm{e}^{-2u}}{1 + \mathrm{e}^{-2u}} du = N \log 2 . \end{aligned} \tag{9.60}$$

This is a surprisingly simple result. From this result and (8.31), the ground-state energy for the one-dimensional antiferromagnet ($S = 1/2$) is obtained as

$$E_g = \frac{NJ}{2}[1 + 2(2\log 2 - 1)] = \frac{NJ}{2}(1 + 2 \times 0.3863) . \tag{9.61}$$

Orbach [9.11] extended the Bethe-Hulthén theory to the anisotropic exchange interaction $[J_z = J, J_x = J_y = (1-\alpha)J]$. One can discuss also this case in a similar way. As a result, (9.51, 52) are generalized as follows:

$$\cot\frac{1}{2}\varphi(x,y) = \frac{\cot\frac{1}{2}f(x) - \cot\frac{1}{2}f(y)}{(2-\alpha) + \alpha\cot\frac{1}{2}f(x)\cdot\cot\frac{1}{2}f(y)} , \tag{9.62}$$

$$2\varepsilon = N\int_0^1 [1 - (1-\alpha)\cos f(x)]dx . \tag{9.63}$$

As in the isotropic case, $\varphi(x,y)$ changes discontinuously from π to $-\pi$ at $x = y$. If we introduce ξ and $g(\xi)$ by (9.53, 54), a similar procedure as before leads to the following integral equation for $g(\xi)$:

$$\frac{2}{1+\xi^2} = \pi g(\xi) + \int_{-\infty}^{\infty} \frac{2 - \alpha + \alpha\eta^2}{[(2-\alpha) + \alpha\xi\eta]^2 + (\xi - \eta)^2} g(\eta)d\eta . \tag{9.64}$$

The ground-state energy ε can be calculated from (9.63) by using this $g(\xi)$. Orbach solved (9.64) numerically and determined the energy as a function of α. The result showed that ε changes smoothly from $\alpha = 0$ corresponding to the isotropic case to $\alpha = 1$ corresponding to the Ising case; no indication of a discontinuous change in the derivative was detected between these limits. The short-range correlation also shows no discontinuity.

One can use, instead of $g(\xi) = -dx/d\xi$, the function

$$A(f) = \frac{dx}{df} = \frac{1}{2}g(\xi)\frac{1}{\sin^2\frac{1}{2}f} \tag{9.65}$$

to obtain from (9.64) the following integral equation for $A(f)$:

$$A(f) = \frac{1}{\pi} - \frac{1}{2\pi}\int_0^{2\pi} df' \frac{(1 - \beta\cos f')A(f')}{1 - \beta(\cos f + \cos f') + \beta^2\cos^2\frac{1}{2}(f + f')} . \tag{9.66}$$

With this $A(f)$, ε can be obtained from

$$\varepsilon = \frac{N}{2}\int_0^{2\pi}(1 - \beta\cos f)A(f)df , \tag{9.67}$$

where $\beta = 1 - \alpha$.

Walker [9.12] solved (9.66) by Fourier transformation. New variables λ ($\lambda > 0$) and ψ are introduced by $\beta = \text{sech}\lambda$ and

$$e^{i\psi} = \frac{e^{(1/2)(\lambda - if)} - e^{(-1/2)(\lambda - if)}}{e^{(-1/2)(\lambda + if)} - e^{(1/2)(\lambda + if)}} . \tag{9.68}$$

Then $B(\psi)$, which is defined by

$$B(\psi)(e^{i\psi} - e^{\lambda})(e^{i\psi} - e^{-\lambda}) = A(f) , \tag{9.69}$$

is introduced instead of $A(f)$. The integral equation for $B(\psi)$ is given by

$$B(\psi) = [\pi(e^{i\psi} - e^{\lambda})(e^{i\psi} - e^{-\lambda})]^{-1} \\ - \frac{e^{2\lambda} - e^{-2\lambda}}{2\pi} \int_0^{2\pi} \frac{e^{2i\psi'} B(\psi') d\psi'}{(e^{-\lambda + i\psi'} - e^{\lambda + i\psi})(e^{-\lambda + i\psi} - e^{\lambda + i\psi'})} . \tag{9.70}$$

Using the Fourier-expansion formulae

$$\frac{\sin\psi}{\cosh\lambda - \cos\psi} = 2\sum_{n=1}^{\infty} e^{-n\lambda} \sin n\psi ,$$

$$\frac{\sinh\lambda}{\cosh\lambda - \cos\psi} = 1 + 2\sum_{n=1}^{\infty} e^{-n\lambda} \cos n\psi ,$$

we expand the first term on the right-hand side of (9.70) as

$$\frac{e^{-i\psi}}{\cosh\lambda - \cos\psi} = \frac{1 - \tanh\lambda}{\tanh\lambda} + \frac{1 - \tanh\lambda}{\tanh\lambda} \sum_{n=1}^{\infty} e^{-n\lambda} e^{in\psi} \\ + \frac{1 + \tanh\lambda}{\tanh\lambda} \sum_{n=1}^{\infty} e^{-n\lambda} e^{-in\psi} . \tag{9.71}$$

The integration kernel is also expanded in terms of $e^{in(\psi - \psi')}$. Then $B(\psi)$ is obtained as an expansion in $e^{i\psi}$. This procedure results in

$$B(\psi) = -\frac{1}{4\pi} \frac{1}{\sinh\lambda} \sum_{m=-\infty}^{\infty} \frac{e^{im\psi}}{\cosh(m+1)\lambda} , \tag{9.72}$$

$$A(\psi) = -\frac{1}{2\pi} \frac{\cos\psi - \cosh\lambda}{\sinh\lambda} \sum_{m=-\infty}^{\infty} \frac{e^{im\psi}}{\cosh m\lambda} . \tag{9.73}$$

Using (9.73) in (9.67), we obtain the energy of the ground state as

$$E_g = \frac{NJ}{2} \left[2\tanh\lambda \left(1 + \sum_{n=1}^{\infty} \frac{4}{e^{2n\lambda} + 1} \right) - 1 \right] . \tag{9.74}$$

This expression for E_g agrees with (9.61) for $\alpha \to 0$, $\beta \to 1$, $\lambda \to 0$ and with the value of the Ising model $(1/2)NJ$ for $\alpha \to 1$, $\beta \to 0$, $\lambda \to \infty$. There is no singular point between these two limiting cases as a function of λ. λ becomes pure imaginary for $\beta > 1$. In this case (9.74) has a pole for any small value of λ. Therefore $\alpha = 0$ (i.e., $\beta = 1$) is a singular point in the one-dimesional antiferromagnet. In the Ising model the system can be divided into two sublattices with + and − spins; the state having such sublattice moments (i.e., long-range order) is realized up to the isotropic case $\alpha \to 0, \beta \to 1$. On the other hand, the long range order vanishes for the case $\beta > 1$ in which the transverse component is larger than the z component. The boundary which separates these two cases is believed to be $\alpha = 0, \beta = 1$. *Yang* and *Yang* [9.13] extended the above treatment for $\beta \leq 1$ to the region of $\beta > 1$ and discussed the analyticity of the ground-state energy as a function of β. The $\beta \to \infty$ case corresponds to purely transverse coupling, i.e., the XY model. The XY model was studied by *Lieb* et al. [9.14] and *Katsura* [9.15].

The calculation of the excitation energy for the one-dimensional antiferromagnet is fairly complicated. *des Cloizeaux* and *Pearson* [9.16] derived the spin-wave excitation energy by extending the above method due to Bethe and Hulthén. Although we do not go into the details of the calculation here, they obtained

$$\Delta E = -J\pi|\sin q| \tag{9.75}$$

as the excitation energy of the spin wave having wavenumber q. This should be compared with the result of the Anderson's spin-wave theory ($S = 1/2$)

$$\Delta E_{\mathrm{A}} = -2J|\sin q| \ .$$

The coefficient of $|\sin q|$ is larger by $\pi/2$ in the exact solution.

The calculation of the ground-state energy for the one-dimensional antiferromagnet was extended by *Griffiths* [9.17] and *Yang* and *Yang* [9.18] to the case of finite magnetic fields. According to them the susceptibilty at $T = 0$ (for the case of isotropic exchange interaction) is given by $(1/2\pi^2)g^2\mu_{\mathrm{B}}^2/|J|$. *Bonner* and *Fisher* [9.19] carried out numerical calculations of the specific heat and the susceptibility for finite chains ($N = 2 \sim 11$), which cover the ratio of the transverse and longitudinal exchange interactions, β, from 1 to 0.

9.4 Exact Solution of One-Dimensional Hubbard Model

As shown in (4.16) of Part I, the antiferromagnetic Heisenberg Hamiltonian is derived from the so-called Hubbard Hamiltonian (3.1) by taking the large-U limit of the Coulomb integral between electrons on the same site. Namely, the Hamiltonian (3.1)

$$\mathcal{H}_{\mathrm{Hubbard}} = \sum_{nn's} b_{n'-n} a_{n's}^{\dagger} a_{ns} + U \sum_{n} a_{n\uparrow}^{\dagger} a_{n\uparrow} a_{n\downarrow}^{\dagger} a_{n\downarrow} \tag{3.1}$$

reduces in the limit $U \to \infty$ to

$$\mathcal{H}_{\mathrm{Heisenberg}} = -\sum_{nn'} \frac{|b_{n'-n}|^2}{U} \left(\frac{1}{2} - 2\boldsymbol{S}_n \cdot \boldsymbol{S}_{n'} \right) , \tag{4.16}$$

in the case that the electron number N_e is equal to the number of lattice points N. The ground-state energy of (4.16) for the one-dimensional case can be obtained from (9.61) as

$$E_g = 2NJ \log 2 \,, \tag{9.76}$$

where J is defined by

$$J = -2\frac{b^2}{U} \,. \tag{9.77}$$

It is assumed here that $b_{n'-n}$ has a finite value only between nearest-neighbor lattice points. Thus the exact solution of the one-dimensional Hubbard model includes the Bethe-Hulthén solution as the limit $U \to \infty$. In this sense, as well as in connection with the problem of the metal-insulator transition (Mott transition), the study of the one-dimensional Hubbard model is believed to be important.

The exact solution of the one-dimensional Hubbard model was obtained in the studies of *Yang* [9.20] and *Lieb* and *Wu* [9.21]. First, Yang obtained the exact solution of the ground state of the one-dimensional electron gas with the δ-function-type Coulomb repulsion. Lieb and Wu applied Yang's theory to the one-dimensional Hubbard model and succeeded in finding its exact solution. The derivation is so complicated that we simply show some results; readers interested in the details are referred to the original papers.

Let M and N_e be the electron number with the reversed spin and the total number of electrons, respectively. Without losing generality we can assume that $S_{z\mathrm{tot}} \equiv (1/2)(N_e - 2M)$ and $N_e < N$. Introducing two densities of states $\sigma(\Lambda)$ and $\rho(k)$, we express M and N_e as

$$\int_{-Q}^{Q} \rho(k) dk = \frac{N_e}{N} \,, \tag{9.78}$$

$$\int_{-B}^{B} \sigma(\Lambda) d\Lambda = \frac{M}{N} \,. \tag{9.79}$$

Then $\rho(k)$ and $\sigma(\Lambda)$ are given by the solution of the following integral equations:

$$2\pi\rho(k) = 1 + \cos k \int_{-B}^{B} \frac{8U\sigma(\Lambda)d\Lambda}{U^2 + 16(\sin k - \Lambda)^2} \,, \tag{9.80}$$

$$\int_{-Q}^{Q} \frac{8U\rho(k)dk}{U^2 + 16(\Lambda - \sin k)^2} = 2\pi\sigma(\Lambda) + \int_{-B}^{B} \frac{4U\sigma(\Lambda')d\Lambda'}{U^2 + 4(\Lambda - \Lambda')^2} \,, \tag{9.81}$$

where Q and B must satisfy $Q \leq \pi$ and $B \leq \infty$. In terms of $\rho(k)$, the ground-state energy is given as

$$E = -2N \int_{-Q}^{Q} \rho(k) \cos k \, dk \,. \tag{9.82}$$

Both M and N_e increase monotonically with increasing B and Q. $B = \infty$ is realized at $M = N_e/2$, i.e., $S_{z\text{tot}} = 0$; $Q = \pi$ holds for $N_e = N$ corresponding to the half-filled band. Let us discuss the case for the electron density of one electron per site and $S_{z\text{tot}} = 0$, for which $Q = \pi$ and $B = \infty$. In this case the solution of (9.80, 81) is obtained as

$$\sigma(\Lambda) = (2\pi)^{-1} \int_0^\infty \operatorname{sech}\left(\frac{1}{4}\omega U\right) \cos(\omega\Lambda)\, \mathrm{J}_0(\omega)\, d\omega\,, \tag{9.83}$$

$$\rho(k) = (2\pi)^{-1} + \pi^{-1} \cos k \int_0^\infty \frac{\cos(\omega \sin k)\mathrm{J}_0(\omega)}{1 + \exp\left(\frac{1}{2}\omega U\right)}\, d\omega\,. \tag{9.84}$$

From (9.82) the energy is given by

$$E = -4N \int_0^\infty \frac{\mathrm{J}_0(\omega)\mathrm{J}_1(\omega)d\omega}{\omega\left[1 + \exp\left(\frac{1}{2}\omega U\right)\right]}\,, \tag{9.85}$$

where J_0 and J_1 are Bessel functions of the zeroth and first orders, respectively, and for simplicity $b = -1$ has been assumed. The expression (9.85) for the energy shows clearly that $U = 0$ is a singular point. In fact, if U is finite, the minimum energy which is needed to change the electron number by one is always finite, suggesting that the one-dimensional electron system has an insulating property in this case. On the other hand, for $U = 0$ this energy becomes zero and the system exhibits metallic properties. Therefore the Mott transition occurs at $U = 0$ in one dimension. In other words, the Mott transition does not occur in the region of finite U. The system is always metallic for $N_e \neq N$.

Using an indefinite integral of $\mathrm{J}_0(\omega)\mathrm{J}_1(\omega)/\omega$

$$I(\omega) = \omega\left[\mathrm{J}_1^2(\omega) - \mathrm{J}_2(\omega)\mathrm{J}_0(\omega)\right] + \mathrm{J}_0(\omega)\mathrm{J}_1(\omega)\,, \tag{9.86}$$

we can rewrite (9.85) as

$$E = -\frac{1}{2}NU \int_0^\infty I(\omega)\operatorname{sech}^2\left(\frac{1}{4}\omega U\right) d\omega\,. \tag{9.87}$$

Since $\operatorname{sech}^2(\omega U/4)$ behaves as the δ function in the limit $U \to \infty$, (9.87) may be approximated in this limit as

$$\begin{aligned} E &\sim -\frac{1}{2}NU \int_0^\infty \frac{\omega}{2}\operatorname{sech}^2\left(\frac{1}{4}\omega U\right) d\omega = -N\frac{4}{U}\int_0^\infty x \operatorname{sech}^2 x\, dx \\ &= -4N\frac{1}{U}\log 2\,. \end{aligned} \tag{9.88}$$

This value agrees with the exact solution due to Bethe and Hulthén, (9.76) with J defined by (9.77). Higher-order terms in $1/U$ can be calculated by expanding $I(\omega)$ in terms of ω and integrating term by term. The term proportional to $1/U^3$ is obtained as

$$E^{(3)} = 12N\frac{1}{U^3}\sum_{n=1}^\infty (-1)^{n-1}\frac{1}{n^3} = 12N\frac{1}{U^3}\cdot\frac{3}{4}\times\frac{\pi^3}{25.79\ldots}\,. \tag{9.89}$$

10. Two-Dimensional XY Model – Kosterlitz-Thouless Transition

According to the spin-wave theory for the ferromagnetic Heisenberg model, the number of excited spin waves at finite temperatures diverges in one- and two-dimensional systems. This implies that the phase transition from the paramagnetic state to the ferromagnetic state does not occur in these low-dimensional systems. In two dimensions, the ferromagnetic phase transition becomes possible only when the anisotropy energy is taken into account.

Let us consider the two-dimensional XY model. If spins are treated classically, the exchange interaction $\mathcal{H}$ between neighboring spins only can be written as

$$\mathcal{H} = -J \sum_{\langle n,m \rangle} \cos(\theta_n - \theta_m) = -\frac{1}{2} J \sum_{\langle n,m \rangle} \left[\mathrm{e}^{\mathrm{i}(\theta_n - \theta_m)} + \mathrm{e}^{-\mathrm{i}(\theta_n - \theta_m)} \right] , \quad (10.1)$$

where $\langle n, m \rangle$ denotes the summation over nearest-neighbor spin pairs on the two-dimensional square lattice. θ_n represents the angle between the n-th spin and a particular axis. In this case the correlation function between spins is given by

$$\langle \mathrm{e}^{\mathrm{i}\theta_0} \mathrm{e}^{-\mathrm{i}\theta_n} \rangle = \int \prod_m d\theta_m \mathrm{e}^{\mathrm{i}(\theta_0 - \theta_n)} \exp\left[\frac{J}{kT} \sum_{\langle n,m \rangle} \cos(\theta_n - \theta_m) \right] \Big/ Z , \quad (10.2)$$

where Z is the partition function.

At high temperatures this correlation function can be evaluated by means of the high-temperature expansion. Noting that

$$\int_0^{2\pi} d\theta_m = 2\pi , \qquad \int_0^{2\pi} d\theta_m \mathrm{e}^{\mathrm{i}\theta_m} = 0 ,$$

one finds that the first nonvanishing contribution in the high-temperature expansion of (10.2) is the term proportional to $(J/kT)^{|n|}$. Thus we have

$$\langle \mathrm{e}^{\mathrm{i}\theta_0} \mathrm{e}^{-\mathrm{i}\theta_n} \rangle \simeq (J/kT)^{|n|} = \exp\left[-|n| \log(kT/J) \right] . \quad (10.3)$$

In this way the correlation function at high temperatures decreases exponentially with the distance between spins.

At low temperatures, thermal fluctuations are not so large and the change of spin direction is gradual. Therefore we can expand the cosine function in (10.1) and approximate it by

$$\mathcal{H} = \frac{1}{2} J \sum_{n,i} [\nabla_i \theta(\boldsymbol{n})]^2 , \tag{10.4}$$

keeping the terms to second order. Here the abbreviation

$$\nabla_i \theta(\boldsymbol{n}) = \theta_{\boldsymbol{n}} - \theta_{\boldsymbol{n}+i} \tag{10.5}$$

has been used. In this case the correlation function is calculated as

$$\langle \exp[\mathrm{i}(\theta_0 - \theta_{\boldsymbol{n}})] \rangle \simeq \int \prod_m d\theta_m \mathrm{e}^{\mathrm{i}(\theta_0 - \theta_{\boldsymbol{n}})} \exp\left[-\frac{J}{2kT} \sum (\nabla\theta)^2 \right] \Big/ Z . \tag{10.6}$$

The integral in (10.6) has the form of a Gaussian integral for each θ; therefore the integral can be easily carried out, resulting in

$$\langle \exp[\mathrm{i}(\theta_0 - \theta_{\boldsymbol{n}})] \rangle \sim \exp\left[\frac{kT}{J} \Delta(\boldsymbol{n}) \right] , \tag{10.7}$$

where $\Delta(\boldsymbol{n})$ is the lattice Green's function for the square lattice. For large $|\boldsymbol{n}|$, $\Delta(\boldsymbol{n})$ can be approximated as

$$\Delta(\boldsymbol{n}) \simeq -\frac{1}{2\pi} \log |\boldsymbol{n}| \quad (|\boldsymbol{n}| \gg 1) . \tag{10.8}$$

By using this expression in (10.7), one obtains the correlation function as

$$\langle \exp[\mathrm{i}(\theta_0 - \theta_{\boldsymbol{n}})] \rangle \simeq \left(\frac{1}{|\boldsymbol{n}|} \right)^{kT/2\pi J} . \tag{10.9}$$

In the limit $|\boldsymbol{n}| \to \infty$ the expectation value of the product of two spins approaches the product of averages; thus we have

$$\lim_{|\boldsymbol{n}| \to \infty} \langle \mathrm{e}^{\mathrm{i}(\theta_0 - \theta_{\boldsymbol{n}})} \rangle = \langle \mathrm{e}^{\mathrm{i}\theta_0} \rangle \langle \mathrm{e}^{-\mathrm{i}\theta_{\boldsymbol{n}}} \rangle \to 0 . \tag{10.10}$$

Thus the result of (10.9) implies that the two-dimensional classical XY spin system does not show ferromagnetism even at low temperatures.

As mentioned in Sect. 7.4, the Fourier component $G(k)$ of the spin correlation function can be expressed at the critical point $T = T_c$ as $k^{-2+\eta}$ with the critical exponent η; see (7.85). By using the inverse Fourier transformation in two dimension, the spin correlation function is obtained as

$$\langle \boldsymbol{S}(0) \cdot \boldsymbol{S}(\boldsymbol{n}) \rangle = \left(\frac{1}{|\boldsymbol{n}|} \right)^{\eta} . \tag{10.11}$$

In the two-dimensional XY system, (10.9) (which was derived in the spin-wave approximation for the low-temperature region) has the same form as (10.11), if we take

$$\eta = \frac{kT}{2\pi J} . \tag{10.12}$$

Therefore this system has a critical line instead of a critical point. Corresponding to this fact, the spin correlation length ξ is infinite below a certain temperature and the susceptibility diverges there.

The above argument shows that the two-dimensional XY model has two distinct phases, the low-temperature phase and the high-temperature phase, which are topologically different from each other. Kosterlitz and Thouless clarified the nature of the phase transition between these two phases [10.1–3]. They noted the importance of vortex states, as shown in Fig. 10.1. From (10.4), the excitation energy of such a vortex state is obtained as

$$\mathcal{H} \simeq \pi J \log(R/a) , \tag{10.13}$$

where a is the distance between lattice points and R is the radius of the system. In the limit $R \to \infty$ this energy diverges. However, the center of the vortex can be anywhere in the crystal so that the entropy due to the vortex is given by

$$k \log(R/a)^2 . \tag{10.14}$$

Thus the free energy of one vortex takes the form

$$F = (\pi J - 2kT) \log(R/a) . \tag{10.15}$$

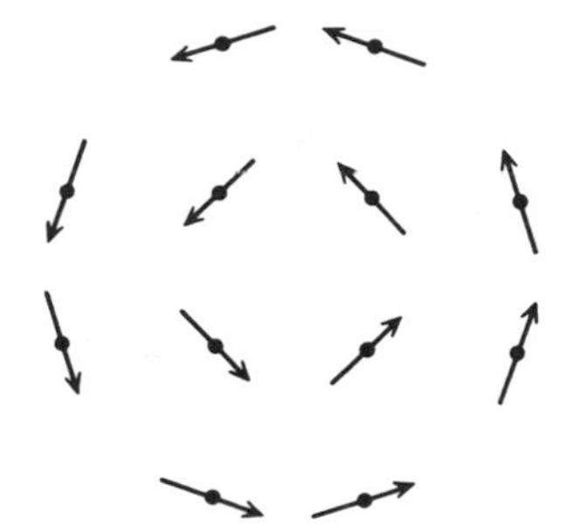

Fig. 10.1. A vortex state of spins

As the temperature increases, the free energy becomes zero at T_c given by

$$T_c = \frac{\pi J}{2k} . \tag{10.16}$$

Above this temperature many vortices are generated and condensation of vortices occurs. The temperature T_c is the transition temperature between the two phases.

Kosterlitz and *Thouless* [10.1] defined the vorticity q of a vortex by

$$\oint d\theta(\boldsymbol{r}) = 2\pi q . \tag{10.17}$$

Here the same $\boldsymbol{r}$ as a continuous variable is used instead of $\boldsymbol{n}$ to represent the coordinates of the lattice points. The integral in (10.17) denotes a summation over peripheral lattice points surrounding an area. Let us write $\theta(\boldsymbol{r})$ as

$$\theta(\boldsymbol{r}) = \psi(\boldsymbol{r}) + \bar{\theta}(\boldsymbol{r}) \tag{10.18}$$

in terms of the local average $\bar{\theta}(\boldsymbol{r})$ and the deviation from it $\psi(\boldsymbol{r})$. The condition for the absolute minimum of the ferromagnetic Hamiltonian is, of course, $\bar{\theta}(\boldsymbol{r}) = \text{constant}$.

Since $\bar{\theta}(\boldsymbol{r})$ obeys Laplace's equation except for the singular point at the center of vortex, we can introduce the function $\bar{\theta}'(\boldsymbol{r})$, the conjugate function of $\bar{\theta}(\boldsymbol{r})$, by

$$f(z) = \bar{\theta}(\boldsymbol{r}) + \mathrm{i}\bar{\theta}'(\boldsymbol{r}) \, . \tag{10.19}$$

Here we have used $z = x + \mathrm{i}y$ and $\boldsymbol{r} = (x, y)$. Rewriting (10.17) as an integral over an area, we try to obtain a differential equation for $\bar{\theta}(\boldsymbol{r})$. We note at this stage that there is a singular point corresponding to the center of the vortex within this area. To avoid this difficulty we use the Cauchy-Riemann relation and replace the derivative of $\bar{\theta}(\boldsymbol{r})$ by the derivative of $\bar{\theta}'(\boldsymbol{r})$. Then one can derive a differential equation for $\bar{\theta}'(\boldsymbol{r})$:

$$\nabla^2 \bar{\theta}'(\boldsymbol{r}) = -2\pi\rho(\boldsymbol{r}) \, , \tag{10.20}$$

$$\rho(\boldsymbol{r}) = \sum_i q_i \delta(\boldsymbol{r} - \boldsymbol{r}_i) \, , \tag{10.21}$$

where $\rho(\boldsymbol{r})$ is the distribution function of the vortex centers. The Laplacian symbol ∇^2 represents actually the discrete version of the second derivative, i.e.,

$$\nabla^2 = \nabla_x^2 + \nabla_y^2 \, , \tag{10.22}$$

$$\nabla_x^2 f(\boldsymbol{r}) = f(\boldsymbol{r} + x) + f(\boldsymbol{r} - x) - 2f(\boldsymbol{r}) \, . \tag{10.23}$$

The solution of (10.20) for the two-dimensional lattice is given by

$$\bar{\theta}'(\boldsymbol{r}) = -2\pi \int d\boldsymbol{r}' \rho(\boldsymbol{r}') g(\boldsymbol{r} - \boldsymbol{r}') + O\left[\log \frac{R}{r_0} \int \rho(\boldsymbol{r}) d\boldsymbol{r}\right] \, , \tag{10.24}$$

where $g(\boldsymbol{r})$ is identical to the lattice Green function $\Delta(\boldsymbol{n})$ except for a constant; it satisfies

$$\nabla^2 g(\boldsymbol{r}) = \delta(\boldsymbol{r}) \, . \tag{10.25}$$

The solution of this differential equation is given by

$$g(\boldsymbol{r}) = \int_{-\pi}^{\pi} \frac{dk_x}{2\pi} \int_{-\pi}^{\pi} \frac{dk_y}{2\pi} \frac{1 - \mathrm{e}^{\mathrm{i}\boldsymbol{k}\cdot\boldsymbol{r}}}{(4 - 2\cos k_x - 2\cos k_y)} \, , \tag{10.26}$$

as can be easily confirmed by substituting this solution directly into (10.25). We note that $g(\boldsymbol{r} = 0) = 0$ has been assumed here. It was shown by *Spitzer*

[10.4] that this lattice Green function for $r \gg a$ is accurately approximated by

$$g(\boldsymbol{r}) \sim \frac{1}{2\pi} \log \frac{r}{r_0} = \frac{1}{2\pi} \log \frac{r}{a} + \frac{1}{2\pi} \log \frac{a}{r_0} , \tag{10.27}$$

where $a/r_0 = 2\sqrt{2}e^{\gamma}$. Therefore the second term on the right-hand side of (10.27) is almost equal to 1/4 and this approximation is fairly good even at $r = a$.

One may consider that the second term on the right-hand side of (10.24) has been added to cancel the divergence of the first term for $r \to R$. Since the second term diverges as $\log(R/r_0)$, the condition that the sum be well behaved is

$$\int \rho(\boldsymbol{r}) d\boldsymbol{r} = 0 \quad \text{or} \quad \sum_i q_i = 0 . \tag{10.28}$$

Namely, the total vorticity must be zero due to cancellation of the positive and negative contributions. Because of this condition, $\bar{\theta}'(\boldsymbol{r})$ is given by the first term of (10.24). Since $\bar{\theta}'(\boldsymbol{r})$ is the imaginary part of $f(z)$, the conjugate function $\bar{\theta}(\boldsymbol{r})$ is given by the imaginary part of

$$-\sum_i q_i \log \left(\frac{z - z_i}{r_0}\right) .$$

However,

$$\iint d^2r \left[\nabla \bar{\theta}(\boldsymbol{r})\right]^2 = \iint d^2r \left[\nabla \bar{\theta}'(\boldsymbol{r})\right]^2 \tag{10.29}$$

holds from the Cauchy-Riemann relation. Therefore it is not necessary to go back to $\theta(\boldsymbol{r})$, but we obtain, by using (10.20, 24),

$$\begin{aligned} \mathcal{H}_{2n}(r_1, \dots, r_{2n}) = \frac{1}{2} J \iint d^2r \left[\nabla \psi(\boldsymbol{r})\right]^2 \\ - \pi J \sum_{i \neq j}^{2n} q_i q_j \log \left|\frac{\boldsymbol{r}_i - \boldsymbol{r}_j}{a}\right| + \mu \sum_i q_i^2 \end{aligned} \tag{10.30}$$

as an effective Hamiltonian for the system containing $2n$ vortices. Here

$$2\mu = 2\pi J \log \frac{a}{r_0} \simeq \pi^2 J \tag{10.31}$$

corresponds to the energy of a pair of vortices with opposite signs, which are separated by the lattice constant a. If we use (10.16) for the transition temperature, we obtain $\mu/kT_c \sim \pi$. This value is so large that the density of vortices at T_c is expected to be sufficiently low. The first term of (10.30) is the contribution from spin-wave excitations. In the following we restrict the vorticity to ± 1 for simplicity, since vortices with $|q| > 1$ hardly occur. Let the partition function for the Hamiltonian (10.30) be

$$Z = \mathrm{Tr}\{ \exp(-\beta \mathcal{H}_{2n}) \} . \tag{10.32}$$

Here the trace Tr means

$$\mathrm{Tr} \equiv \int \delta\psi(\boldsymbol{r}) \sum_{n=0}^{\infty} \frac{1}{(n!)^2} \iint_{D_{2n}} d^2 r_{2n} \ldots \iint_{D_1} d^2 r_1 . \tag{10.33}$$

The region of integration over the angle $\psi(\boldsymbol{r})$ can be extended to $(-\infty, \infty)$. The integrals over the positions $\boldsymbol{r}_i$ of the vortices are carried out over the entire plane subject to the restriction $|\boldsymbol{r}_i - \boldsymbol{r}_j| > a$. Because of the presence of n pairs of + and − vortices, the statistical factor $1/(n!)^2$ is included and the summation over n is taken from 0 to ∞. From this partition function one can understand that, despite their presence, the vortices do not affect the spin correlation function much. The reason is that μ/kT_c is so large that the density of vortices is sufficiently low below T_c, and that pairs are formed due to a strong attractive force between + and − vortices, even if they are present. Therefore the spin correlation function is determined mainly by spin-wave excitations and the result (10.9), which does not take into account the presence of vortices, is essentially correct. When the temperature is raised to the vicinity of T_c, pairs of vortices dissociate and dissociated single vortices condense above $T = T_c$.

The partition function (10.32) together with (10.33) is equivalent to the partition function for the *s-d* system, which was derived in the study of the Kondo effect by *Anderson* and *Yuval* [10.5]. The only difference is that in the *s-d* system the integral over r is one-dimensional, whereas it is two-dimensional here. Noting this point, *Kosterlitz* [10.2] derived a scaling law with respect to the lattice constant a, which is similar to *Anderson*, *Yuval* and *Hamann*'s method [10.6] for the *s-d* system, and examined properties of the two-dimensional XY model near its critical point by applying considerations based on the renormalization group. Leaving the details to the original paper, we describe in the following the main results of *Kosterlitz* [10.2].

First, the transition temperature is given by

$$\frac{\pi J}{kT_c} - 2 = \exp\left(-\frac{\pi^2 J}{2kT_c}\right) . \tag{10.34}$$

The difference from the previous value (10.16) is due to pair excitations of vortices. The susceptibility diverges in the entire region below T_c and is finite above T_c. When the temperature is decreased from high temperature to T_c, the correlation function behaves as

$$\begin{aligned} \xi &\sim \exp(bt^{-1/2}) \qquad (t > 0) \\ &= \infty \qquad (t < 0) \end{aligned} \tag{10.35}$$

where $t = (T - T_c)/T_c$ and $b \simeq 1.5$; the susceptibility behaves as

$$\begin{aligned} &\chi \sim \xi^{2-\eta} , \qquad \eta = \frac{1}{4} \quad (t > 0) \\ &\chi = \infty \qquad\qquad\qquad (t < 0) \end{aligned} \tag{10.36}$$

Thus, instead of the usual power law in t, ξ diverges more rapidly so that the critical exponents ν and γ cannot be defined. However, the singular part of the free energy above T_c is given by

$$F \sim \xi^{-2} \qquad (t > 0) , \tag{10.37}$$

and the magnetization under a weak field follows

$$m \sim h^{1/\delta} , \qquad \delta = 15 \quad (t = 0) . \tag{10.38}$$

The susceptibility χ and the specific heat c can be expressed in terms of ξ as $\chi \sim \xi^{\tilde{\gamma}}$ and $c \sim \xi^{\tilde{\alpha}}$. Then we find

$$\begin{aligned} \tilde{\gamma} &= \frac{\gamma}{\nu} = 2 - \eta , \\ \tilde{\alpha} &= \frac{\alpha}{\nu} = -d , \\ \delta &= \frac{d + 2 - \eta}{d - 2 + \eta} , \end{aligned} \tag{10.39}$$

which suggest that the standard scaling relations hold. It is notable that $\eta = 1/4$ and $\delta = 15$ agree with the corresponding values for the two-dimensional Ising system.

In summary, the Kosterlitz-Thouless transition is a phase transition, which is caused by the appearance of vortices; it is a special phase transition between two disordered phases in contrast to the transition from the spin-ordered state to the disordered state in ordinary magnets. There is no symmetry breaking here, and so this type of transition is called a topological phase transition. Real systems showing the Kosterlitz-Thouless transition include the two-dimensional film of superfluid helium [10.7].

Part III

Magnetism of Metals

11. Magnetism of Free Electrons

In Parts I and II we have discussed the cases where the $3d$- or $4f$-electrons are localized on each atom and do not contribute to the conduction. The magnetism of many insulating crystals of iron-group and rare-earth compounds as well as rare-earth metals are well described by spin Hamiltonians. On the other hand, in iron-group transition metals such as Ni, Co and Fe, and also in their alloys, the $3d$-electrons carrying magnetism move about in crystals, like the $4s$-electrons, so that their energy levels broaden to form bands with certain widths, narrower than the $4s$ bands. Thus $3d$-electrons become *itinerant electrons* and participate in the metallic cohesion. However, Coulomb integrals on the same atomic sites are large for the $3d$-electrons of Ni, Co and Fe so that the electrons are strongly correlated although they move about the crystals. In Part III, the subject of our consideration is mainly metallic magnetism with strong correlation; we begin by describing free-electron systems without correlation (that is, the opposite limit to ionic crystals).

11.1 Electronic States of Free Electrons

Let us consider a valence electron in a metal. The electron is affected by the periodic potential due to the ionic lattice as well as the Coulomb potential due to other electrons. The potential due to the other electrons is a function of each electron coordinate; we divide it into two parts: one is the average potential of the electrons and the other is the deviation from the average. For the time being we neglect the latter contribution. This approximation is generally called the *Hartree approximation* and is always employed as a starting approximation to treat many-body problems. In this way a many-body problem is, in the first place, reduced to a one-body problem. The potential due to the average distribution is calculated from the solution of the one-body problem; the process is repeated until the potential thus obtained is equal to the starting potential. In this sense the method is called a *self-consistent method.* The potential determined from the average distribution of the other electrons has also the periodicity of the lattice. The spatially uniform part of the potential cancels the uniformly averaged potential due to the positively charged ions. (Here we neglect the quantity 1 compared to the number of lattice points N.) The cancellation occurs because of the electrical neutrality of the whole crystal. Therefore, in this approximation, the potential influencing one electron has the same periodicity as the lattice.

If we neglect the periodic lattice potential, electrons move freely. The wave function of a free electron in a cube of side length L is a plane wave:

$$\psi_{\boldsymbol{k}}(\boldsymbol{r}) = \frac{1}{\sqrt{V}} \mathrm{e}^{\mathrm{i}\boldsymbol{k}\cdot\boldsymbol{r}} , \tag{11.1}$$

where V is the volume of the cube, $\boldsymbol{r}$ is the coordinate of the electron and $\boldsymbol{k}$ is the wavevector. For the periodic boundary condition of Born-von Kármán, the wavevectors are

$$(k_x, k_y, k_z) = \frac{2\pi}{L}(n_x, n_y, n_z) , \tag{11.2}$$

where n_x, n_y and n_z are positive or negative integers. The energy $\varepsilon_{\boldsymbol{k}}$ of an electron with wavevector $\boldsymbol{k}$ is

$$\varepsilon_{\boldsymbol{k}} = \frac{\hbar^2 k^2}{2m} , \tag{11.3}$$

where m denotes the electron mass and $\hbar$ is Planck constant h divided by 2π.

Let the total number of electrons be N_e. The ground state of the N_e-electron system is the state where up- and down-spin electrons occupy the plane wave states, (11.1), in order of energy, according to the Pauli principle. The highest occupied energy $\hbar^2 k_\mathrm{F}^2/2m$ is called the Fermi energy; the highest wavevector, k_F, is the Fermi wavenumber. The number of electron states n contained inside a sphere of radius k in k-space is given by

$$n(k) = \frac{V}{8\pi^3}\frac{4\pi k^3}{3} = \frac{V}{6\pi^2}k^3 = \frac{V}{6\pi^2}\left(\frac{2m}{\hbar^2}\right)^{3/2}\varepsilon_{\boldsymbol{k}}^{3/2} . \tag{11.4}$$

By differentiating this quantity with respect to the energy $\varepsilon_{\boldsymbol{k}}$, one can calculate the number of states in the energy interval between ε and $\varepsilon + d\varepsilon$ (this is called the density of states):

$$\rho(\varepsilon) = \frac{V}{6\pi^2}\left(\frac{2m}{\hbar^2}\right)^{3/2}\frac{3}{2}\varepsilon^{1/2} = \frac{3n}{2\varepsilon} . \tag{11.5}$$

The sphere of radius k_F in k-space occupied by the electrons is called the Fermi sphere and its surface is called the Fermi surface. The density of states at the Fermi surface for one spin is given as

$$\rho(\varepsilon_\mathrm{F}) = \frac{V}{4\pi^2}\left(\frac{2m}{\hbar^2}\right)^{3/2}\varepsilon_\mathrm{F}^{1/2} = \frac{3N_\mathrm{e}}{4\varepsilon_\mathrm{F}} , \tag{11.6}$$

as obtained by setting ε and n in (11.5) to be ε_F and $N_\mathrm{e}/2$, respectively.

11.2 Conduction-Electron States in Crystals

Conduction electrons in metals experience a periodic potential with the period of the ionic lattice. Suppose that the crystal lattice is generated by repeated translations of three independent primitive vectors, $\boldsymbol{a}_1, \boldsymbol{a}_2, \boldsymbol{a}_3$. A parallel hexahedron composed of these three primitive vectors is called the *unit cell*; its volume v_0 is given by $\boldsymbol{a}_1 \cdot (\boldsymbol{a}_2 \times \boldsymbol{a}_3)$. If each unit cell contains only one atom, the lattice is called a Bravais lattice; in general, several atoms belong to one unit cell. For a periodic potential, $V(\boldsymbol{r})$, the periodicity is expressed by the relation

$$V(\boldsymbol{r}) = V(\boldsymbol{r} + l\boldsymbol{a}_1 + m\boldsymbol{a}_2 + n\boldsymbol{a}_3) \ , \tag{11.7}$$

where l, m, n are arbitrary integers. The real function $V(\boldsymbol{r})$, which has this character, can be expressed as the Fourier series

$$V(\boldsymbol{r}) = \sum_{\boldsymbol{K}} \mathrm{e}^{-\mathrm{i}\boldsymbol{K}\cdot\boldsymbol{r}} V(\boldsymbol{K}) \tag{11.8}$$

$$V(-\boldsymbol{K}) = V^*(\boldsymbol{K})$$

with the following conditions on $\boldsymbol{K}$:

$$\boldsymbol{K} \cdot \boldsymbol{a}_1 = 2\pi n_1 \ , \quad \boldsymbol{K} \cdot \boldsymbol{a}_2 = 2\pi n_2 \ , \quad \boldsymbol{K} \cdot \boldsymbol{a}_3 = 2\pi n_3 \ , \tag{11.9}$$

where n_1, n_2, and n_3 are integers. The condition (11.9) is satisfied if

$$\boldsymbol{K} = n_1\boldsymbol{b}_1 + n_2\boldsymbol{b}_2 + n_3\boldsymbol{b}_3 \ . \tag{11.10}$$

Here $\boldsymbol{b}_1, \boldsymbol{b}_2$ and $\boldsymbol{b}_3$ are the primitive vectors of the reciprocal lattice. They are connected to the primitive vectors of the original lattice by the following relations:

$$\boldsymbol{a}_i \cdot \boldsymbol{b}_j = 2\pi\delta_{ij} \ , \quad \boldsymbol{b}_1 = 2\pi \frac{\boldsymbol{a}_2 \times \boldsymbol{a}_3}{\boldsymbol{a}_1 \cdot (\boldsymbol{a}_2 \times \boldsymbol{a}_3)} \ , \dots \ . \tag{11.11}$$

If the periodic potential $V(\boldsymbol{r})$ is weak, it can be treated as a perturbation. The matrix elements of (11.8) with respect to plane waves, $\psi_{\boldsymbol{k}}$, have nonvanishing values $V(\boldsymbol{K})$ only between two states $\boldsymbol{k}$ and $\boldsymbol{k}'$ satisfying $\boldsymbol{k} - \boldsymbol{k}' = \boldsymbol{K}$.

Thus the perturbed wave function is expressed as a superposition of plane waves $\mathrm{e}^{\mathrm{i}\boldsymbol{k}\cdot\boldsymbol{r}}$ and $\mathrm{e}^{\mathrm{i}(\boldsymbol{k}+\boldsymbol{K})\cdot\boldsymbol{r}}$:

$$\psi_{\boldsymbol{k}}(\boldsymbol{r}) = \frac{1}{\sqrt{V}} \mathrm{e}^{\mathrm{i}\boldsymbol{k}\cdot\boldsymbol{r}} (1 + \sum_{\boldsymbol{K}\neq 0} a_{\boldsymbol{k}+\boldsymbol{K}} \mathrm{e}^{\mathrm{i}\boldsymbol{K}\cdot\boldsymbol{r}}) = \mathrm{e}^{\mathrm{i}\boldsymbol{k}\cdot\boldsymbol{r}} u_{\boldsymbol{k}}(\boldsymbol{r}) \ . \tag{11.12}$$

The function $u_{\boldsymbol{k}}(\boldsymbol{r})$ in (11.12) is obviously a periodic function of $\boldsymbol{r}$ since $\boldsymbol{K}$ is a reciprocal-lattice vector. This result known as Bloch's theorem is one of the fundamental theorems in the theory of electrons in solids.

If one makes a perturbation calculation of the energy $\varepsilon_{\boldsymbol{k}}$ up to second order, one obtains

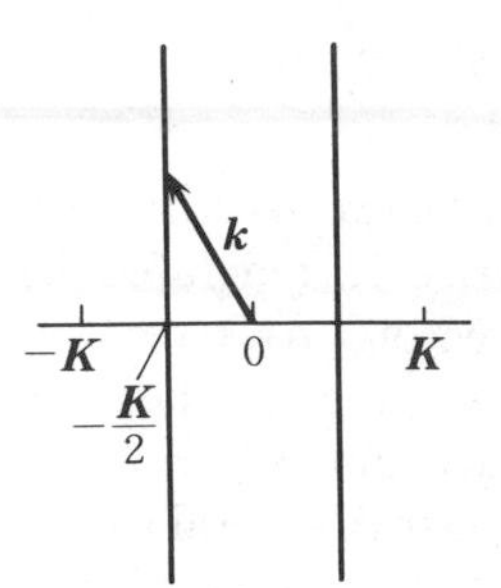

Fig. 11.1. Relation between $\boldsymbol{k}$ and $\boldsymbol{K}$ satisfying (11.14)

$$\varepsilon_{\boldsymbol{k}} = \varepsilon^0_{\boldsymbol{k}} + \sum_{\boldsymbol{K}} \frac{|V(\boldsymbol{K})|^2}{\varepsilon^0_{\boldsymbol{k}} - \varepsilon^0_{\boldsymbol{k}+\boldsymbol{K}}} . \tag{11.13}$$

The energy denominator in the second term becomes very small when $\boldsymbol{k}$ approaches the values satisfying the relation

$$2\boldsymbol{k} \cdot \boldsymbol{K} + \boldsymbol{K}^2 = 0 ; \tag{11.14}$$

the approximation given by (11.13) then fails. The condition of (11.14) is satisfied by all $\boldsymbol{k}$ in the plane bisecting the line connecting the point $-\boldsymbol{K}$ and the origin (Fig. 11.1). For wavevectors $\boldsymbol{k}$ in the plane and nearby, the contribution to the sum over $\boldsymbol{K}$ in (11.13) is expected to be small except for the special $\boldsymbol{K}$ determining the plane. We neglect these small terms and take into account only the matrix elements between the two states $\boldsymbol{k}$ and $\boldsymbol{k} + \boldsymbol{K}$. Then the energy of the conduction electron is calculated by solving the determinantal equation of degree 2:

$$\begin{vmatrix} \varepsilon^0_{\boldsymbol{k}} - \varepsilon & V(\boldsymbol{K}) \\ V(-\boldsymbol{K}) & \varepsilon^0_{\boldsymbol{k}+\boldsymbol{K}} - \varepsilon \end{vmatrix} = 0 ;$$

the solution is

$$\varepsilon_{\boldsymbol{k}} = \varepsilon^0_{\boldsymbol{k}} - \frac{1}{2}\left[(\varepsilon^0_{\boldsymbol{k}} - \varepsilon^0_{\boldsymbol{k}+\boldsymbol{K}}) \pm \sqrt{(\varepsilon^0_{\boldsymbol{k}} - \varepsilon^0_{\boldsymbol{k}+\boldsymbol{K}})^2 + 4|V(\boldsymbol{K})|^2}\right] . \tag{11.15}$$

This approximation is called the *two-wave approximation.* Let us now put $\boldsymbol{k} = -\boldsymbol{K}/2 + \boldsymbol{k}_\parallel + \boldsymbol{k}_\perp$ and use $\varepsilon^0_{\boldsymbol{k}} = (\hbar^2/2m)k^2$. Then (11.15) leads to

$$\varepsilon_{\boldsymbol{k}} = \frac{\hbar^2}{2m}\left[\left(\frac{K}{2}\right)^2 + k_\parallel^2 + k_\perp^2\right] \pm \left[\left(\frac{\hbar^2}{2m}Kk_\parallel\right)^2 + |V(\boldsymbol{K})|^2\right]^{1/2} , \tag{11.16}$$

where $k_\parallel$ is the difference between the parallel component of $\boldsymbol{k}$ along the vector $\boldsymbol{K}$ and $-\boldsymbol{K}/2$, while $k_\perp$ is the component of $\boldsymbol{k}$ perpendicular to $\boldsymbol{K}$. Equation (11.16) shows that the electron energy changes discontinuously by $2|V(\boldsymbol{K})|$ when $\boldsymbol{k}$ crosses the surface $k_\parallel = 0$. The surface defined by $k_\parallel = 0$ is called the boundary surface of the *Brillouin zone.* Suppose we construct such boundary surfaces for all reciprocal-lattice vectors. We call the region in

k-space surrounded by these boundary surfaces the Brillouin zone; in particular, the Brillouin zone including the origin is called the first Brillouin zone. We can bring outer Brillouin zones into the first zone by shifting them by appropriately chosen reciprocal-lattice vectors. In this way the electron energy is specified by the number of the Brillouin zone and the value of $\boldsymbol{k}$ in the first zone. This representation of the energy is called the *reduced-zone representation.* Since the normal components of $d\varepsilon_{\boldsymbol{k}}/d\boldsymbol{k}$ vanish at the boundary surfaces of Brillouin zones, we may formally extend reduced energy bands to outside the first Brillouin zone as periodic functions, $\varepsilon(\boldsymbol{k}+\boldsymbol{K}) = \varepsilon(\boldsymbol{k})$. In the reduced-zone representation, the electron energy as a function of $\boldsymbol{k}$ has a band-like structure separated by forbidden regions; we call these *energy bands.*

The wave functions of conduction electrons in a periodic potential are generally called *Bloch orbitals*; several approximations have been devised to obtain the latter. The simplest one is a perturbation calculation starting from plane waves and is called the *nearly-free-electron approximation.* In the opposite limit, there is a method to approximate a Bloch orbital by a linear combination of atomic orbitals at each lattice site, such as

$$\psi_{\boldsymbol{k}l} = \sum_{nm} a_m \mathrm{e}^{\mathrm{i}\boldsymbol{k}\cdot\boldsymbol{R}_n} \varphi_m(\boldsymbol{r}-\boldsymbol{R}_n) \; ; \tag{11.17}$$

this is called the *tight-binding approximation.* l in (11.17) represents the band index. In case of degenerate atomic orbitals such as d-orbital, $\varphi_m(\boldsymbol{r})$ represents one of five degenerate orbitals. In this method of approximation, the energy levels are divided into s-, p- and d-bands corresponding to the s-, p- and d-orbitals of the atoms; practically, mixing may occur between them. The tight-binding approximation is often used for qualitative discussion but it is not good enough for quantitative calculations of band structures. For this purpose, other methods are used, such as the method of Orthogonalized Plane Waves (OPW), the method of augmented plane waves (APW; used particulally for the 3d-band of the iron group), and also the method utilizing Green functions.

As mentioned above, conduction-electron states in crystals are specified by Bloch orbitals belonging to each band (say, the s-band or the p-band). Electrons occupy these states from the lowest energy state to states at the Fermi energy. Generally, in these cases the Fermi surface deforms from a simple sphere to a complicated surface with a strange shape, and sometimes extends over several bands. Nevertheless, the density of states $\rho(\varepsilon)$, as a function of energy, is still defined as in the case of free electrons. In this situation it is essential for conduction electrons that the region occupied by these electrons is distinguishably separated from the unoccupied region, with the Fermi surface as a boundary.

11.3 Paramagnetic Susceptibility of Conduction Electrons

When a uniform magnetic field, $\boldsymbol{H}$, is applied to the conduction-electron state described above, the energy of electrons with spin direction parallel to the magnetic field decreases by $(1/2)g\mu_{\mathrm{B}}H$, while the energy of electrons with antiparallel spin increases by the same amount. Thus, as shown in Fig. 11.2, a number $(1/2)\rho g\mu_{\mathrm{B}}H$ of electrons with antiparallel spin near the Fermi surface transfer to the parallel spin states; here ρ is the one-spin density of states at the Fermi surface. This change destroys the balance between the numbers of conduction elecrons with spins parallel and antiparallel to the field so that the conduction-electron system becomes magnetized. The susceptibility due to such a process is called the *Pauli paramagnetic susceptibility* and is given by

$$\chi_{\mathrm{Pauli}} = \frac{1}{2}g^2\mu_{\mathrm{B}}^2\rho(\varepsilon_{\mathrm{F}}) \ . \tag{11.18}$$

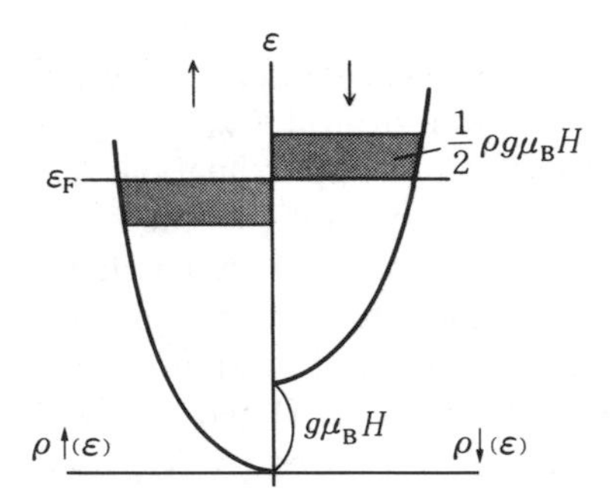

Fig. 11.2. Energy distribution of electrons in the presence of a magnetic field

In addition to the Pauli susceptibility related to electronic spin magnetic moments, a conduction-electron system has a diamagnetic susceptibility; the latter originates from the change in the orbital states caused by the applied magnetic field. This portion is usually called the Landau diamagnetic susceptibility, the value of which is given by (11.18) multiplied by $-(1/3)$ for $g = 2$, as we show in the next section for free-electron systems. For general Bloch electrons the expression for the diamagnetic susceptibility becomes rather complicated, reflecting the shape of the Fermi surface and also because of the contribution from interband transitions.

In the derivation of (11.18), the external magnetic field was assumed to be spatially uniform. To obtain the corresponding expression in the case of a spatially varying external field, we decompose the field into Fourier components as

$$\boldsymbol{H}(\boldsymbol{r}) = \sum_{\boldsymbol{q}} \boldsymbol{H}_{\boldsymbol{q}} \mathrm{e}^{-\mathrm{i}\boldsymbol{q}\cdot\boldsymbol{r}} , \tag{11.19}$$

$$\boldsymbol{H}_{-\boldsymbol{q}} = \boldsymbol{H}_{\boldsymbol{q}}^{*} .$$

We calculate the susceptibility, $\chi_{\boldsymbol{q}}$, in response to one of the components, $\boldsymbol{H}_{\boldsymbol{q}}\mathrm{e}^{-\mathrm{i}\boldsymbol{q}\cdot\boldsymbol{r}}$, or its real part. The total spin-magnetization density is obtained to first order in magnetic field by summing up over $\boldsymbol{q}$ the magnetization induced by each magnetic field component, $\boldsymbol{\sigma}_{\boldsymbol{q}}(\boldsymbol{r})$. That is,

$$\boldsymbol{\sigma}(\boldsymbol{r}) = \frac{1}{V} \sum_{\boldsymbol{q}} \chi_{\boldsymbol{q}} \boldsymbol{H}_{\boldsymbol{q}} \mathrm{e}^{-\mathrm{i}\boldsymbol{q}\cdot\boldsymbol{r}} . \tag{11.20}$$

In calculating $\chi_{\boldsymbol{q}}$, it is essential to know the $\boldsymbol{k}$-dependence of the conduction-electron energy, as well as the density of states $\rho(\varepsilon)$. For simplicity we adopt the free-electron model. Taking the z-axis in the direction of the magnetic field, we treat the Zeeman energy

$$\mathcal{H}_{\mathrm{Z}} = -\frac{1}{2} g\mu_{\mathrm{B}} \sum_{i} s_{iz} (H_{\boldsymbol{q}} \mathrm{e}^{-\mathrm{i}\boldsymbol{q}\cdot\boldsymbol{r}_i} + H_{\boldsymbol{q}}^{*} \mathrm{e}^{\mathrm{i}\boldsymbol{q}\cdot\boldsymbol{r}_i}) \tag{11.21}$$

as the perturbation. Here s_{iz} is the z-component of the spin of the ith electron. In first order, the electron state described by a plane wave with wavevector $\boldsymbol{k}$ and spin $+$ or $-$ is perturbed to

$$\psi_{\boldsymbol{k}\pm} = \frac{1}{\sqrt{V}} \mathrm{e}^{\mathrm{i}\boldsymbol{k}\cdot\boldsymbol{r}} \left[1 \pm \frac{m}{2\hbar^2} g\mu_{\mathrm{B}} \left(\frac{H_{\boldsymbol{q}} \mathrm{e}^{-\mathrm{i}\boldsymbol{q}\cdot\boldsymbol{r}}}{(\boldsymbol{k}-\boldsymbol{q})^2 - k^2} + \frac{H_{\boldsymbol{q}}^{*} \mathrm{e}^{\mathrm{i}\boldsymbol{q}\cdot\boldsymbol{r}}}{(\boldsymbol{k}+\boldsymbol{q})^2 - k^2} \right) \right] . \tag{11.22}$$

The number density of electrons with wavevector $\boldsymbol{k}$ and spin $+$ or $-$ is found by taking the square of the absolute value of (11.22):

$$\rho_{\boldsymbol{k}\pm} = \frac{1}{V} \left[1 \pm \frac{m}{2\hbar^2} g\mu_{\mathrm{B}} \left(\frac{1}{(\boldsymbol{k}-\boldsymbol{q})^2 - k^2} + \frac{1}{(\boldsymbol{k}+\boldsymbol{q})^2 - k^2} \right) (H_{\boldsymbol{q}} \mathrm{e}^{-\mathrm{i}\boldsymbol{q}\cdot\boldsymbol{r}} + H_{\boldsymbol{q}}^{*} \mathrm{e}^{\mathrm{i}\boldsymbol{q}\cdot\boldsymbol{r}}) \right] . \tag{11.23}$$

The spatial density of the spin magnetic moment is obtained by multiplying (11.23) by $\pm(1/2)g\mu_{\mathrm{B}}$, adding both expressions for $+$ and $-$ spins, and then summing over $\boldsymbol{k}$ within the Fermi sphere:

$$\begin{aligned} \sigma(\boldsymbol{r}) &= \frac{1}{V} \chi_q \frac{1}{2} (H_{\boldsymbol{q}} \mathrm{e}^{-\mathrm{i}\boldsymbol{q}\cdot\boldsymbol{r}} + H_{\boldsymbol{q}}^{*} \mathrm{e}^{\mathrm{i}\boldsymbol{q}\cdot\boldsymbol{r}}) \\ &= \frac{3}{16} \frac{N_{\mathrm{e}}}{V} \frac{g^2 \mu_{\mathrm{B}}^2}{\varepsilon_{\mathrm{F}}} \left(1 + \frac{4k_{\mathrm{F}}^2 - q^2}{4k_{\mathrm{F}} q} \log \left| \frac{2k_{\mathrm{F}} + q}{2k_{\mathrm{F}} - q} \right| \right) \\ &\quad \times \frac{1}{2} (H_{\boldsymbol{q}} \mathrm{e}^{-\mathrm{i}\boldsymbol{q}\cdot\boldsymbol{r}} + H_{\boldsymbol{q}}^{*} \mathrm{e}^{\mathrm{i}\boldsymbol{q}\cdot\boldsymbol{r}}) . \end{aligned} \tag{11.24}$$

Therefore $\chi_{\boldsymbol{q}}$ is found to be

$$\chi_q = \frac{3}{16} N_{\mathrm{e}} \frac{g^2 \mu_{\mathrm{B}}^2}{\varepsilon_{\mathrm{F}}} f\left(\frac{q}{2k_{\mathrm{F}}} \right) = \chi_{\mathrm{Pauli}} \frac{1}{2} f\left(\frac{q}{2k_{\mathrm{F}}} \right) , \tag{11.25}$$

where the function $f(x)$ is defined by

$$f(x) = 1 + \frac{1-x^2}{2x} \log \left| \frac{1+x}{1-x} \right| . \tag{11.26}$$

As illustrated in Fig. 11.3, $f(x)$ has a singularity at $x = 1$, where its derivative diverges to $-\infty$. The function decreases monotonically starting from 2, the value at $x = 0$, and approaches zero as x^{-2} for $x \gg 1$. Accordingly, χ_q is identical to the Pauli paramagnetic susceptibility at $q = 0$ and decreases monotonically with increasing q. The singularity of $f(q/2k_\mathrm{F})$ at $q = 2k_\mathrm{F}$ reflects the existence of the Fermi surface (because $2k_\mathrm{F}$ is the diameter of the Fermi surface).

If the external magnetic field varies not only in space but also in time, we must take Fourier components also in time. In terms of the susceptibility $\chi(\boldsymbol{q}, \omega)$ of each component, we can obtain the variation of the spin magnetic moment in time and space. Details of this dynamical susceptibility are described in a later chapter.

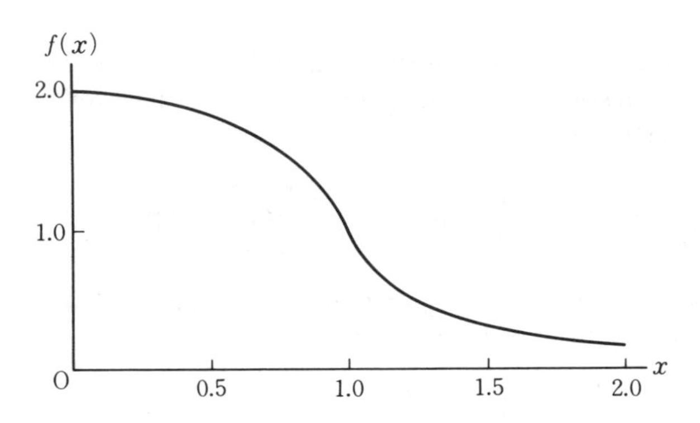

Fig. 11.3. $f(x)$ as a function of x

11.4 Diamagnetic Susceptibility of Conduction Electrons

Conduction-electron systems have both a diamagnetic susceptibility (originating in diamagnetic motion) and the Pauli paramagnetic susceptibility (originating in the magnetic moment associated with the spin angular momentum). The diamagnetism arises according to Lenz's law of electromagnetism; the conduction electrons respond to an externally applied magnetic field so as to cancel it. It is sometimes called the *Landau diamagnetism* because *Landau* was the first to calculate the diamagnetic susceptibility of free electrons [11.1]. The calculation of the diamagnetism of conduction electrons is not a trivial task and many others have devised different methods. The result is that the magnetization includes an oscillatory function of the reciprocal of the external magnetic field; the period is

$$\Delta\left(\frac{1}{H}\right) = \frac{2\pi e}{\hbar c S} , \tag{11.27}$$

where S is the stationary value (the maximum or the minimum) of the section of the Fermi surface cut perpendicular to the direction of the magnetic field. Here the electric charge of electron is chosen as $-e$. The oscillation is known as the *de Haas-van Alphen effect* and has been a powerful experimental method to determine the structure of the Fermi surface; it is observed in strong magnetic field. We can find the diamagnetic magnetization caused by weak magnetic fields most simply by a perturbational method, without worrying about the surface of the metal.

Using the method of second quantization, we can express the Hamiltonian describing the interaction between the conduction electrons and the external magnetic field as

$$H_{\mathrm{I}} = \int d\tau \psi^{\dagger}(\boldsymbol{r})\left[-\frac{\mathrm{i}e\hbar}{2mc}(\boldsymbol{A}\cdot\nabla + \nabla\cdot\boldsymbol{A}) + \frac{e^2}{2mc^2}A^2(\boldsymbol{r})\right]\psi(\boldsymbol{r}) . \tag{11.28}$$

Here $\psi^{\dagger}(\boldsymbol{r})$ and $\psi(\boldsymbol{r})$ are Fermi operators describing the conduction electron fields and represent creation and annihilation for an electron at space coordinate $\boldsymbol{r}$, respectively. $\boldsymbol{A}(\boldsymbol{r})$ is the vector potential and the integration over $d\tau$ is a volume integration together with a sum over spin states. We expand $\psi^{\dagger}(\boldsymbol{r})$ and $\psi(\boldsymbol{r})$ in plane waves as follows:

$$\psi(\boldsymbol{r}) = \frac{1}{\Omega^{1/2}}\sum_{\boldsymbol{k}\sigma} c_{\boldsymbol{k}\sigma}u_{\sigma}\mathrm{e}^{\mathrm{i}\boldsymbol{k}\cdot\boldsymbol{r}} , \tag{11.29}$$

$$\psi^{\dagger}(\boldsymbol{r}) = \frac{1}{\Omega^{1/2}}\sum_{\boldsymbol{k}\sigma} c^{\dagger}_{\boldsymbol{k}\sigma}u_{\sigma}\mathrm{e}^{-\mathrm{i}\boldsymbol{k}\cdot\boldsymbol{r}} , \tag{11.30}$$

where $c^{\dagger}_{\boldsymbol{k}\sigma}$ and $c_{\boldsymbol{k}\sigma}$ are, respectively, creation and annihilation operators for an electron with wavevector $\boldsymbol{k}$ and spin σ, while u_{σ} is a spin eigenfunction. Ω stands for the total volume. In a similar way, we expand the vector potential in Fourier series:

$$\boldsymbol{A}(\boldsymbol{r}) = \frac{(2\pi)^{3/2}}{\Omega}\sum_{\boldsymbol{q}} \boldsymbol{a}(\boldsymbol{q})\mathrm{e}^{\mathrm{i}\boldsymbol{q}\cdot\boldsymbol{r}} , \tag{11.31}$$

$$\boldsymbol{a}(\boldsymbol{q}) = \left(\frac{1}{2\pi}\right)^{3/2}\int d\tau \boldsymbol{A}(\boldsymbol{r})\mathrm{e}^{-\mathrm{i}\boldsymbol{q}\cdot\boldsymbol{r}} . \tag{11.32}$$

The gauge of the vector potential is fixed by

$$\mathrm{div}\boldsymbol{A}(\boldsymbol{r}) = 0 , \quad \boldsymbol{q}\cdot\boldsymbol{a}(\boldsymbol{q}) = 0 . \tag{11.33}$$

By using (11.29–31), we can rewrite $H_{\rm I}$ given by (11.28) as

$$H_{\rm I} = \frac{e\hbar}{mc}\frac{(2\pi)^{3/2}}{\Omega}\sum_{kq\sigma} c^\dagger_{k+q,\sigma}c_{k\sigma}\boldsymbol{a}(\boldsymbol{q})\cdot\boldsymbol{k} \ . \tag{11.34}$$

Here we retain terms to first order in $\boldsymbol{a}$.

Let us first calculate the electric current density induced by the vector potential. The current-density operator is expressed as

$$\begin{aligned} \boldsymbol{J}(\boldsymbol{r}) &= \frac{\mathrm{i}e\hbar}{2m}(\psi^\dagger\nabla\psi - \text{h.c.}) - \frac{e^2}{mc}\psi^\dagger\boldsymbol{A}(\boldsymbol{r})\psi \\ &= \boldsymbol{J}_{\rm P}(\boldsymbol{r}) + \boldsymbol{J}_{\rm D}(\boldsymbol{r}) \ , \end{aligned} \tag{11.35}$$

where h.c. means the Hermitian conjugate. The first term represents the paramagnetic current, and the second the diamagnetic current. The currents in (11.35) are also represented as

$$\boldsymbol{J}_{\rm P}(\boldsymbol{r}) = -\frac{e\hbar}{2m\Omega}\sum_{kq\sigma} c^\dagger_{k+q,\sigma}c_{k\sigma}\mathrm{e}^{-\mathrm{i}\boldsymbol{q}\cdot\boldsymbol{r}}(2\boldsymbol{k}+\boldsymbol{q}) \ , \tag{11.36}$$

$$\boldsymbol{J}_{\rm D}(\boldsymbol{r}) = -\frac{e^2}{mc}\frac{1}{\Omega}\sum_{kq\sigma} c^\dagger_{k+q,\sigma}c_{k\sigma}\mathrm{e}^{-\mathrm{i}\boldsymbol{q}\cdot\boldsymbol{r}}\boldsymbol{A}(\boldsymbol{r}) \ , \tag{11.37}$$

in terms of $c^\dagger_{k,\sigma}$ and $c_{k,\sigma}$ as in (11.34).

The wave function ψ_0 of the ground state, i.e., the Fermi sphere, is perturbed by the interaction (11.34). The change in first order, Φ_1, is obtained as

$$\Phi_1 = \sum_{i\neq 0}\frac{\langle\psi_i|H_{\rm I}|\psi_0\rangle}{E_0 - E_i}\psi_i \ . \tag{11.38}$$

With this Φ_1, the expectation value of the current density $\boldsymbol{j}(\boldsymbol{r})$ is given by

$$\begin{aligned} \boldsymbol{j}(\boldsymbol{r}) &= \langle\Phi_1|\boldsymbol{J}_{\rm P}(\boldsymbol{r})|\psi_0\rangle + \langle\psi_0|\boldsymbol{J}_{\rm P}(\boldsymbol{r})|\Phi_1\rangle \\ &\quad + \langle\psi_0|\boldsymbol{J}_{\rm D}(\boldsymbol{r})|\psi_0\rangle \end{aligned} \tag{11.39}$$

to first order in the vector potential. Using (11.34, 36–38), we can express the expectation value of the paramagnetic current, $\boldsymbol{j}_{\rm P}(\boldsymbol{r})$, as

$$\begin{aligned} \boldsymbol{j}_{\rm P}(\boldsymbol{r}) = &-\frac{e^2\hbar^2(2\pi)^{3/2}}{2m^2c\Omega^2}\sum_{i\neq 0}\sum_{kq\sigma}\sum_{k'q'\sigma'}(2\boldsymbol{k}+\boldsymbol{q})\boldsymbol{k}'\cdot\Big[\boldsymbol{a}(\boldsymbol{q}')\mathrm{e}^{-\mathrm{i}\boldsymbol{q}\cdot\boldsymbol{r}} \\ &\times\langle\psi_0|c^\dagger_{k+q,\sigma}c_{k\sigma}|\psi_i\rangle\langle\psi_i|c^\dagger_{k'+q',\sigma'}c_{k'\sigma'}|\psi_0\rangle\frac{1}{E_0-E_i} + \text{c.c.}\Big] \end{aligned} \tag{11.40}$$

and the diamagnetic current, $\boldsymbol{j}_{\rm D}(\boldsymbol{r})$, as

$$\boldsymbol{j}_{\rm D}(\boldsymbol{r}) = -\frac{ne^2}{mc}\boldsymbol{A}(\boldsymbol{r}) \ , \tag{11.41}$$

where n is the electron number per unit volume. Since the excited state i in (11.40) is the state where an electron is excited from a one-electron state $(\boldsymbol{k}', \sigma')$ to a state $(\boldsymbol{k}' + \boldsymbol{q}', \sigma')$, the energy difference $E_0 - E_i$ is given by the difference of the one-electron energies $\varepsilon_{\boldsymbol{k}',\sigma'} - \varepsilon_{\boldsymbol{k}'+\boldsymbol{q}',\sigma'}$. To return to the initial state from the excited state by operators $c^{\dagger}_{\boldsymbol{k}+\boldsymbol{q},\sigma} c_{\boldsymbol{k},\sigma}$ the relations, $\boldsymbol{k} + \boldsymbol{q} = \boldsymbol{k}'$, $\boldsymbol{k} = \boldsymbol{k}' + \boldsymbol{q}'$ and $\sigma = \sigma'$ must hold. Thus the $\boldsymbol{q}$-component in (11.40) is found to be

$$\boldsymbol{j}_{\mathrm{P}\boldsymbol{q}}(\boldsymbol{r}) = \frac{4e^2(2\pi)^{3/2}}{mc\Omega^2} \mathrm{e}^{\mathrm{i}\boldsymbol{q}\cdot\boldsymbol{r}} \sum_{\boldsymbol{k}} \frac{(2\boldsymbol{k} + \boldsymbol{q})\boldsymbol{k} \cdot \boldsymbol{a}(\boldsymbol{q})}{(\boldsymbol{k} + \boldsymbol{q})^2 - \boldsymbol{k}^2} f(\varepsilon_{\boldsymbol{k}}) \,, \tag{11.42}$$

where $f(\varepsilon_{\boldsymbol{k}})$ is the Fermi distribution function and the summation over $\boldsymbol{k}$ is to be taken inside the Fermi sphere for $T = 0$. If one chooses the z-axis in the direction of $\boldsymbol{q}$ vector and the x-axis in the direction of $\boldsymbol{a}(\boldsymbol{q})$, only the x-component remains in $\boldsymbol{j}_{\mathrm{P}\boldsymbol{q}}(\boldsymbol{r})$; the y- and z- components vanish because of symmetry. In other words, $\boldsymbol{j}_{\mathrm{P}\boldsymbol{q}}(\boldsymbol{r})$ is proportional to $\boldsymbol{a}(\boldsymbol{q})$. After performing the summation over $\boldsymbol{k}$ in (11.42), one obtains

$$\begin{aligned} \boldsymbol{j}_{\mathrm{P}\boldsymbol{q}}(\boldsymbol{r}) &= \frac{e^2(2\pi)^{3/2}}{mc\Omega} \frac{1}{24\pi^2} \boldsymbol{a}(\boldsymbol{q}) \mathrm{e}^{\mathrm{i}\boldsymbol{q}\cdot\boldsymbol{r}} \\ &\times k_{\mathrm{F}}^3 \left\{ 5 - 3\left(\frac{q}{2k_{\mathrm{F}}}\right)^2 + \frac{3}{2}\left(\frac{2k_{\mathrm{F}}}{q}\right)\left[\left(\frac{q}{2k_{\mathrm{F}}}\right)^2 - 1\right]^2 \log\left|\frac{1 + q/2k_{\mathrm{F}}}{1 - q/2k_{\mathrm{F}}}\right| \right\} . \end{aligned} \tag{11.43}$$

On the other hand, the diamagnetic current is expressed as the $\boldsymbol{q}$-component of (11.41):

$$\boldsymbol{j}_{\mathrm{D}\boldsymbol{q}}(\boldsymbol{r}) = -\frac{e^2(2\pi)^{3/2}}{mc\Omega} \frac{1}{3\pi^2} k_{\mathrm{F}}^3 \boldsymbol{a}(\boldsymbol{q}) \mathrm{e}^{\mathrm{i}\boldsymbol{q}\cdot\boldsymbol{r}} \,. \tag{11.44}$$

Consequently, the $\boldsymbol{q}$-component of the total current density is given by the sum of (11.43, 44):

$$\boldsymbol{j}_{\boldsymbol{q}}(\boldsymbol{r}) = -\frac{e^2}{mc} \frac{(2\pi)^{3/2}}{\Omega} \frac{k_{\mathrm{F}}^3}{8\pi^2} L_j\left(\frac{q}{2k_{\mathrm{F}}}\right) \boldsymbol{a}(\boldsymbol{q}) \mathrm{e}^{\mathrm{i}\boldsymbol{q}\cdot\boldsymbol{r}} \,, \tag{11.45}$$

$$L_j(x) = 1 + x^2 - \frac{1}{2x}(1 - x^2)^2 \log\left|\frac{1 + x}{1 - x}\right| . \tag{11.46}$$

Since $L_j(x)$ vanishes as $x \to 0$, the paramagnetic current and the diamagnetic one cancel perfectly in the limit of $q \to 0$, and thus the total current is zero. This is a well-known result, as shown, for example, in the famous paper by *Bardeen-Cooper-Schrieffer* on their theory of superconductivity [11.2]. In a superconductor, an energy gap emerges at the Fermi surface of the conduction electrons because of the formation of Cooper pairs; consequently, the paramagnetic part of the current vanishes for $q = 0$ and the diamagnetic current is unaffected.

Next we calculate the magnetization by using a relation between the current and the magnetization $\boldsymbol{m}$:

$$\boldsymbol{j}(\boldsymbol{r}) = c \text{ curl } \boldsymbol{m} \ . \tag{11.47}$$

Since the magnetic field is given by curl$\boldsymbol{A}(\boldsymbol{r})$, the $\boldsymbol{q}$-component of the magnetic field is

$$\boldsymbol{H}_q = \mathrm{i}\frac{(2\pi)^{3/2}}{\Omega}\boldsymbol{q} \times \boldsymbol{a}(\boldsymbol{q})\mathrm{e}^{\mathrm{i}\boldsymbol{q}\cdot\boldsymbol{r}} \ . \tag{11.48}$$

Introducing the susceptibility χ_q by

$$\boldsymbol{m}_q = \chi_q \boldsymbol{H}_q \ , \tag{11.49}$$

we have

$$\begin{aligned} \text{curl } \boldsymbol{m}_q &= \frac{1}{c}\boldsymbol{j}_q(\boldsymbol{r}) = -\frac{(2\pi)^{3/2}}{\Omega}\boldsymbol{q} \times [\boldsymbol{q} \times \boldsymbol{a}(\boldsymbol{q})]\mathrm{e}^{\mathrm{i}\boldsymbol{q}\cdot\boldsymbol{r}}\chi_q \\ &= \frac{(2\pi)^{3/2}}{\Omega}q^2\boldsymbol{a}(\boldsymbol{q})\mathrm{e}^{\mathrm{i}\boldsymbol{q}\cdot\boldsymbol{r}}\chi_q \ . \end{aligned} \tag{11.50}$$

On using the expression (11.45) for $\boldsymbol{j}_q(\boldsymbol{r})$ in (11.50), we obtain the diamagnetic susceptibility responding to the $\boldsymbol{q}$-component of the field as

$$\chi_q = -\frac{e^2}{mc^2}\frac{k_\mathrm{F}}{12\pi^2}L\left(\frac{q}{2k_\mathrm{F}}\right) \ . \tag{11.51}$$

In the above equation $L(x)$ is given by

$$\begin{aligned} L(x) &= \frac{3}{8x^2}L_j(x) \\ &= 1 - \frac{1}{5}x^2 + \ldots \quad (x \ll 1) \ . \end{aligned} \tag{11.52}$$

Thus, in the limit $q \to 0$, we obtain the diamagnetic susceptibility χ_d:

$$\chi_\mathrm{d} = -\frac{e^2}{mc^2}\frac{k_\mathrm{F}}{12\pi^2} = -\frac{1}{3}\chi_\mathrm{Pauli} \ . \tag{11.53}$$

This is just equal to $-1/3$ of the Pauli paramagnetic suscceptibility. The absolute value of the diamagnetic susceptibility $\chi_\mathrm{d}(q)$ decreases proportionally to q^2 as q increases from 0; its value at $q = 2k_\mathrm{F}$ is 3/4 of its value at $q = 0$, so that the ratio $|\chi_\mathrm{d}|/\chi_\mathrm{para}$ is 1/2 [where $\chi_\mathrm{para} = \chi_q$ given in (11.25) with $q = 2k_\mathrm{F}$]. We note that $L(x)$ has a singularity at $x = 1$ reflecting the existence of the Fermi surface; it decreases as $1/x^2$ for $x \gg 1$. Figure 11.4 illustrates the x-dependence of $L(x)$.

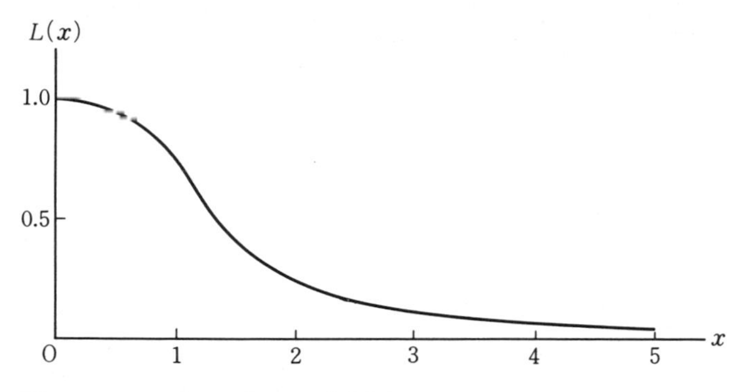

Fig. 11.4. x-dependence of $L(x)$

12. Coulomb Interaction Between Electrons

Conduction electrons in metals do not move about completely free; rather they move under the influences of their mutual Coulomb interaction and the periodic potential of the ions. For the present we neglect the influence of the periodic potential and consider exclusively the effect of the Coulomb interaction.

12.1 First-Order Perturbation and Exchange Interactions

The repulsive Coulomb interaction between electrons is e^2/r_{ij}, where r_{ij} denotes the interelectron distance. This interaction can be expressed in Fourier-series expansion as

$$\frac{e^2}{r_{ij}} = \frac{1}{V}\sum_{q\neq 0}\frac{4\pi e^2}{q^2}\mathrm{e}^{\mathrm{i}\boldsymbol{q}\cdot(\boldsymbol{r}_i-\boldsymbol{r}_j)} \tag{12.1}$$

with $\boldsymbol{r}_i$ and $\boldsymbol{r}_j$ the positions of the electrons. The term with $\boldsymbol{q}=0$ is omitted from the summation over $\boldsymbol{q}$; it cancells with a term due to the average (uniform) positive charge of the ionic lattice, by charge neutrality of the whole crystal. In the formalism of second quantization, with plane waves (11.1) as the basis, the Coulomb interaction (12.1) can be written as

$$\mathcal{H}_{\mathrm{Coulomb}} = \frac{1}{V}\sum_{q\neq 0}\frac{2\pi e^2}{q^2}\sum_{\boldsymbol{k}_1\boldsymbol{k}_2\sigma_1\sigma_2} a^\dagger_{\boldsymbol{k}_1+\boldsymbol{q},\sigma_1}a^\dagger_{\boldsymbol{k}_2-\boldsymbol{q},\sigma_2}a_{\boldsymbol{k}_2\sigma_2}a_{\boldsymbol{k}_1\sigma_1}\,, \tag{12.2}$$

where we have introduced $a^\dagger_{\boldsymbol{k}\sigma}$ and $a_{\boldsymbol{k}\sigma}$, the Fermi creation and annihilation operators for electrons with wavevector $\boldsymbol{k}$ and spin σ, respectively. The sums over σ_1 and σ_2 are taken for the two components of the electron spin. Equation (12.2) can also be derived directly from the Coulomb integral

$$\frac{1}{2}\int\varphi^*(\boldsymbol{r}_1)\varphi^*(\boldsymbol{r}_2)\frac{e^2}{r_{12}}\varphi(\boldsymbol{r}_2)\varphi(\boldsymbol{r}_1)d\tau_1 d\tau_2$$

if we use the expansion of the electron wave function $\varphi(\boldsymbol{r})$ in terms of plane waves (11.1):

$$\varphi(\boldsymbol{r}) = \sum_{\boldsymbol{k}\sigma} a_{\boldsymbol{k}\sigma}\psi_{\boldsymbol{k}\sigma}(\boldsymbol{r}) \ .$$

Let us now treat the Coulomb interaction as a perturbation to the kinetic energy

$$\sum_{\boldsymbol{k}\sigma} \varepsilon_{\boldsymbol{k}} a^{\dagger}_{\boldsymbol{k}\sigma} a_{\boldsymbol{k}\sigma} \tag{12.3}$$

of free electrons. The first-order term in the perturbation is given by the expectation value with respect to the ground state of the free electrons. If one takes such two pairs as, $a^{\dagger}_{\boldsymbol{k}_2-\boldsymbol{q},\sigma_2}$, $a_{\boldsymbol{k}_2,\sigma_2}$ and $a^{\dagger}_{\boldsymbol{k}_1+\boldsymbol{q},\sigma_1}$, $a_{\boldsymbol{k}_1,\sigma_1}$, then the diagonal element with respect to the ground state is nonvanishing only for $\boldsymbol{q} = 0$. This term is, however, excluded from (12.2) and so one does not need to consider it. Therefore, for the diagonal element of (12.2), we need consider only the cases satisfiying $\boldsymbol{k}_2 = \boldsymbol{k}_1+\boldsymbol{q}$ and $\sigma_1 = \sigma_2$ to make pairs with $a^{\dagger}_{\boldsymbol{k}_1+\boldsymbol{q},\sigma_1}$ and $a_{\boldsymbol{k}_2,\sigma_2}$ and with $a^{\dagger}_{\boldsymbol{k}_2-\boldsymbol{q},\sigma_2}$ and $a_{\boldsymbol{k}_1,\sigma_1}$. Then the perturbation energy in first order is calculated as

$$V_{\text{exch}} = -\frac{1}{V}\sum_{\boldsymbol{k}_1\boldsymbol{k}_2\sigma} \frac{2\pi e^2}{|\boldsymbol{k}_1-\boldsymbol{k}_2|^2} n_{\boldsymbol{k}_1\sigma} n_{\boldsymbol{k}_2\sigma} \ . \tag{12.4}$$

Since the contribution to (12.4) comes from the mutual exchange of $a_{\boldsymbol{k}_2\sigma_2}$ and $a_{\boldsymbol{k}_1\sigma_1}$ in (12.2), (12.4) is called the exchange interaction; it necessarily has a negative sign. In (12.4), $n_{\boldsymbol{k}\sigma}$ denotes the occupation number of electrons in the state of wavevector $\boldsymbol{k}$ and spin σ. In the ground state $n_{\boldsymbol{k}\sigma} = 1$ for $|\boldsymbol{k}| < k_{\text{F}}$ and $n_{\boldsymbol{k}\sigma} = 0$ for $|\boldsymbol{k}| > k_{\text{F}}$.

Keeping only terms with $\boldsymbol{k}_1 + \boldsymbol{q} = \boldsymbol{k}_2$ in (12.2), we obtain

$$\begin{aligned}
\mathcal{H}_{\text{exch}} &= -\frac{1}{V}\sum_{\boldsymbol{k}_1\boldsymbol{k}_2} \frac{2\pi e^2}{|\boldsymbol{k}_1-\boldsymbol{k}_2|^2} \sum_{\sigma_1\sigma_2} a^{\dagger}_{\boldsymbol{k}_2\sigma_1} a_{\boldsymbol{k}_2\sigma_2} a^{\dagger}_{\boldsymbol{k}_1\sigma_2} a_{\boldsymbol{k}_1\sigma_1} \\
&= -\frac{1}{V}\sum_{\boldsymbol{k}_1\boldsymbol{k}_2} \frac{2\pi e^2}{|\boldsymbol{k}_1-\boldsymbol{k}_2|^2} \Big[\frac{1}{2}(a^{\dagger}_{\boldsymbol{k}_1\uparrow}a_{\boldsymbol{k}_1\uparrow} + a^{\dagger}_{\boldsymbol{k}_1\downarrow}a_{\boldsymbol{k}_1\downarrow})(a^{\dagger}_{\boldsymbol{k}_2\uparrow}a_{\boldsymbol{k}_2\uparrow} + a^{\dagger}_{\boldsymbol{k}_2\downarrow}a_{\boldsymbol{k}_2\downarrow}) \\
&\quad + \frac{1}{2}(a^{\dagger}_{\boldsymbol{k}_1\uparrow}a_{\boldsymbol{k}_1\uparrow} - a^{\dagger}_{\boldsymbol{k}_1\downarrow}a_{\boldsymbol{k}_1\downarrow})(a^{\dagger}_{\boldsymbol{k}_2\uparrow}a_{\boldsymbol{k}_2\uparrow} - a^{\dagger}_{\boldsymbol{k}_2\downarrow}a_{\boldsymbol{k}_2\downarrow}) \\
&\quad + a^{\dagger}_{\boldsymbol{k}_1\uparrow}a_{\boldsymbol{k}_1\downarrow}a^{\dagger}_{\boldsymbol{k}_2\downarrow}a_{\boldsymbol{k}_2\uparrow} + a^{\dagger}_{\boldsymbol{k}_1\downarrow}a_{\boldsymbol{k}_1\uparrow}a^{\dagger}_{\boldsymbol{k}_2\uparrow}a_{\boldsymbol{k}_2\downarrow}\Big] \ ,
\end{aligned} \tag{12.5}$$

which reduces to (12.4) on taking the diagonal element with respect to the spin σ. The second, third and fourth terms in the square bracket can be rewritten in the familiar form of an exchange interaction, $2(\boldsymbol{s}_{\boldsymbol{k}_1}\cdot\boldsymbol{s}_{\boldsymbol{k}_2})$ by using the spin operators, $\boldsymbol{s}_{\boldsymbol{k}_1}$ and $\boldsymbol{s}_{\boldsymbol{k}_2}$, of electrons with $\boldsymbol{k}_1$ and $\boldsymbol{k}_2$. The first term gives a constant, 1/2, when the states of $\boldsymbol{k}_1$ and $\boldsymbol{k}_2$ are singly occupied by electrons; together with $2(\boldsymbol{s}_1\cdot\boldsymbol{s}_2)$ it forms an operator which interchanges spin 1 and spin 2.

Equation (12.4) shows that the exchange interaction acts between electrons with parallel spin, and reduces their energy; there is no reduction in energy for electrons with antiparallel spin. Therefore free electrons tend to align their spins parallel to each other due to the exchange interaction. *Bloch* discussed the ferromagnetism of free electrons based on the exchange interaction, (12.4), in the following way [12.1]. The ground state of free electrons is, as well known, the state in which electrons with up and down spin occupy equally the orbitals of wavevector $\boldsymbol{k}$ inside the Fermi surface. Anticipating the spin alignment, we set the electron numbers with spin $\pm$ to be N_+ and N_-, and the corresponding radii of the Fermi spheres to be k_{F+} and k_{F-}. The kinetic energy, (12.3), and the exchange energy, (12.4), can be easily expressed as functions of N_+ and N_-; the results are

$$E_{\text{kin}} = E_{\text{kin}+} + E_{\text{kin}-} = V\frac{\hbar^2}{2m}\frac{6}{5}\pi\left(\frac{9\pi}{2}\right)^{1/3}(n_+^{5/3} + n_-^{5/3}) , \tag{12.6}$$

$$E_{\text{exch}} = -V\frac{9}{4}e^2\left(\frac{2}{9\pi}\right)^{1/3}(n_+^{4/3} + n_-^{4/3}) , \tag{12.7}$$

where $n_\pm$ is the number density of electrons with spin $\pm$; that is, $n_\pm = N_\pm/V$. The total energy is found, by adding the above two equations, as

$$\frac{E_{\text{T}}}{V} = \frac{3}{5}\pi\frac{\hbar^2}{m}\left(\frac{9\pi}{2}\right)^{1/3}\left[(n_+^{5/3} + n_-^{5/3}) - \alpha(n_+^{4/3} + n_-^{4/3})\right] \tag{12.8}$$

with

$$\alpha = \frac{5}{6\pi^2}\frac{e^2 m}{\hbar^2}\left(\frac{9\pi}{2}\right)^{1/3} .$$

Let the square bracket on the right-hand side of (12.8) be $g(n_+)$. If one differentiates it with respect to n_+ under the condition, $n = n_+ + n_- = N_e/V = \text{constant}$, one obtains

$$\frac{dg}{dn_+} = \frac{5}{3}(n_+^{1/3} - n_-^{1/3})\left[(n_+^{1/3} + n_-^{1/3}) - \frac{4}{5}\alpha\right] .$$

Thus extrema are present at two points, i.e., at $n_+ = n_-$ and at $n_+^{1/3} + n_-^{1/3} = (4/5)\alpha$. On differentiating the above equation and setting $n_+ = n_-$ we have

$$\left.\frac{d^2 g}{dn_+^2}\right|_{n_+=n_-=n/2} = \frac{20}{9}\left(\frac{n}{2}\right)^{-2/3}\left[\left(\frac{n}{2}\right)^{1/3} - \frac{2}{5}\alpha\right] .$$

Therefore one finds the following for the first extremum:

$$\text{minimum at } n_+ = n_- = n/2 \quad \text{for } \left(\frac{n}{2}\right)^{1/3} > \frac{2}{5}\alpha ,$$

$$\text{maximum at } n_+ = n_- = n/2 \quad \text{for } \left(\frac{n}{2}\right)^{1/3} < \frac{2}{5}\alpha .$$

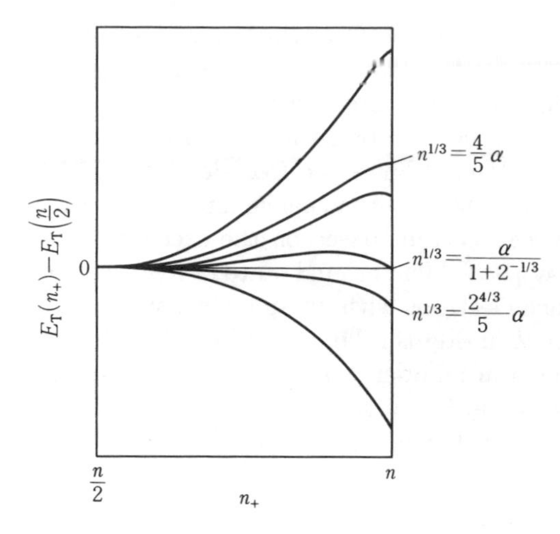

Fig. 12.1. Relation between E_T and n_+

The second extremum is realized only when $(4/5)\alpha$ falls between the maximum of $n_+^{1/3} + n_-^{1/3}$, $2^{2/3}n^{1/3}$, and the minimum, $n^{1/3}$. These situations are illustrated in Fig. 12.1.

As seen from Fig. 12.1, the condition for ferromagnetism is that the energy $E_T(n_+ = n)$ be lower than $E_T(n_+ = n/2)$; explicitly, it is

$$n^{1/3} < \frac{1}{1 + 2^{-1/3}}\alpha \, .$$

Let r_0 be the radius of the sphere with volume equal to the volume per electron in the metal; its ratio to Bohr radius, $a_0 = \hbar^2/(e^2 m)$, is denoted by r_s. Then the above condition for ferromagnetism becomes

$$r_s = \frac{r_0}{a_0} > \frac{2\pi}{5}(1 + 2^{1/3})\left(\frac{9\pi}{4}\right)^{1/3} = 5.4531 \, . \tag{12.9}$$

This condition is satisfied at low density of conduction electrons. Notice that imperfect ferromagnetism, where both n_+ and n_- are finite, does not occur in this model.

12.2 Second-Order Perturbation Theory and Correlation Energy

Some of the correlation between electrons with parallel spin is naturally taken into account by the Pauli principle, but this is not the case for antiparallel spins, within first-order perturbation theory. Since the repulsive Coulomb interaction acts for all electrons, electrons with antiparallel spin also tend to keep away from each other. This tendency arises from the change in the wave functions, and appears only in the second- or higher-order terms in the perturbed energy. It would tend to reduce the exchange effect in (12.4). In this sense the first-order perturbed energy is insufficient to discuss the ferromagnetism of the electron gas; so we go back to the Coulomb interaction, (12.2), and examine the effect of the second-order perturbation on the Fermi sphere.

The Coulomb interaction (12.2) excites two electrons occupying the states, $(\boldsymbol{k}_1\sigma_1)$ and $(\boldsymbol{k}_2\sigma_2)$, inside the Fermi sphere to the states, $(\boldsymbol{k}_1 + \boldsymbol{q}\sigma_1)$ and $(\boldsymbol{k}_2 - \boldsymbol{q}\sigma_2)$, outside the sphere. The excitation energy in the intermediate state is

$$\Delta E = \varepsilon_{\boldsymbol{k}_1+\boldsymbol{q}} + \varepsilon_{\boldsymbol{k}_2-\boldsymbol{q}} - \varepsilon_{\boldsymbol{k}_1} - \varepsilon_{\boldsymbol{k}_2} \ .$$

The second application of the perturbation returns the electrons to inside the Fermi sphere. Figure 12.2 shows the four possible cases; the relations for the wavevectors and the spins are:

Process (1)

$$\boldsymbol{k}_1' = \boldsymbol{k}_1 + \boldsymbol{q} \ , \quad \boldsymbol{k}_2' = \boldsymbol{k}_2 - \boldsymbol{q} \ , \quad \boldsymbol{q}' = -\boldsymbol{q} \ , \quad \sigma_1' = \sigma_1 \ , \quad \sigma_2' = \sigma_2 \ ,$$

Process (2)

$$\boldsymbol{k}_1' = \boldsymbol{k}_1 + \boldsymbol{q} \ , \quad \boldsymbol{k}_2' = \boldsymbol{k}_2 - \boldsymbol{q} \ , \quad \boldsymbol{q}' = \boldsymbol{k}_2 - \boldsymbol{k}_1 - \boldsymbol{q} \ , \quad \sigma_1' = \sigma_1 = \sigma_2' = \sigma_2 \ ,$$

Process (3)

$$\boldsymbol{k}_1' = \boldsymbol{k}_2 - \boldsymbol{q} \ , \quad \boldsymbol{k}_2' = \boldsymbol{k}_1 + \boldsymbol{q} \ , \quad \boldsymbol{q}' = -\boldsymbol{k}_2 + \boldsymbol{k}_1 + \boldsymbol{q} \ , \quad \sigma_1' = \sigma_1 = \sigma_2' = \sigma_2 \ ,$$

Process (4)

$$\boldsymbol{k}_1' = \boldsymbol{k}_2 - \boldsymbol{q} \ , \quad \boldsymbol{k}_2' = \boldsymbol{k}_1 + \boldsymbol{q} \ , \quad \boldsymbol{q}' = \boldsymbol{q} \ , \quad \sigma_1' = \sigma_2 \ , \quad \sigma_2' = \sigma_1 \ .$$

The relations for the prcesses (3) and (4) are identical to those for (2) and (1), respectively, if one interchanges $\boldsymbol{k}_1'$ with $\boldsymbol{k}_2'$ and σ_1' with σ_2' and also reverses the sign of $\boldsymbol{q}'$. Accordingly, the second-order perturbed energy is given by the sum of the two terms corresponding to these two processes:

$$E^{(2)} = E_{\mathrm{a}}^{(2)} + E_{\mathrm{b}}^{(2)} , \tag{12.10}$$

$$E_{\mathrm{a}}^{(2)} = -8\left(\frac{2\pi e^2}{V}\right)^2 \frac{m}{\hbar^2} \sum_{\boldsymbol{q}} \frac{1}{q^4} \sum_{\boldsymbol{k}_1} \sum_{\boldsymbol{k}_2} \frac{1}{q^2 + \boldsymbol{q}\cdot(\boldsymbol{k}_1 + \boldsymbol{k}_2)} , \tag{12.11}$$

$$E_{\mathrm{b}}^{(2)} = 4\left(\frac{2\pi e^2}{V}\right)^2 \frac{m}{\hbar^2} \sum_{\boldsymbol{q}} \frac{1}{q^2} \sum_{\boldsymbol{k}_1} \sum_{\boldsymbol{k}_2} \frac{1}{(\boldsymbol{k}_1 + \boldsymbol{k}_2 + \boldsymbol{q})^2} \frac{1}{q^2 + \boldsymbol{q}\cdot(\boldsymbol{k}_1 + \boldsymbol{k}_2)} , \tag{12.12}$$

where the factors 8 and 4 on the right-hand sides come from the sums over spins as well as the sum over the two processes (1) and (4), and (2) and (3).

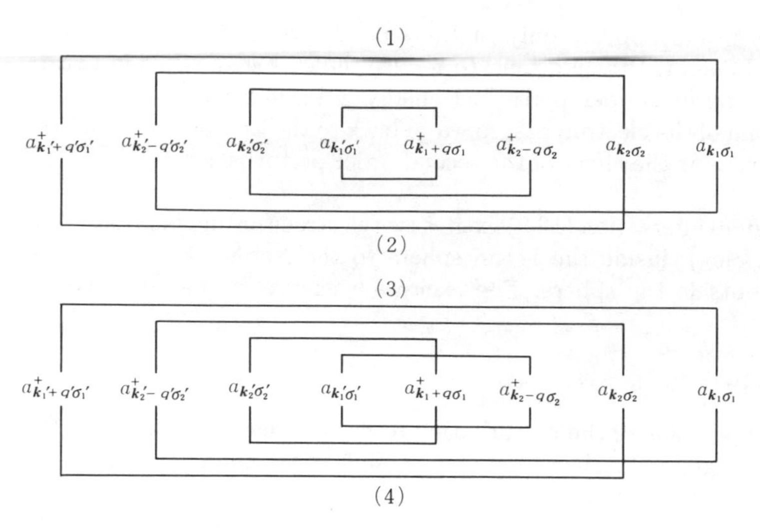

Fig. 12.2. Four possible processes of the perturbation

The sums over $\boldsymbol{k}_1$ and $\boldsymbol{k}_2$ are taken in the regions satisfying the conditions $|\boldsymbol{k}_1+\boldsymbol{q}|>k_{\mathrm{F}}$, $|\boldsymbol{k}_1|<k_{\mathrm{F}}$, $|\boldsymbol{k}_2+\boldsymbol{q}|>k_{\mathrm{F}}$, $|\boldsymbol{k}_2|<k_{\mathrm{F}}$. We have here assumed equal numbers of electrons with + and − spin. By converting the sums into integrals and representing $\boldsymbol{q}$, $\boldsymbol{k}_1$ and $\boldsymbol{k}_2$ in units of k_{F}, radius of the Fermi sphere, and also by choosing the Rydberg energy, $e^4m/(2\hbar^2)$, as the unit of energy, we can express (12.11, 12) as

$$\begin{aligned}\varepsilon_{\mathrm{a}}^{(2)} &= \frac{E_{\mathrm{a}}^{(2)}}{N_{\mathrm{e}}}\left(\frac{e^4m}{2\hbar^2}\right)^{-1}\\ &= -\frac{3}{8\pi^5}\int\frac{d^3q}{q^4}\int_{k_1<1,|\boldsymbol{k}_1+\boldsymbol{q}|>1}d^3k_1\int_{k_2<1,|\boldsymbol{k}_2+\boldsymbol{q}|>1}d^3k_2\frac{1}{q^2+\boldsymbol{q}\cdot(\boldsymbol{k}_1+\boldsymbol{k}_2)},\end{aligned} \tag{12.13}$$

$$\begin{aligned}\varepsilon_{\mathrm{b}}^{(2)} = \frac{E_{\mathrm{b}}^{(2)}}{N_{\mathrm{e}}}\left(\frac{e^4m}{2\hbar^2}\right)^{-1} &= \frac{3}{16\pi^5}\int\frac{d^3q}{q^2}\int_{k_1<1,|\boldsymbol{k}_1+\boldsymbol{q}|>1}d^3k_1\int_{k_2<1,|\boldsymbol{k}_2+\boldsymbol{q}|>1}d^3k_2\\ &\times\frac{1}{(\boldsymbol{k}_1+\boldsymbol{k}_2+\boldsymbol{q})^2}\frac{1}{q^2+\boldsymbol{q}\cdot(\boldsymbol{k}_1+\boldsymbol{k}_2)}.\end{aligned} \tag{12.14}$$

Using the same units, one can express the kinetic energy, (12.6), and the exchange interaction energy, (12.7), calculated in first order; the results for $n_+=n_-=N_{\mathrm{e}}/(2V)$ are, respectively,

$$\varepsilon_{\mathrm{kin}} = \frac{E_{\mathrm{kin}}}{N_{\mathrm{e}}}\left(\frac{e^4m}{2\hbar^2}\right)^{-1} = \frac{3}{5}\left(\frac{9\pi}{4}\right)^{2/3}r_{\mathrm{s}}^{-2}, \tag{12.15}$$

$$\varepsilon_{\mathrm{exch}} = \frac{E_{\mathrm{exch}}}{N_{\mathrm{e}}}\left(\frac{e^4m}{2\hbar^2}\right)^{-1} = -\frac{3}{2\pi}\left(\frac{9\pi}{4}\right)^{1/3}r_{\mathrm{s}}^{-1}. \tag{12.16}$$

From these equations, together with the results in second order, (12.13, 14), we see that the perturbation expansion appears to be a power-series expansion in r_s. That is, the second-order term is proportional to r_s^0-term and the third-order term is proportional to r_s. But inspection shows that (12.13) is logarithmically divergent at the lower limit of q-integration because of $1/q^4$ in the integrand. On the other hand, in (12.14), a factor $1/q^2$ out of $1/q^4$ has been replaced by $1/(\boldsymbol{q}+\boldsymbol{k}_1+\boldsymbol{k}_2)^2$ (because of the exchange of electrons) and so there is no divergence in $\varepsilon_{\mathrm{b}}^{(2)}$. Let us examine the divergence in (12.13) more closely. Since the divergence arises at $q \to 0$ of the integrand, we perform the integration over $\boldsymbol{k}_1$ and $\boldsymbol{k}_2$ by assuming $q \ll 1$. If the z-axis is taken in the direction of $\boldsymbol{q}$ and the direction cosine of $\boldsymbol{k}_1$ with respect to the z-axis is denoted by $\cos\theta = x_1$, the integral becomes

$$\iiint d^3k_1 = 2\pi \int dx_1 \int k_1^2 dk_1 \ .$$

In the case $q \ll 1$, the conditions $k < 1$, $|\boldsymbol{k}+\boldsymbol{q}| > 1$ reduce to $0 < x < 1$ and $1 - qx < k < 1$. Then the regions of integration in (12.13) are from 0 to 1 for x and from $1 - qx$ to 1 for k. Thus we have

$$\begin{aligned}\iiint \frac{d^3k_1}{q^2 + \boldsymbol{q}\cdot(\boldsymbol{k}_1+\boldsymbol{k}_2)} &= 2\pi \int_0^1 dx_1 \int_{1-qx_1}^1 k_1^2 dk_1 \frac{1}{qk_2x_2 + qk_1x_1} \\ &= 2\pi q \int_0^1 x_1 dx_1 \frac{1}{qk_2x_2 + qx_1} \\ &= 2\pi \left[1 - k_2x_2 \log\left(\frac{1+k_2x_2}{k_2x_2}\right)\right] \ .\end{aligned}$$

The integral over k_2 can be carried out similarly, and so (12.13) diverges as $\log q$:

$$\varepsilon_{\mathrm{a}}^{(2)} \sim -\frac{4}{\pi^2}(1 - \log 2) \int_0 \frac{dq}{q} \ \to \ \frac{4}{\pi^2}(1 - \log 2) \lim_{q\to 0} \log q \ . \qquad (12.17)$$

12.3 Higher-Order Perturbation – Gell-Mann–Brueckner Theory

The investigation of the correlation energy due to Coulomb interactions (this usually means contributions from the second- and higher-order terms) for electron systems has a long history; it was started by *Wigner* [12.2] and continued by *Macke* [12.3], *Bohm* and *Pines* [12.4], *Pines* [12.5], *Gell-Mann* and *Brueckner* [12.6], *Sawada* [12.7] and *Sawada* et al. [12.8]. According to the results of these studies, all terms higher than second order diverge; the most strongly divergent terms in the nth order perturbation are proportional to

$$\varepsilon^{(n)} \sim r_s^{n-2} \int \frac{dq}{q^{2n-3}} . \tag{12.18}$$

Gell-Mann and *Brueckner* collected all the most strongly divergent terms (called the most divergent terms) before integrating over q and cxpressed them in a closed form. Then carrying out the integral over q, they found a finite value for the energy; there are three kinds of terms : terms proportional to $\log r_s$, terms independent of r_s and terms which vanish for $r_s \to 0$. The result is interpreted as follows; to collect all the most divergent terms has ultimately the effect of introducing a cutoff $q_{\min}$ at long wavelength, so that the Coulomb interaction is reduced to the screened Coulomb type. With this cutoff, the sum of the most divergent terms has the form

$$\varepsilon \sim \frac{4}{\pi^2}(1 - \log 2) \log q_{\min} + c_1\left(\frac{r_s}{q_{\min}^2}\right) + c_2\left(\frac{r_s}{q_{\min}^2}\right)^2 + \dots . \tag{12.18$'$}$$

The coefficient c_1 and c_2 are constants fixed by actual perturbation calculations to third and fourth orders, respectively. The result of Gell-Mann and Brueckner implies that the cutoff, $q_{\min}$, is proportional to $r_s^{1/2}$; if one assumes that

$$q_{\min} = c r_s^{1/2}, \tag{12.19}$$

then one finds that the right-hand side of (12.18′) is the sum of a term proportional to $\log r_s$ and a constant term independent of r_s. Previous to Gell-Mann and Brueckner, Bohm and Pines had studied the electron gas by separating out the plasma oscillation, a collective motion of the electrons. Based on the physical consideration that the separation should lead to a cutoff in q in the long-wavelength region, they estimated the two terms in the correlation energy, $\log r_s$-term plus constant term.

Now let us turn to explain the essential point of Gell-Mann- Brueckner theory. It is not very difficult to get the most divergent terms of third and fourth order in a perturbative calculation of energy by a similar method used to get the second-order term. The results are as follows:

$$\begin{aligned} \varepsilon^{(3)} &= 2\left(\frac{\alpha r_s}{\pi^2}\right)\left(\frac{3}{8\pi^5}\right)\int \frac{d\boldsymbol{q}}{q^6}\int d\boldsymbol{k}_1 \int d\boldsymbol{k}_2 \int d\boldsymbol{k}_3 \\ &\quad \times \frac{1}{q^2 + \boldsymbol{q}\cdot(\boldsymbol{k}_1 + \boldsymbol{k}_2)} \frac{1}{q^2 + \boldsymbol{q}\cdot(\boldsymbol{k}_1 + \boldsymbol{k}_3)} , \end{aligned} \tag{12.20}$$

$$\begin{aligned} \varepsilon^{(4)} &= -2\left(\frac{\alpha r_s}{\pi^2}\right)^2\left(\frac{3}{8\pi^5}\right)\int \frac{d\boldsymbol{q}}{q^8}\int d\boldsymbol{k}_1 \int d\boldsymbol{k}_2 \int d\boldsymbol{k}_3 \int d\boldsymbol{k}_4 \\ &\quad \times \left[\frac{1}{q^2 + \boldsymbol{q}\cdot(\boldsymbol{k}_1 + \boldsymbol{k}_2)} \frac{1}{q^2 + \boldsymbol{q}\cdot(\boldsymbol{k}_1 + \boldsymbol{k}_3)} \frac{1}{q^2 + \boldsymbol{q}\cdot(\boldsymbol{k}_1 + \boldsymbol{k}_4)}\right. \\ &\quad + \frac{1}{q^2 + \boldsymbol{q}\cdot(\boldsymbol{k}_1 + \boldsymbol{k}_2)} \frac{1}{q^2 + \boldsymbol{q}\cdot(\boldsymbol{k}_1 + \boldsymbol{k}_3)} \frac{1}{q^2 + \boldsymbol{q}\cdot(\boldsymbol{k}_3 + \boldsymbol{k}_4)} \\ &\quad \left. + \frac{1}{q^2 + \boldsymbol{q}\cdot(\boldsymbol{k}_1+\boldsymbol{k}_2)} \frac{1}{2q^2 + \boldsymbol{q}\cdot(\boldsymbol{k}_1+\boldsymbol{k}_2+\boldsymbol{k}_3+\boldsymbol{k}_4)} \frac{1}{q^2 + \boldsymbol{q}\cdot(\boldsymbol{k}_1+\boldsymbol{k}_3)}\right] , \end{aligned} \tag{12.21}$$

where $(4/9\pi)^{1/3}$ has been replaced by α [this is different from the α used in (12.8)]. At this stage we introduce a time-dependent function

$$F_q(t) = \int d\boldsymbol{k} \exp\Big[-|t|\Big(\frac{1}{2}q^2 + \boldsymbol{q}\cdot\boldsymbol{k}\Big)\Big] \qquad (-\infty < t < \infty) \tag{12.22}$$

and calculate the quantity

$$A_2 \equiv \frac{1}{2}\int_{-\infty}^{\infty} dt_1 \int_{-\infty}^{\infty} dt_2 F_q(t_1)F_q(t_2)\delta(t_1+t_2) \ . \tag{12.23}$$

The integrations over $\boldsymbol{k}$ in (12.20–22) are carried out under the conditions $|\boldsymbol{k}+\boldsymbol{q}| > 1$, $|\boldsymbol{k}| < 1$. On using (12.22), we evaluate (12.23) as

$$\begin{aligned} A_2 &= \int d\boldsymbol{k}_1 \int d\boldsymbol{k}_2 \int_0^{\infty} dt \exp\Big[-\Big(\frac{1}{2}q^2 + \boldsymbol{q}\cdot\boldsymbol{k}_1\Big)t + \Big(\frac{1}{2}q^2 + \boldsymbol{q}\cdot\boldsymbol{k}_2\Big)(-t)\Big] \\ &= \int d\boldsymbol{k}_1 \int d\boldsymbol{k}_2 \frac{1}{q^2 + \boldsymbol{q}\cdot(\boldsymbol{k}_1+\boldsymbol{k}_2)} \ ; \end{aligned} \tag{12.24}$$

this is just the integral occurring in (12.13) for $\varepsilon_{\mathrm{a}}^{(2)}$. In third order, by calculating

$$A_3 \equiv \frac{1}{3}\int_{-\infty}^{\infty} dt_1 \int_{-\infty}^{\infty} dt_2 \int_{-\infty}^{\infty} dt_3 F_q(t_1)F_q(t_2)F_q(t_3)\delta(t_1+t_2+t_3) \ , \tag{12.25}$$

we obtain

$$\begin{aligned} A_3 &= 2\int d\boldsymbol{k}_1 \int d\boldsymbol{k}_2 \int d\boldsymbol{k}_3 \int_0^{\infty} dt_1 \int_{-t_1}^{0} dt_2 \\ &\quad \times \exp\Big[-\Big(\frac{1}{2}q^2 + \boldsymbol{q}\cdot\boldsymbol{k}_1\Big)t_1 + \Big(\frac{1}{2}q^2 + \boldsymbol{q}\cdot\boldsymbol{k}_2\Big)t_2 \\ &\quad + \Big(\frac{1}{2}q^2 + \boldsymbol{q}\cdot\boldsymbol{k}_3\Big)(-t_1-t_2)\Big] \\ &= 2\int d\boldsymbol{k}_1 \int d\boldsymbol{k}_2 \int d\boldsymbol{k}_3 \frac{1}{q^2 + \boldsymbol{q}\cdot(\boldsymbol{k}_1+\boldsymbol{k}_2)} \frac{1}{q^2 + \boldsymbol{q}\cdot(\boldsymbol{k}_1+\boldsymbol{k}_3)} \ . \end{aligned} \tag{12.26}$$

This integral occurs in (12.20) for $\varepsilon^{(3)}$. The correspondence betweeen the integrals A_n and the most divergent terms $\varepsilon^{(n)}$ holds in general.

Now by making use of the Fourier transform of the δ-function

$$\delta(t) = \frac{1}{2\pi}\int_{-\infty}^{\infty} \mathrm{e}^{itu} du \ ,$$

we can express A_n as

$$A_n = \frac{q}{2\pi n}\int_{-\infty}^{\infty} du \Big[Q_q(u)\Big]^n \ , \tag{12.27}$$

where $Q_q(u)$ is given by

$$Q_q(u) = \int d\boldsymbol{k} \int_{-\infty}^{\infty} dt \mathrm{e}^{\mathrm{i}tuq} \exp\Big[-|t|\Big(\frac{1}{2}q^2 + \boldsymbol{q}\cdot\boldsymbol{k}\Big)\Big] . \tag{12.28}$$

Consequently, except for $\varepsilon_{\mathrm{b}}^{(2)}$, the correlation energy is now brought together as

$$\begin{aligned}\varepsilon' &= \varepsilon_{\mathrm{a}}^{(2)} + \varepsilon^{(3)} + \varepsilon^{(4)} + \dots \\ &= -\frac{3}{8\pi^5}\int \frac{d\boldsymbol{q}}{q^3}\frac{1}{2\pi}\int_{-\infty}^{\infty} du \sum_{n=2}^{\infty}\frac{(-1)^n}{n}[Q_q(u)]^n\Big(\frac{\alpha r_{\mathrm{s}}}{\pi^2 q^2}\Big)^{n-2} .\end{aligned} \tag{12.29}$$

In the integration over $\boldsymbol{q}$ for ε higher than and equal to second order, the contribution from the region near $q = 0$ is the most important. So we keep the exact behavior in the integrand for $q \to 0$, but take arbitrarily $q = 1$ for the upper limit. For small q, as we have done in second order, we may transform the integral over $\boldsymbol{k}$ to an integral over x (the direction cosine of $\boldsymbol{k}$ with respect to $\boldsymbol{q}$) from 0 to 1 and an integral over k from $1 - qx$ to 1. Thus we obtain the approximation

$$\begin{aligned}Q_q(u) &\simeq 2\pi q \int_0^1 x dx \int_{-\infty}^{\infty} dt \mathrm{e}^{\mathrm{i}tuq}\mathrm{e}^{-|t|qx} \\ &= 2\pi \int_0^1 x dx \int_{-\infty}^{\infty} ds \mathrm{e}^{\mathrm{i}us}\mathrm{e}^{-|s|x} \\ &= 4\pi[1 - u\tan^{-1}(u^{-1})] = 4\pi R(u) ;\end{aligned} \tag{12.30}$$

for $q > 1$ we let $Q_q(u)$ be zero. For $\varepsilon_{\mathrm{a}}^{(2)}$, the difference between the exact value and the approximate value obtained by using (12.30) is a constant independent of r_{s}. Then, taking into account the difference δ we obtain

$$\varepsilon' = -\frac{12}{\pi^3}\int_{-\infty}^{\infty} du \int_0^1 \frac{dq}{q} \sum_{n=2}^{\infty} \frac{(-1)^n}{n}[R(u)]^n \Big(\frac{4\alpha r_{\mathrm{s}}}{\pi q^2}\Big)^{n-2} + \delta , \tag{12.31}$$

where δ is given by

$$\delta \equiv \varepsilon_{\mathrm{a}}^{(2)} - \Big[-\frac{12}{\pi^3}\int_{-\infty}^{\infty} du \int_0^1 \frac{dq}{q}\frac{1}{2}R^2\Big] . \tag{12.32}$$

The series in (12.31) converges well for large q but diverges for small q; we assume that the result obtained for large q can be extrapolated to the region of small q. The series in (12.31) can be summed by using the expansion of logarithmic function. After integrating over q, and neglecting terms which vanish in the limit of $r_{\mathrm{s}} \to 0$, we obtain

$$\begin{aligned}\varepsilon' &= \frac{3}{\pi^3}\int_{-\infty}^{\infty} du [R(u)]^2\Big[\log\Big(\frac{4\alpha r_{\mathrm{s}}}{\pi}\Big) + \log R(u) - \frac{1}{2}\Big] + \delta \\ &= \frac{2}{\pi^2}(1 - \log 2)\Big[\log\Big(\frac{4\alpha r_{\mathrm{s}}}{\pi}\Big) + \langle \log R(u)\rangle_{\mathrm{av}} - \frac{1}{2}\Big] + \delta ,\end{aligned} \tag{12.33}$$

where

$$\langle \log R(u) \rangle_{\mathrm{av}} = \frac{\int_{-\infty}^{\infty} du R^2 \log R}{\int_{-\infty}^{\infty} du R^2} . \tag{12.34}$$

From (12.33) for ε', we find that the coefficient of the term of $\log r_{\mathrm{s}}$-form is equal to $(2/\pi^2)(1 - \log 2)$. The constant term c, which is independent of r_{s}, can be written as

$$c = \frac{2}{\pi^2}(1 - \log 2)\left\{ \log\left[\frac{4}{\pi}\left(\frac{4}{9\pi}\right)^{1/3}\right] - \frac{1}{2} + \langle \log R \rangle_{\mathrm{av}} \right\} + \delta . \tag{12.35}$$

Numerical integration gives the values

$$\langle \log R \rangle_{\mathrm{av}} = -0.551 \ , \quad \delta = -0.0508 \ , \quad [12.5] \ , \quad \varepsilon_{\mathrm{b}}^{(2)} = 0.046 \pm 0.002 \ .$$

Using these, we are finally led to the correlation energy of the high-density electron gas ($r_{\mathrm{s}} \to 0$):

$$\varepsilon_{\mathrm{cor}} = 0.0622 \ \log r_{\mathrm{s}} - 0.096 \ . \tag{12.36}$$

The result (12.33) shows that the lower limit of the q integration term by term in (12.31) is in fact cut off at a certain value proportional to $r_{\mathrm{s}}^{1/2}$. This result by Gell-Mann and Brueckner can also been derived by means of the equation-of-motion method in the RPA (Random Phase Approximation), as shown by *Sawada* [12.7] and *Sawada* et al. [12.8]. The discussion based on the RPA will be postponed to a later Chapter.

Can the electron gas become ferromagnetic when the correlation energy is taken into account? The answer to this question is provided by the calculation of paramagnetic susceptibility, carried out by *Pines* [12.9] and also by *Brueckner* and *Sawada* [12.10]. If we define ζ, the spin polarization, in terms of the numbers N_+ and N_- of electrons with + and – spins by

$$N_\pm = \frac{1}{2} N_{\mathrm{e}}(1 \pm \zeta) \ , \tag{12.37}$$

the change in energy (per electron) due to the spin polarization is expressed as follows:

$$\Delta E = \frac{1}{4}\zeta^2(\alpha_{\mathrm{F}} + \alpha_{\mathrm{exch}} + \alpha_{\mathrm{c}}) \ ; \tag{12.38}$$

the terms are the change in the kinetic energy, α_{F}, the change in the exchange interaction energy, α_{exch}, and the change in correlation energy, α_{c}, respectively. In Rydberg units, they are

$$\alpha_{\mathrm{F}} = \frac{4.91}{r_{\mathrm{s}}^2} \ , \quad \alpha_{\mathrm{exch}} = -\frac{0.814}{r_{\mathrm{s}}} \ ,$$

$$\alpha_{\mathrm{c}} = \frac{2}{3\pi^2}\left[1 - \log\frac{\alpha r_{\mathrm{s}}}{\pi} - \langle \log R \rangle_{\mathrm{av}}\right] \ ; \tag{12.39}$$

α_c is the value calculated by *Brueckner* and *Sawada* [12.10]. The correlation energy acts to inhibit completely the ferromagnetism found in the Hartree-Fock approximation.

Nearly twenty years later, *Shastry* [12.11] revised the calculations by Brueckner and Sawada to derive the exact value for α_c. According to Shastry, $1 - \langle \log R \rangle_{av} = 1.551$ in (12.39) is corrected as 0.306.

13. Theory of Strong Electron Correlation

In this chapter, we treat the problem of electron correlation between Bloch electrons under the crystal potential. Starting with the considerations on the two-electron system in the Bloch bands, we describe two standard theories to treat the strong electron correlation, i.e., the t-matrix theory by Kanamori and the Gutzwiller's variational theory.

13.1 Exchange Interaction of Electrons in the Tight-Binding Approximation – Slater Theory

Iron-group metals such as Ni, Co and Fe become ferromagnetic. The ferromagnetism of these metals appears superficially to be understood by the band theory, that is, within the one-body approximation of Hartree-Fock theory. In particular, for the ground states of Ni and Fe, band calculations give not only satisfactory values of the spontaneous magnetization but also acceptable results for the details of Fermi surfaces observed in experiment. On the other hand, for free electrons the ferromagnetism cannot be discussed by considering only the exchange interaction (i.e., the diagonal element of the Coulomb interaction); we must include also the correlation energy, through an infinite-order perturbation calculation. Therefore it seems unlikely that the Hartree-Fock approximation can correctly describe the ferromagnetism of iron-group metals, in particular at finite temperatures.

An accurate treatment of the exchange potential is a difficult problem even within a band theory based on the Hartree-Fock approximation. Usually we adopt the mean value of the exchange potential for free electrons as obtained by Slater, or that further multiplied by a reduction factor $\alpha \cong 2/3$. The basis for choosing $\alpha = 2/3$ was discussed by *Kohn* and *Sham*, as follows [13.1]: Slater took the mean value of the exchange potential used in the Schrödinger equation derived by taking the variation of the expectation value of total energy with respect to the Bloch function $u_i(\boldsymbol{r})$. *Kohn* and *Sham* took the other way arround: reversing two operations of averaging and variation, they started with the spatial integral of the exchange interaction energy as calulated for free electrons, (12.7):

$$E_{\text{exch}+} = -\frac{3}{2}e^2\left(\frac{3}{4\pi}\right)^{1/3}\int d\tau n_+^{4/3}$$
$$= -\frac{3}{2}e^2\left(\frac{3}{4\pi}\right)^{1/3}\int d\tau\Big[\sum_i u_{i+}^*(\boldsymbol{r})u_{i+}(\boldsymbol{r})\Big]^{4/3} .$$

Then they took the variation with respect to $u_{i+}(\boldsymbol{r})$ and obtained

$$-2e^2\left(\frac{3}{4\pi}\right)^{1/3} n_+^{1/3} \tag{13.1}$$

as the exchange potential. This value is just 2/3 that obtained by Slater.

A second method to evaluate the exchange potential uses a calculation based on the tight-binding approximation for electrons. The calculation was in fact carried out long time ago by *Slater* [13.2]. *Slater* evaluated the average value of the exchange integral between two electrons with parallel spins in Bloch orbital states $(\boldsymbol{k}_i m)$ and $(\boldsymbol{k}_j m')$ of 3d bands:

$$I = \iint u_{\boldsymbol{k}_i m}^*(\boldsymbol{r}_1)u_{\boldsymbol{k}_j m'}^*(\boldsymbol{r}_2)\frac{e^2}{r_{12}}u_{\boldsymbol{k}_j m'}(\boldsymbol{r}_1)u_{\boldsymbol{k}_i m}(\boldsymbol{r}_2)d\tau_1 d\tau_2 , \tag{13.2}$$

where $u_{\boldsymbol{k}m}(\boldsymbol{r})$ stands for the wave function of the Bloch orbital with wavevector, $\boldsymbol{k}$, in the first Brillouin zone and with d-band number, m ($m = 1, 2, ...5$). Since $u_{\boldsymbol{k}m}(\boldsymbol{r})$ satisfies the Bloch condition

$$u_{\boldsymbol{k}m}(\boldsymbol{r}) = \mathrm{e}^{\mathrm{i}\boldsymbol{k}\cdot\boldsymbol{R}_n}u_{\boldsymbol{k}m}(\boldsymbol{r}-\boldsymbol{R}_n) , \tag{13.3}$$

(13.2) can be rewritten as

$$I = \sum_p\sum_q \mathrm{e}^{\mathrm{i}(\boldsymbol{k}_j-\boldsymbol{k}_i)\cdot(\boldsymbol{R}_p-\boldsymbol{R}_q)}\int_{p\text{cell}} d\tau_1 \int_{q\text{cell}} d\tau_2$$
$$\times\, u_{\boldsymbol{k}_i m}^*(\boldsymbol{r}_1-\boldsymbol{R}_p)u_{\boldsymbol{k}_j m'}^*(\boldsymbol{r}_2-\boldsymbol{R}_q)\frac{e^2}{r_{12}}u_{\boldsymbol{k}_j m'}(\boldsymbol{r}_1-\boldsymbol{R}_p)u_{\boldsymbol{k}_i m}(\boldsymbol{r}_2-\boldsymbol{R}_q) . \tag{13.4}$$

Here the integrations over $\boldsymbol{r}_1$ and $\boldsymbol{r}_2$ are carried out within the pth and qth unit cells, respectively. The largest contribution occurs when p and q refer to the same cell. Since electron wave functions in a cell are considered as similar to atomic orbitals, we may make the approximation $u_{\boldsymbol{k}m}(\boldsymbol{r}-\boldsymbol{R}_n) \rightarrow N^{-1/2}\varphi_m^a(\boldsymbol{r}-\boldsymbol{R}_n)$, where $\varphi_m^a(\boldsymbol{r}-\boldsymbol{R}_n)$ is the atomic orbital function. Thus we can evaluate the contribution from the $p = q$ term in (13.4) as $(1/N)J_{mm'}$, where $J_{mm'}$ is the exchange integral for a free atom.

Next let us examine the case $p \neq q$. For $m = m'$ we may now make the approximation

$$\int_{p\text{cell}} d\tau u_{\boldsymbol{k}_i m}^*(\boldsymbol{r}-\boldsymbol{R}_p)u_{\boldsymbol{k}_j m}(\boldsymbol{r}-\boldsymbol{R}_p) \sim \frac{1}{N}$$

to obtain

$$\sum_{q \neq p} \frac{1}{N^2} \frac{e^2}{|\boldsymbol{R}_p - \boldsymbol{R}_q|} \mathrm{e}^{\mathrm{i}(\boldsymbol{k}_j - \boldsymbol{k}_i)\cdot(\boldsymbol{R}_p - \boldsymbol{R}_q)} = \frac{1}{N} \sum_{\boldsymbol{R}_p \neq 0} \frac{e^2}{R_p} \mathrm{e}^{\mathrm{i}(\boldsymbol{k}_j - \boldsymbol{k}_i)\cdot \boldsymbol{R}_p} .$$

Next, for $m \neq m'$, the atomic orbitals belonging to the same atom are orthogonal to each other, so that the integrals in (13.4) are regarded as interactions between multipoles separated by $|\boldsymbol{R}_p - \boldsymbol{R}_q|$, and are neglected compared to the $m = m'$ terms. Then (13.2) is approximated as

$$I \sim \frac{J_{mm'}}{N} + \frac{1}{N} \sum_{\boldsymbol{R}_p \neq 0} \frac{e^2}{R_p} \mathrm{e}^{\mathrm{i}(\boldsymbol{k}_j - \boldsymbol{k}_i)\cdot \boldsymbol{R}_p} \delta_{mm'} . \tag{13.5}$$

We can estimate the average of the second term with respect to $\boldsymbol{k}_i$ and $\boldsymbol{k}_j$ inside the Fermi sphere by assuming the Fermi surface to be of a sphere of radius k_{F}. The result is so small compared with the first term that we neglect it. Consequently all the exchange energy is approximated in terms of the mean value of exchange integrals for atomic orbitals, as follows:

$$I \sim \frac{\bar{J}}{N} = \frac{1}{25} \frac{1}{N} \sum_{mm'} J_{mm'} . \tag{13.6}$$

The exchange integrals $J_{mm'}$ of the atomic orbitals can be written as sum of products of angular and radial integrals:

$$J_{mm'} = \sum_k b^k(lm, lm') F^k_{nl} , \quad (l = 2, n = 3) \tag{13.7}$$

$$b^k(lm, lm') = \frac{(k - |m - m'|)!}{(k + |m - m'|)!} \frac{(2l+1)^2 (l - |m|)!(l - |m'|)!}{(l + |m|)!(l + |m'|)!} \times \left[\int_0^{\pi} P_l^{|m|}(\cos\theta) P_l^{|m'|}(\cos\theta) P_k^{|m-m'|}(\cos\theta) \frac{\sin\theta}{2} d\theta \right]^2 , \quad (k \geq |m - m'|) \tag{13.8}$$

$$F^k_{nl} = e^2 (4\pi^2) \int_0^{\infty} \int_0^{\infty} R^2_{nl}(r) R^2_{nl}(r') \frac{r(a)^k}{r(b)^{k+1}} r^2 r'^2 dr dr' , \tag{13.9}$$

where $R_{nl}(r)$ is the radial part of the wave function in the nl orbital. The distance $r(a)$ denotes the smaller of r and r', while $r(b)$ denotes the larger. The numerical values of the angular part, (13.8), as calculated by *Slater* [13.3] are given in Table 13.1.

The numerical values in the table give the average of (13.6) over m and m' as

$$\bar{J} = \frac{1}{5} F^0_{\mathrm{3d}} + \frac{2}{35} (F^2_{\mathrm{3d}} + F^4_{\mathrm{3d}}) \tag{13.10}$$

in terms of three integrals F^0, F^2 and F^4. Slater evaluated these from the observed multiplet splitting of the Ni atom, $(3d)^8(4s)^2$, as $F^2_{\mathrm{3d}} = 80\,000\ \mathrm{cm}^{-1}$, $F^4_{\mathrm{3d}} = 50\,000\ \mathrm{cm}^{-1}$ and obtained the value $7430\ \mathrm{cm}^{-1}$ for the contribution

Table 13.1. $b^k(lm, lm'), l = 2$

m	m'	k = 0	1	2	3	4
±2	±2	1	0	4/49	0	1/441
±2	±1	0	0	6/49	0	5/441
±2	0	0	0	4/49	0	15/441
±2	∓1	0	0	0	0	35/441
±2	∓2	0	0	0	0	70/441
±1	±1	1	0	1/49	0	16/441
±1	0	0	0	1/49	0	30/441
±1	∓1	0	0	6/49	0	40/441
0	0	1	0	4/49	0	36/441

from F^2 and F^4 to (13.10). F^0, the contribution from the Coulomb integral of intra-atomic $3d$ electrons, is much larger (about 15 eV) than that from the exchange integrals involving F^2.

Now let the number of atoms be N, the number of $3d$-electrons per atom be n, and the difference in number of electrons with ± spin be $n_+ - n_- = \delta n$. The difference in the exchange integrals per electron caused by the imbalance of $n_\pm$ is expressed as

$$\frac{\bar{J}}{N} \cdot N(n_+ - n_-) = \bar{J}\delta n \ . \tag{13.11}$$

This is the average energy shift of $3d$-electrons with up spin relative to down spin, and is called the *exchange splitting.* For the Ni band it takes the value 0.05 Ryd and J_{eff} is roughly 0.08 Ryd≅ 1 eV, for $\delta n = 0.6$. This value of the exchange splitting is believed to be approximately correct. If the number of $3d$-electrons is 9.4, the exchange potential given by (13.1) is estimated as

$$-2e^2\left(\frac{3}{4\pi v_0}\right)^{1/3}\left(\frac{9.4}{2}\right)^{1/3} \simeq -1 \text{ Ryd} = -13.6 \text{ eV} \ ,$$

which is a very large value. However, the exchange splitting is much reduced (by factor $\sim 1/20$) due to the 1/3 power dependence of $n_\pm$ and takes the much smaller value of about 0.05 Ryd.

From the point of view based on the tight-binding approximation for electrons, the exchange interaction energy between electrons is approximately independent of the electron wavevectors, and is determined by the exchange interaction within an atom. Thus the exchange potential is proportional to $n_\pm$, so that the exchange splitting is proportional to $\bar{J}$ itself. If the contribution from F^0 is included, the splitting would take a value much larger than the expected value (about 1 eV). To interpret this result the following argument is offered: the exchange splitting of (13.11) was obtained within the

Hartree-Fock approximation, but the effect of electron correlations tends to weaken the Hartree-Fock results. Particularly the F^0 term originating from intra-atomic Coulomb integrals is considerably reduced, so that one obtains a value of about 1 eV for the effective J_{eff} including F^0. Therefore, in this sense, we must recognize that electron systems are very strongly correlated.

13.2 Slater-Hubbard Hamiltonian and Two-Electron Problems

Let us now examine how correlation effect between electrons reduces in practice the effect of the intra-atomic exchange interaction, in particular, the effect of the Coulomb integral in atoms. To this end, we simplify the problem following *Slater* et al. [13.4]; we treat two electrons in a single band, and examine the effect of the Coulomb interaction between them in the tight-binding approximation. We consider the following linear combination of the atomic orbitals $\varphi_i(\boldsymbol{r} - \boldsymbol{R}_n)$ for an electron:

$$v_i(\boldsymbol{k}, \boldsymbol{r}) = N^{-1/2} \sum_m \mathrm{e}^{\mathrm{i}\boldsymbol{k}\cdot\boldsymbol{R}_m} \varphi_i(\boldsymbol{r} - \boldsymbol{R}_m) , \tag{13.12}$$

where the suffix i designates the atomic orbital ($i = 1, 2, ..., 5$ for the d-band) and $\boldsymbol{R}_n$ is the site vector for the atoms. The wave functions constructed in this way are deficient since the atomic orbitals belonging to different sites are not mutually orthogonal. As a result, the wave functions $v_i(\boldsymbol{k}, \boldsymbol{r})$ are not properly normalized; the integral of the absolute square of each function is not unity. Moreover, two states described by (13.12) with same $\boldsymbol{k}$ but different i are not orthogonal to each other. Thus the Bloch orbitals are given by linear combinations of (13.12) with different i, when we take several atomic orbitals into account. Let us represent such a Bloch orbital by $u_i(\boldsymbol{k}, \boldsymbol{r})$; here i refers to the ith energy band. In order to avoid the inconvenience that the atomic orbitals for different sites are not orthogonal, we prefer to use the following system of functions in place of the atomic orbitals: In the new system, the orbital functions are localized on each site and further are orthogonal to each other; that is

$$a_i(\boldsymbol{r} - \boldsymbol{R}_m) = N^{-1/2} \sum_{\boldsymbol{k}} \mathrm{e}^{-\mathrm{i}\boldsymbol{k}\cdot\boldsymbol{R}_m} u_i(\boldsymbol{k}, \boldsymbol{r}) . \tag{13.13}$$

The functions thus defined are called Wannier functions. In case of degenerate atomic orbitals, we point out that the Wannier functions defined by (13.13) differ considerably from the original atomic functions; care is needed in practical calculations using them. By the inverse of (13.13), Bloch functions are given as linear combinations of Wannier functions:

$$u_i(\boldsymbol{k}, \boldsymbol{r}) = N^{-1/2} \sum_m \mathrm{e}^{\mathrm{i}\boldsymbol{k}\cdot\boldsymbol{R}_m} a_i(\boldsymbol{r} - \boldsymbol{R}_m) . \tag{13.14}$$

Let the energy of the Bloch orbital be $\varepsilon_i(\boldsymbol{k})$. The matrix elements of the one-electron Hamiltonian $\mathcal{H}_1$ with respect to the Wannier functions of (13.13) are

$$\int a_i^*(\boldsymbol{r})\mathcal{H}_1 a_j(\boldsymbol{r}-\boldsymbol{R}_m)d\tau = \delta_{ij}\varepsilon_i(-\boldsymbol{R}_m) , \tag{13.15}$$

where $\varepsilon_i(\boldsymbol{R}_m)$ is defined by

$$\varepsilon_i(\boldsymbol{R}_m) = N^{-1}\sum_{\boldsymbol{k}} \varepsilon_i(\boldsymbol{k})\mathrm{e}^{\mathrm{i}\boldsymbol{k}\cdot\boldsymbol{R}_m} ; \tag{13.16}$$

this $\varepsilon(\boldsymbol{R}_m)$ is represented by b_m in (3.1) of Part I.

The Hamiltonian of the two-electron system is given by sum of the two one-electron Hamiltonians, $\mathcal{H}_1$ and $\mathcal{H}_2$, and the Coulomb interaction $g_{12} = e^2/r_{12}$:

$$\mathcal{H} = \mathcal{H}_1 + \mathcal{H}_2 + g_{12} . \tag{13.17}$$

The wave function of the system is generally expressed as

$$\psi(\boldsymbol{r}_1,\boldsymbol{r}_2) = \sum_{ij}\sum_{\boldsymbol{k}_1,\boldsymbol{k}_2} \Gamma_{ij}(\boldsymbol{k}_1,\boldsymbol{k}_2)u_i(\boldsymbol{k}_1,\boldsymbol{r}_1)u_j(\boldsymbol{k}_2,\boldsymbol{r}_2) \tag{13.18}$$

in the basis of Bloch orbitals. One does not need to write down the spin singlet and triplet functions explicitly. Instead, one may invoke symmetric or antisymmetric conditions on the wave function $\psi(\boldsymbol{r}_1,\boldsymbol{r}_2)$ with respect to the interchange of position coordinates $\boldsymbol{r}_1$ and $\boldsymbol{r}_2$ for spin-singlet or triplet states. The conditions are

$$\Gamma_{ji}(\boldsymbol{k}_2,\boldsymbol{k}_1) = \pm\Gamma_{ij}(\boldsymbol{k}_1,\boldsymbol{k}_2) , \tag{13.19}$$

where the + sign corresponds to the spin-singlet state and the − sign to the triplet state.

On substituting (13.18) into Schrödinger equation for the Hamiltonian (13.17), we obtain an equation for $\Gamma_{ij}(\boldsymbol{k}_1,\boldsymbol{k}_2)$:

$$\begin{aligned}
&\Big[\varepsilon_i(\boldsymbol{k}_1) + \varepsilon_j(\boldsymbol{k}_2)\Big]\Gamma_{ij}(\boldsymbol{k}_1,\boldsymbol{k}_2) \\
&\quad + N^{-2}\sum_{\boldsymbol{k}_1',\boldsymbol{k}_2'}\sum_{i'j'}\Gamma_{i'j'}(\boldsymbol{k}_1',\boldsymbol{k}_2')\sum_{mnrs}\mathrm{e}^{-\mathrm{i}(\boldsymbol{k}_1\cdot\boldsymbol{R}_m+\boldsymbol{k}_2\cdot\boldsymbol{R}_n-\boldsymbol{k}_1'\cdot\boldsymbol{R}_r-\boldsymbol{k}_2'\cdot\boldsymbol{R}_s)} \\
&\quad \times\iint a_i^*(\boldsymbol{r}_1-\boldsymbol{R}_m)a_j^*(\boldsymbol{r}_2-\boldsymbol{R}_n)g_{12}a_{i'}(\boldsymbol{r}_1-\boldsymbol{R}_r)a_{j'}(\boldsymbol{r}_2-\boldsymbol{R}_s)d\tau_1 d\tau_2 \\
&= E\Gamma_{ij}(\boldsymbol{k}_1,\boldsymbol{k}_2) .
\end{aligned} \tag{13.20}$$

The integral on the left-hand side of the equation is a generalized version of the integral in (13.4), so that we can use the same argument as before to neglect integrals involving different atoms. However, there remains the question whether we may neglect completely the Coulomb integrals with

$\boldsymbol{R}_m = \boldsymbol{R}_r$, $\boldsymbol{R}_n = \boldsymbol{R}_s$, $i = i'$ and $j = j'$, although they are reduced by the screening effects of the 4s-electrons. If we consider only the integral related to a single atom, we keep only terms with $\boldsymbol{R}_m - \boldsymbol{R}_n = \boldsymbol{R}_r = \boldsymbol{R}_s$; then (13.20) becomes

$$\Big[\varepsilon_i(\boldsymbol{k}_1) + \varepsilon_j(\boldsymbol{k}_2) - E\Big]\Gamma_{ij}(\boldsymbol{k}_1, \boldsymbol{k}_2) + N^{-1}\sum_{i'j'}\langle ij|g|i'j'\rangle \sum_{\boldsymbol{k}'}\Gamma_{i'j'}\Big(\frac{\boldsymbol{K}}{2} + \boldsymbol{k}', \frac{\boldsymbol{K}}{2} - \boldsymbol{k}'\Big) = 0 , \tag{13.21}$$

where $\boldsymbol{K} = \boldsymbol{k}_1 + \boldsymbol{k}_2$. If $\boldsymbol{K}/2 \pm \boldsymbol{k}$ falls outside the first Brillouin zone, it is to be reduced into the first zone by adding a proper reciprocal-lattice vector.

The two-electron wave function (13.18) can also be written in the Wannier-function basis:

$$\psi(\boldsymbol{r}_1, \boldsymbol{r}_2) = \sum_{ij}\sum_{mn}\Gamma_{ij}(\boldsymbol{R}_m, \boldsymbol{R}_n)a_i(\boldsymbol{r} - \boldsymbol{R}_m)a_j(\boldsymbol{r} - \boldsymbol{R}_n) ; \tag{13.22}$$

in this case, the conditions (13.19) for the spin-singlet and -triplet states become

$$\Gamma_{ji}(\boldsymbol{R}_n, \boldsymbol{R}_m) = \pm\Gamma_{ij}(\boldsymbol{R}_m, \boldsymbol{R}_n) . \tag{13.23}$$

Furthermore, since the Hamiltonian (13.17) is invariant under the translation of both $\boldsymbol{r}_1$ and $\boldsymbol{r}_2$ by a lattice vector $\boldsymbol{R}_s$, $\Gamma_{ij}(\boldsymbol{R}_m, \boldsymbol{R}_n)$ must satisfy the following Bloch condition:

$$\Gamma(\boldsymbol{R}_m + \boldsymbol{R}_s, \boldsymbol{R}_n + \boldsymbol{R}_s) = \mathrm{e}^{\mathrm{i}\boldsymbol{K}\cdot\boldsymbol{R}_s}\Gamma(\boldsymbol{R}_m, \boldsymbol{R}_n) .$$

Therefore $\Gamma_{ij}(\boldsymbol{R}_m, \boldsymbol{R}_n)$ factors as

$$\Gamma_{ij}(\boldsymbol{R}_m, \boldsymbol{R}_n) = \mathrm{e}^{\mathrm{i}\boldsymbol{K}\cdot(\boldsymbol{R}_m + \boldsymbol{R}_n)/2}F_{ij}(\boldsymbol{R}_m - \boldsymbol{R}_n) , \tag{13.24}$$

which is a product of functions of the center-of-mass and relative coordinates; here $\hbar\boldsymbol{K}$ is the total momentum. The conditions for the spin singlet and the spin triplet in terms of F_{ij} are

$$F_{ji}(-\boldsymbol{R}) = \pm F_{ij}(\boldsymbol{R}) . \tag{13.25}$$

The equation for $\Gamma_{ij}(\boldsymbol{R}_m, \boldsymbol{R}_n)$ can be derived directly from the Hamiltonian (13.17) or from (13.21) with use of the relation

$$\Gamma_{ij}(\boldsymbol{k}_1, \boldsymbol{k}_2) = N^{-1}\sum_{mn}\mathrm{e}^{-\mathrm{i}(\boldsymbol{k}_1\cdot\boldsymbol{R}_m + \boldsymbol{k}_2\cdot\boldsymbol{R}_n)}\Gamma_{ij}(\boldsymbol{R}_m, \boldsymbol{R}_n) . \tag{13.26}$$

The result is written down as

$$\sum_s\Big[\varepsilon_i(\boldsymbol{R}_s)\mathrm{e}^{-\mathrm{i}\boldsymbol{K}\cdot\boldsymbol{R}_s/2} + \varepsilon_j(\boldsymbol{R}_s)\mathrm{e}^{\mathrm{i}\boldsymbol{K}\cdot\boldsymbol{R}_s/2}\Big]F_{ij}(\boldsymbol{R}_p - \boldsymbol{R}_s) + \sum_{kl}\langle ij|g|kl\rangle\delta(\boldsymbol{R}_p)F_{kl}(\boldsymbol{R}_p) = EF_{ij}(\boldsymbol{R}_p) . \tag{13.27}$$

For a single band with nondegenerate atomic orbitals, the above equation can be simplified. Further, by putting $\boldsymbol{K} = 0$, we have

$$2\sum_{s} \varepsilon(\boldsymbol{R}_s)F(\boldsymbol{R}_p - \boldsymbol{R}_s) + U\delta(\boldsymbol{R}_p)F(\boldsymbol{R}_p) = EF(\boldsymbol{R}_p) , \tag{13.28}$$

where U is the intra-atomic Coulomb integral $\langle ii|g|ii\rangle$. *Slater* et al. [13.4] derived a solution of the above difference equation for a one-dimensional lattice in the following way. Let us take $2N+1$ lattice points ($n = 0, \pm 1, \ldots, \pm N$), and adopt the boundary condition that F vanishes at both ends $R_{\pm N}$. We simplify the notation by representing R_n by n. Further, regarding the transfer energy $\varepsilon(R)$, we consider only nearest-neighbor tranfer; that is, $\varepsilon(\pm 1) = \varepsilon_a < 0$ and the others are 0.

Then (13.28) separates into the following two equations:

$$\begin{aligned} &2\varepsilon_a[F(n+1) + F(n-1)] = EF(n) && \text{for } n \neq 0 , \\ &2\varepsilon_a[F(1) + F(-1)] + UF(0) = EF(0) && \text{for } n = 0 . \end{aligned} \tag{13.29}$$

The solution of the equation for $n \neq 0$ is given immediately as

$$\begin{aligned} &F(n) = \sin[\alpha(N-n)] \qquad (0 < n \leq N) , \\ &F(-n) = \pm F(n) ; \end{aligned} \tag{13.30}$$

the + and − signs correspond to the spin-singlet and spin-triplet states, respectively. Substitution of (13.30) into the $n \neq 0$ part of (13.29) gives the energy eigenvalue E as

$$E = 4\varepsilon_a \cos\alpha , \tag{13.31}$$

which has a minimum at $\alpha = 0$ since $\varepsilon_a < 0$. The value of α itself is determined from the value of $F(0)$: for the spin-triplet state, a solution $F(0) = 0$ leads to $\sin\alpha N = 0$, and thus

$$\alpha = \frac{m\pi}{N} \quad (m = 1, 2, \ldots, N-1) . \tag{13.32}$$

For the spin-singlet state, α is determined by the transcendental equation

$$-\cot\alpha N \sin\alpha = -\frac{U}{4\varepsilon_a} . \tag{13.33}$$

To solve (13.33), we plot $-\cot\alpha N \sin\alpha$ as a function of α as shown in Fig. 13.1, and look for intersection points with the horizontal line $-U/4\varepsilon_a > 0$. In the case $-U/4\varepsilon_a > 1/N$, there exist, in all, $N-1$ solutions for α, each of which is near to, but slightly smaller than, the solution for the spin-triplet, $m\pi/N$. When $U = 0$, the solutions for the spin singlet are given by the zeros of $-\cot\alpha N \sin\alpha$, that is, $\alpha = (\pi/N)(m+1/2)$ ($m = 0, 1, 2, \ldots, N-1$), numbering N in all. The minimum energy of the spin-singlet state corresponds to $\alpha = \pi/(2N)$. As U increases, this α becomes larger; it merges with the value

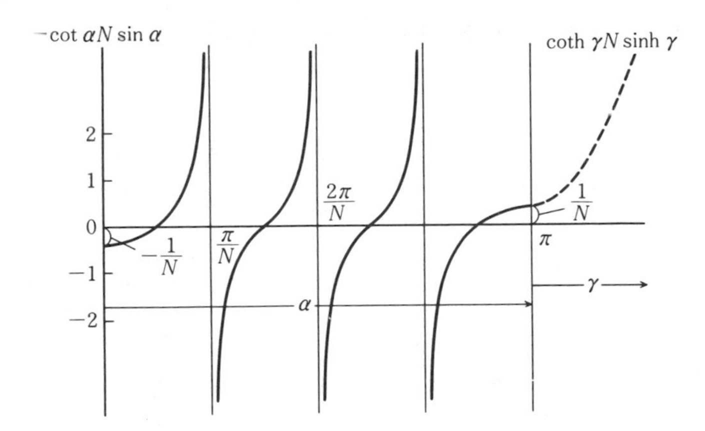

Fig. 13.1. Relations between α and $-\cot\alpha N\sin\alpha$ and between γ and $\coth\gamma N\sinh\gamma$

for the lowest-energy state of the spin triplet, $\alpha=\pi/N$, in the limit $U\to\infty$. Therefore *the ground state is always the spin singlet*; the spin-triplet state never has lower energy.

At this point, we remark that the spin-singlet solution for the highest energy is no longer obtained as a sinusoidal wave in (13.30), if $-U/4\varepsilon_a$ exceeds $1/N$. In this case, the solution is given by

$$\begin{aligned} F(n) &= (-1)^n \sinh\gamma(N-n)\ , \\ F(-n) &= F(n)\ , \end{aligned} \tag{13.34}$$

instead of (13.30). On substituting this solution into the first of (13.29), we have

$$E=-4\varepsilon_a\cosh\gamma\ ; \tag{13.35}$$

substituting this relation into the second of (13.29) for $n=0$, we have a relation to determine γ

$$\coth(\gamma N)\sinh\gamma=-\frac{U}{4\varepsilon_a}\ . \tag{13.36}$$

For $\gamma=0$, Equation (13.36) reduces to $-U/4\varepsilon_a=1/N$, just the condition that the highest-energy sinusoidal solution vanishes. The left-hand side of (13.36) increases monotonically starting from $\gamma=0$ as $\sinh\gamma$, since the approximation $\coth\gamma N\sim 1$ is valid for large N. Thus for $-U/4\varepsilon_a>1/N$ there exists one solution localized at the origin $n=0$; its energy lies above the top of the band, $-4\varepsilon_a$. The bound state thus obtained corresponds to that of Cooper pairs in superconductors, where, due to an attractive interaction dominating over the Coulomb repulsive force, the bound state appears at an energy level lower than the bottom of the band.

The above results apply to a single band in one dimension. *Slater* et al. [13.4] further considered the one-dimensional case, as above, but with a degenerate band (p-band). In this case, they concluded that the spin-triplet state has lower energy than the spin-singlet state, by examining the magnitudes of several intra-atomic interaction integrals. Generally, spin-triplet states tend to have higher energy than spin-singlet states, because electrons must be lifted to higher-energy states to orient their spin directions. Thus the above, contrary conclusion suggests that the energy cost can be rather small for electrons in degenerate orbitals. However, it is premature to conclude from the above argument that the ferromagnetism of three-dimensional electron systems is due to band degeneracies.

Let us return to (13.21) and consider again the lowest energies of spin-singlet and spin-triplet states from a more general point of view. If we assume again the case of a single band (labelled a) and denote the intra-atomic Coulomb integral $\langle aa|g|aa\rangle$ by U, (13.21) is rewritten as

$$\Big[\varepsilon_a(\boldsymbol{k}_1)+\varepsilon_a(\boldsymbol{k}_2)-E\Big]\Gamma(\boldsymbol{k}_1,\boldsymbol{k}_2)+N^{-1}U\sum_{\boldsymbol{k}'}\Gamma\Big(\frac{\boldsymbol{K}}{2}+\boldsymbol{k}',\frac{\boldsymbol{K}}{2}-\boldsymbol{k}'\Big)=0\,. \quad (13.37)$$

For spin-triplet states, the second term vanishes due to the asymmetry of Γ when the summation is carried out. Therefore the Coulomb interaction is ineffective for spin-triplet states; the energy eigenvalues are given by $\varepsilon_a(\boldsymbol{k}_1)+\varepsilon_a(\boldsymbol{k}_2)$. On the other hand, the energy eigenvalues of spin-singlet states are given by the solutions of

$$N^{-1}\sum_{\boldsymbol{k}}\frac{1}{\varepsilon_a(\boldsymbol{K}/2+\boldsymbol{k})+\varepsilon_a(\boldsymbol{K}/2-\boldsymbol{k})-E}=-\frac{1}{U}\,, \quad (13.38)$$

which is obtained from (13.37) by dividing by $\varepsilon_a(\boldsymbol{k}_1)+\varepsilon_a(\boldsymbol{k}_2)-E$ and summing over $\boldsymbol{k}$.

If the lowest energy of the one-electron band $\varepsilon_a(\boldsymbol{k})$ is nondegenerate and has a minimum at $\boldsymbol{k}=0$, we may set $\boldsymbol{K}=0$ in (13.38) to obtain

$$N^{-1}\sum_{\boldsymbol{k}}\frac{1}{\varepsilon_a(\boldsymbol{k})+\varepsilon_a(-\boldsymbol{k})-E}=-\frac{1}{U}\,. \quad (13.39)$$

The left-hand side of this equation is positive if E is smaller than the lowest energy $2\varepsilon_a(0)$; when E approaches $2\varepsilon_a(0)$, it increases and diverges to infinity at $2\varepsilon_a(0)$. Beyond this point it changes to $-\infty$ and increases monotonically to $+\infty$ as E approaches the next energy $2\varepsilon_a(\boldsymbol{k}_1)$ [Fig. 13.2(a)]. Therefore the lowest-energy solution of (13.39) is found between $2\varepsilon_a(0)$ and $\varepsilon_a(0)+\varepsilon_a(\boldsymbol{k}_1)$. Since the lowest-energy solution for the spin-triplet states is $\varepsilon_a(0)+\varepsilon_a(\boldsymbol{k}_1)$, this means that the lowest energy of spin-singlet state is always lower than that of spin-triplet state. Only if the lowest states at the bottom of the one-electron band are degenerate, it is possible that the spin-triplet state becomes lower. Incidentally, if E in (13.39) exceeds twice the energy of the top of the

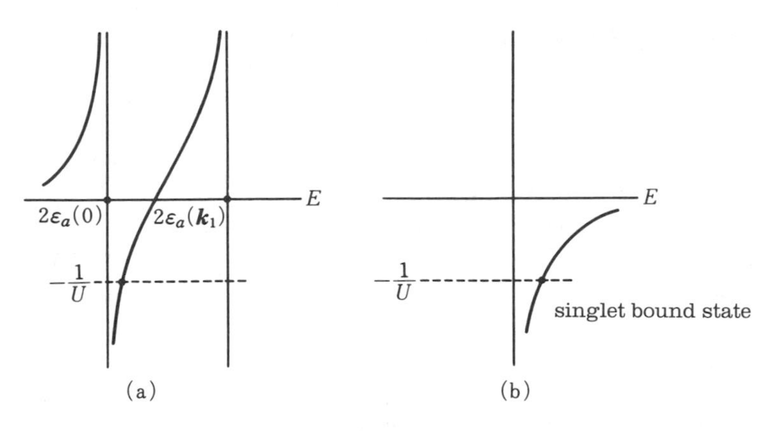

Fig. 13.2. Graphical solution of (13.39)

band, a solution emerges outside the continuum of the energy band, because the left-hand side of (13.39) increases with E from $-\infty$ and approaches 0 for $E \to \infty$ [Fig. 13.2(b)]. This solution is nothing but the bound state of the spin singlet, which was discussed previously for the one-dimensional case.

Now let us consider two electrons with general wavevectors $\boldsymbol{k}_1$ and $\boldsymbol{k}_2$. If the two electrons are in the spin-triplet state, the energy $\varepsilon(\boldsymbol{k}_1) + \varepsilon(\boldsymbol{k}_2)$ is not affected by the interaction g_{12}. On the other hand, the energy of the spin-singlet state shifts by an amount of order of $1/N$, except for the bound state. Let the energy shift be $\Delta E(\boldsymbol{k}_1, \boldsymbol{k}_2)$. Treating separately the two terms with $\boldsymbol{K}/2 \pm \boldsymbol{k} = \boldsymbol{k}_1$ and $\boldsymbol{K}/2 \mp \boldsymbol{k} = \boldsymbol{k}_2$ in the sum over $\boldsymbol{k}$ on the left-hand side of (13.38), we obtain

$$\Delta E(\boldsymbol{k}_1 + \boldsymbol{k}_2) = 2N^{-1} \frac{U}{1 + UG(\boldsymbol{k}_1 + \boldsymbol{k}_2)} , \tag{13.40}$$

$$G(\boldsymbol{k}_1 + \boldsymbol{k}_2) = N^{-1} \sum_{\boldsymbol{q}}{}'' \frac{1}{\varepsilon(\boldsymbol{k}_1 + \boldsymbol{q}) + \varepsilon(\boldsymbol{k}_2 - \boldsymbol{q}) - \varepsilon(\boldsymbol{k}_1) - \varepsilon(\boldsymbol{k}_2)} . \tag{13.41}$$

Here we have neglected ΔE in the denominator of $G(\boldsymbol{k}_1, \boldsymbol{k}_2)$ and omitted the terms with $\boldsymbol{q} = 0$ and $\boldsymbol{q} = \boldsymbol{k}_2 - \boldsymbol{k}_1$ in the summation. The relation (13.40) has been generalized to many-electron cases by Kanamori, who used it particularly to discuss the ferromagnetism of nickel. We describe *Kanamori's theory* [13.5] in the next section.

13.3 Kanamori Theory on Electron Correlation

In first-order perturbation theory in U, that is in the Hartree-Fock approximation, the energy increase of an antiparallel spin pair (measured relative to the energy of a parallel spin pair) is $N^{-1}U$. This quantity is, however, reduced by a factor $1/(1+UG)$ as in (13.40) which includes multiple scattering effects. Thus the effect of correlation (the second-order and higher-order terms) reduces the intra-atomic Coulomb integrals. Roughly speaking, G is given by the inverse of the band-width W, so that U_{eff} is to be limited to the band-width when U becomes large. Now let us consider conduction electrons in the paramagnetic state; the state density at the Fermi surface is denoted by $\rho(\varepsilon_{\mathrm{F}})$. When $\delta n/2$ electrons with $-$ spin near the Fermi surface change orientation to $+$ spin states, giving a polarization $\mu_{\mathrm{B}}\delta n$, the increase in the band energy is given by

$$\Delta E = \frac{1}{4\rho(\varepsilon_{\mathrm{F}})}\delta n^2 \; ; \tag{13.42}$$

the decrease in the exchange interaction energy is

$$\Delta E = -\frac{N^{-1}U_{\mathrm{eff}}}{4}\delta n^2 \; , \tag{13.43}$$

the effective exchange integral being U_{eff}. Thus the instability condition for the paramagnetic state becomes

$$N^{-1}\rho(\varepsilon_{\mathrm{F}})U_{\mathrm{eff}} = \alpha_{\mathrm{eff}} > 1 \; . \tag{13.44}$$

The condition $\alpha_{\mathrm{eff}} = 1$ is that for the divergence of the paramagnetic susceptibility. If one sets here $U_{\mathrm{eff}} = U$, then (13.44) reduces to the condition for appearance of ferromagnetism in the Hartree-Fock approximation to the Hubbard model. For energy bands of usual and not special form, one anticipates that the density of states is approximately given by the inverse of the band-width. Then the condition (13.44) does not hold however large the intra-atomic Coulomb repulsion is, since U_{eff} cannot exceed the band-width W, as seen from (13.40). Therefore the ferromagnetism is not expected for energy bands of simple form. The situation is not greatly modified even on taking the degeneracy of bands into account, as shown below.

Among the intra-atomic Coulomb integrals, $\langle ij|g|kl\rangle$, we neglect those comprising more than two different orbitals, and keep the following four kinds of integrals:

$$\langle ii|g|ii\rangle = U \qquad \langle ij|g|ij\rangle = U' \; , \tag{13.45}$$

$$\langle ij|g|ji\rangle = J \qquad \langle ii|g|jj\rangle = K \; ; \tag{13.46}$$

each is assumed independent of i and j. By extending the method used to derive (13.40) from (13.37), we obtain the energy shifts for each spin state of the two electrons in bands i and j; for the triplet state we have

$$\Delta E_{ij}^{(t)}(\boldsymbol{k}_1, \boldsymbol{k}_2) = N^{-1}\frac{U' - J}{1 + (U' - J)G'} \cong \frac{N^{-1}U'}{1 + U'G'} - \frac{N^{-1}J}{(1 + U'G')^2} \; , \tag{13.47}$$

and for the singlet state

$$\Delta E_{ij}^{(s)}(\boldsymbol{k}_1,\boldsymbol{k}_2) = N^{-1}\frac{U'+J}{1+(U'+J)G'} \cong \frac{N^{-1}U'}{1+U'G'} + \frac{N^{-1}J}{(1+U'G')^2}\,, \tag{13.48}$$

with

$$G' = N^{-1}{\sum_{\boldsymbol{q}}}' \frac{1}{\varepsilon_i(\boldsymbol{k}_1+\boldsymbol{q})+\varepsilon_j(\boldsymbol{k}_2-\boldsymbol{q})-\varepsilon_i(\boldsymbol{k}_1)-\varepsilon_j(\boldsymbol{k}_2)}\,; \tag{13.49}$$

the term $\boldsymbol{q}=0$ is omitted from the summation over $\boldsymbol{q}$.

As seen from the above, if J is smaller than U', then the energy difference between the singlet and triplet states becomes a small quantity; that is, J is further reduced by the factor $(1+U'G')^{-2}$. Incidentally, the two-electron energy for the spin-singlet state in a same band i is also modified as follows:

$$\Delta E_{ii}^{(s)}(\boldsymbol{k}_1,\boldsymbol{k}_2) = \frac{2N^{-1}U^*}{1+U^*G}\,, \tag{13.50}$$

$$U^* = U - \frac{pK^2G''}{1+[U+(p-1)K]G''}\,, \tag{13.51}$$

$$G'' = N^{-1}\sum_{\boldsymbol{q}} \frac{1}{\varepsilon_j(\boldsymbol{k}_1+\boldsymbol{q})+\varepsilon_j(\boldsymbol{k}_2-\boldsymbol{q})-\varepsilon_i(\boldsymbol{k}_1)-\varepsilon_i(\boldsymbol{k}_2)-\Delta E}\,; \tag{13.52}$$

where $p+1$ is the degree of orbital degeneracy. Thus in case of degenerate orbitals, one finds that the interaction between two electrons in different bands is not strongly dependent on the spin state.

So far the argument has been based on the results obtained for two-electron systems. For many-electron systems, the similar argument is expected to be valid qualitatively, if the electron or hole density is low. For many-electron systems, the summation over $\boldsymbol{q}$ in the expression for the function $G(\boldsymbol{k}_1,\boldsymbol{k}_2)$ must be limited to states outside the Fermi surface which are unoccupied by other electrons. Furthermore, we note that two electrons occupying $\boldsymbol{k}_1$ and $\boldsymbol{k}_2$ cannot form a singlet state because of the presence of other electrons; so that one must replace the singlet pair energy by the averaged energy of spin-singlet and spin-triplet states with antiparalel spins $+$ and $-$. The one-electron energy $\varepsilon(\boldsymbol{k})$ is also modified by such interactions; the energy $\varepsilon(\boldsymbol{k})$ in $G(\boldsymbol{k}_1,\boldsymbol{k}_2)$ should be regarded as the band energy modified in this way.

The theory, in which the consideration for two-electron systems is modified in this manner, corresponds to *Brueckner theory* developed for nuclear matter and ^{3}He [13.6, 7]. *Galitskii* has shown that this theory gives a good approximation for low-density fermion systems interacting by short-range forces [13.8].

Kanamori applied the above theory to Ni metal and showed, indeed, that it can become ferromagnetic. The 3d band of Ni has 9.4 electrons per atom, so 0.6 hole is present in 10 electronic states. Therefore, in this case the electron system can be regarded as being of low density. As previously discussed in general, the instability condition for paramagnetism, (13.44), is a rather qualitative one since the actual U_{eff} depends on $\boldsymbol{k}_1$ and $\boldsymbol{k}_2$. To be more quantitative, one needs to calculate the energy change from paramagnetic state, which is caused by the imbalance $n_+ - n_- = \delta n$ of + and − spin electron numbers. Kanamori examined the condition for the instability of paramagnetism of hole bands by taking three degenerate $d\varepsilon$ orbitals as important ones near the top of the $3d$ band of Ni. Following the previous argument, he neglected the interband effect and considered 0.2 hole per atom, which occupy one of the $d\varepsilon$ bands. In this case one can use in (13.44) an approximate value for U_{eff} which is estimated for $\boldsymbol{k}_1$ and $\boldsymbol{k}_2$ at the top of the band. The condition for ferromagnetism turns out to be rather delicate; ferromagnetism is expected only for a band with large band-width and also with large density of states $\rho(\varepsilon_{\text{F}})$ at the Fermi surface, for which U is not strongly reduced. Typical band shapes are illustrated in Fig. 13.3. In nickel these conditions are well satisfied.

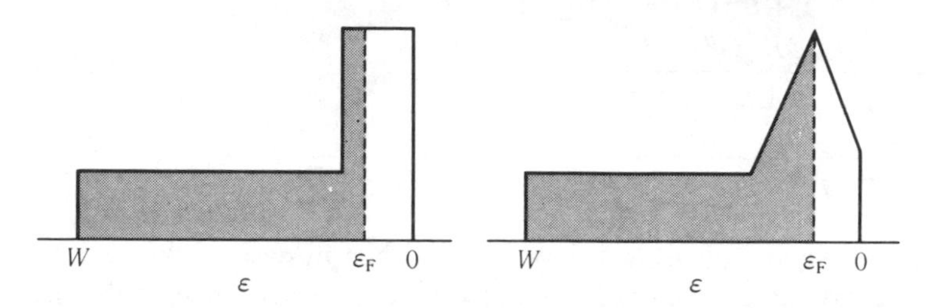

Fig. 13.3. Two examples of the electron bands satisfying the ferromagnetic condition. The ordinate shows the density of states

We can summarize the conclusion of Kanamori's theory as follows: For interacting electron systems with short-range forces (here intra-atomic interactions), the exchange interaction between conduction electrons within the Hartree approximation is given by the Coulomb integral of atomic orbitals or Wannier functions within an atom, if two electrons belong to the same band. When they belong to different bands, the exchange interaction is given by the exchange integral. If we take into account the effect of electron correlation by multiple-scattering theory based on the two-electron approximation, then, roughly speaking, the intra-atomic Coulomb integral is reduced by the factor $1/(1 + U/W)$ and the intra-atomic exchange integral by the factor $1/(1 + U/W)^2$. So exchange interactions between two electrons in different bands cannot make an important contribution to ferromagnetism, compared with that of two electrons in the same band. On the other hand, the exchange interaction of two electrons in the same band cannot become larger than the band-width W even for large U. Therefore ferromagnetism is not expected for a simple band structure, in this mechanism. However, Ni has a band structure with both large band-width ($\sim$ 4eV) and large density of states near the Fermi surface ($\sim$1/eV). Such a band structure makes the ferromagnetism of Ni possible.

Other theories of electron correlation for the Slater-Hubbard Hamiltonian with a single band include *Hubbard's theory* using the two-time Green's function method [13.9] and *Gutzwiller's theory* using a variational method [13.10]. These theories lead to qualitatively the same conclusion for the appearance of ferromagnetism as Kanamori's theory.

13.4 Gutzwiller Variational Method

Gutzwiller applied a variational method [13.10] to study the electron correlation at $T = 0$ for the systems described by the following Hamiltonian which was used above and is usually called Hubbard Hamiltonian,

$$\mathcal{H} = \sum_{ij} t_{ij} c_{i\sigma}^{\dagger} c_{j\sigma} + \frac{1}{2} U \sum_{i\sigma} n_{i\sigma} n_{i-\sigma} , \tag{13.53}$$

where i and j denote the lattice sites and σ specifies the spin state. Let L be the total number of lattice sites, $N_{\uparrow}$ and $N_{\downarrow}$ be the up- and down-spin electron numbers, respectively, and D be the number of lattice sites doubly occupied by electrons with up and down spins. We define the following ratios: $n_{\uparrow} = N_{\uparrow}/L$, $n_{\downarrow} = N_{\downarrow}/L$ and $d = D/L$. The ground state of the system without electron correlation ($U = 0$) is the Fermi sphere where up- and down-spin electrons occupy the levels up to the highest energies, $\varepsilon_{\mathrm{F}\uparrow}$ and $\varepsilon_{\mathrm{F}\downarrow}$, respectively. We represent the ground state wave function of this electron system with $U = 0$ by ψ_0. In this case, the mean value D_0 of the number of lattice sites, which are doubly occupied by electrons, is $D_0 = n_{\uparrow} n_{\downarrow} L$. The number D decreases from D_0, as the repulsive interaction U on the same site sets in, because up- and down-spin electrons tend to keep away from each other.

Gutzwiller adopted the following variational wave function for systems with electron correlation:

$$\psi = \prod_i [1 - (1 - g) n_{i\uparrow} n_{i\downarrow}] \psi_0 . \tag{13.54}$$

Then the ground state energy E_{g} is given by

$$E_{\mathrm{g}} = \frac{\langle \psi | \mathcal{H} | \psi \rangle}{\langle \psi | \psi \rangle} = \left[\left\langle \psi \middle| \sum_{ij\sigma} t_{ij} c_{i\sigma}^{\dagger} c_{j\sigma} \middle| \psi \right\rangle + \left\langle \psi \middle| U \sum_i n_{i\uparrow} n_{i\downarrow} \middle| \psi \right\rangle \right] \Big/ \langle \psi | \psi \rangle . \tag{13.55}$$

Now the problem is how we can evaluate (13.55). To this end Gutzwiller introduced several assumptions and evaluated the energy expectation value E_{g} *in a statistical way. Ogawa* et al. [13.11] derived the Gutzwiller's result by a simpler method. The state ψ_0 is a Slater determinant consisting of Bloch functions with wavevector $\boldsymbol{k}$; if we use the lattice representation, it is a linear combination of states where up- and down-spin electrons are distributed in various ways. We classify those states with various distributions according to the number D of doubly occupied lattice points. For given L, $N_\uparrow$ and $N_\downarrow$, the number N_D of electron configurations with a fixed value of D is given by

$$N_D(L, N_\uparrow, N_\downarrow) = \frac{L!}{(N_\uparrow - D)!(N_\downarrow - D)!D!(L - N_\uparrow - N_\downarrow + D)!} , \tag{13.56}$$

if the number of combinations in probability theory is used. Moreover, we assume the probability of a configuration for N_σ electrons with spin σ to be

$$P(L, N_\sigma) = n_\sigma^{N_\sigma}(1 - n_\sigma)^{L-N_\sigma} , \tag{13.57}$$

where the spatial correlation has been neglected.

Then the normalization integral of ψ is evaluated as

$$\langle\psi|\psi\rangle = \sum_D g^{2D} N_D(L, N_\uparrow, N_\downarrow) P(L, N_\uparrow) P(L, N_\downarrow) . \tag{13.58}$$

Here the summation over the doubly occupied lattice number D can be replaced by the largest term in the thermodynamic limit ($L \to \infty$). This condition leads to the following relation between g and D:

$$g^2 = \frac{d(1 - n_\uparrow - n_\downarrow + d)}{(n_\uparrow - d)(n_\downarrow - d)} . \tag{13.59}$$

According to the above argument, one can think that ψ has a definite value of D. Then the second term in (13.55), corresponding to the Coulomb repulsion, takes evidently the expectation value Ud.

In the case $U = 0$, the kinetic energy per lattice site of spin-σ electrons is given by

$$\frac{1}{L} \sum_{k<k_{F\sigma}} \varepsilon(\boldsymbol{k}) = \bar{\varepsilon}_\sigma < 0 , \tag{13.60}$$

where $\boldsymbol{k}_F$ is the Fermi wavevector. In the presence of U, we suppose that the distribution of electrons (of spin σ) in k-space decreases uniformly to $1 - \alpha_\sigma$ for $k < k_F$, and increases to $1 - \alpha_\sigma - q_\sigma$ for $k > k_F$, as illustrated in Fig. 13.4. Here q_σ represents the discontinuity of the occupation probability at k_F.

Then the kinetic energy is given by

$$E_{\text{kin}} = \sum_\sigma \left[(1 - \alpha_\sigma) \sum_{k<k_{F\sigma}} \varepsilon(\boldsymbol{k}) + (1 - \alpha_\sigma - q_\sigma) \sum_{k>k_{F\sigma}} \varepsilon(\boldsymbol{k})\right] . \tag{13.61}$$

If we adjust the origin of the electron energy by

$$\sum_k \varepsilon(\boldsymbol{k}) = \sum_{k<k_F} \varepsilon(\boldsymbol{k}) + \sum_{k>k_F} \varepsilon(\boldsymbol{k}) = 0 ,$$

then the kinetic energy (13.61) per site is given by

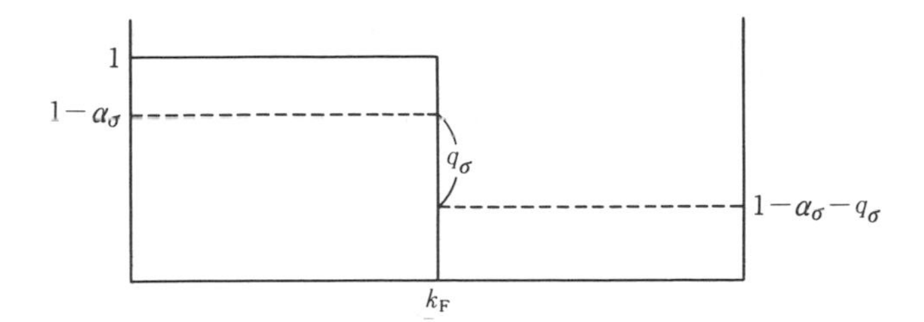

Fig. 13.4. Occupation probability of σ-electron in k-space

$$\frac{1}{L}E_{\mathrm{kin}} = q_\uparrow \bar{\varepsilon}_\uparrow + q_\downarrow \bar{\varepsilon}_\downarrow \ . \tag{13.62}$$

We can also evaluate the expectation value of the kinetic energy, the first term on the right-hand side of (13.55), by means of probability theory as we did in obtaining $\langle\psi|\psi\rangle$. Comparing the result with (13.62), we can determine the jump in the electron distribution function at the Fermi surface:

$$q_\sigma = \frac{\{[(n_\sigma - d)(1 - n_\sigma - n_{-\sigma} + d)]^{1/2} + [(n_{-\sigma} - d)d]^{1/2}\}^2}{n_\sigma(1 - n_\sigma)} \ . \tag{13.63}$$

Regarding the details of the derivation of this formula, readers are recommended to refer to the original papers by *Ogawa* et al. [13.11] and by *Vollhardt* [13.12]. Finally the desired ground state energy is expressed as

$$\frac{E_{\mathrm{g}}}{L} = q_\uparrow(d, n_\uparrow, n_\downarrow)\bar{\varepsilon}_\uparrow + q_\downarrow(d, n_\uparrow, n_\downarrow)\bar{\varepsilon}_\downarrow + Ud \ . \tag{13.64}$$

Using the above result, Gutzwiller studied the condition for the appearence of ferromagnetism. Qualitatively speaking, the condition for ferromagnetism is severe as in Kanamori's theory; the origin of the ferromagnetism is considered to come from the spectrum of $\varepsilon(\boldsymbol{k})$, that is, the special characteristics of the band structure.

In one-dimensional case *Metzner* and *Vollhardt* [13.13] have obtained an exact expression for Gutzwiller's variational wave function and several correlation functions have been analytically evaluated [13.14].

An important result of Gutzwiller's variational theory is the occurrence of the metal-insulator transition in the case that the electron number is one per lattice site, as remarked by *Brinkman* and *Rice* [13.15]. For

$$n = \frac{N}{L} = 1 \ , \quad n_\uparrow = n_\downarrow = \frac{1}{2} \ ,$$

q in (13.63) becomes

$$q = q_\uparrow = q_\downarrow = 8d(1 - 2d) \ . \tag{13.65}$$

On substituting this value into (13.64), we find the value of d which minimizes the energy to be

$$d = \frac{1}{4}\Big(1 - \frac{U}{U_c}\Big) , \quad U_c = 8|\bar{\varepsilon}_0| , \tag{13.66}$$

where $\bar{\varepsilon}_0$ is defined by $\bar{\varepsilon}_\uparrow + \bar{\varepsilon}_\downarrow$. Using this value of d, we find for q and E_g the results

$$q = 1 - \Big(\frac{U}{U_c}\Big)^2 , \tag{13.67}$$

$$\frac{E_g}{L} = -|\bar{\varepsilon}_0|\Big(1 - \frac{U}{U_c}\Big)^2 . \tag{13.68}$$

Equation (13.66) implies that, when U increases from zero, d decreases linearly with U from $d = 1/4$, and vanishes at $U = U_c$. Correspondingly, the discontinuity q of the momentum distribution at the Fermi surface decreases from one, and vanishes at $U = U_c$. Since $q = 0$ means that the Fermi surface does not exist, we conclude that the metal changes to an insulator at $U = U_c$. In the insulator region, Anderson's superexchange interaction will take place between the localized spins occupying each lattice site, so that the electron system becomes antiferromagnetic.

By the way, the exactly treated Gutzwiller's variation does not give rise to the metal-insulator transition in one-dimensional case, whereas the Gutzwiller's approximation becomes exact in the infinite dimension [13.13, 14].

The discontinuity of the momentum distribution at the Fermi surface is associated with the effective mass m^* of electron by

$$q = \Big(1 - \frac{\partial \Sigma}{\partial \omega}\Big|_{\omega=\varepsilon_F}\Big)^{-1} = \Big(\frac{m^*}{m}\Big)^{-1} , \tag{13.69}$$

where $\Sigma(\omega)$ is the electron self-energy and m is the free-electron mass. Thus, as seen from the following expression:

$$\frac{m^*}{m} = q^{-1} = \Big[1 - \Big(\frac{U}{U_c}\Big)^2\Big]^{-1} , \tag{13.70}$$

the effective mass m^* increases with U and diverges to infinity at $U = U_c$.

Let us now find the spin susceptibility. For a given imbalance of up and down spins, we set

$$\begin{aligned} n_\uparrow - n_\downarrow &= m , & n_\uparrow &= \frac{1}{2}(1+m) , \\ n_\uparrow + n_\downarrow &= n = 1 , & n_\downarrow &= \frac{1}{2}(1-m) , \end{aligned} \tag{13.71}$$

where we used the same notation m as the free-electron mass. Then by finding q_σ to order m^2, we obtain

$$\begin{aligned} q_\sigma &= q_{-\sigma} = qf , \\ f &= 1 + \Big[1 - \frac{1}{4}\frac{1}{(1-2d)^2}\Big]m^2 , \end{aligned} \tag{13.72}$$

where q is given by (13.65). The ground-state energy is calculated from (13.64) as

$$\frac{E_{\mathrm{g}}}{L} = qf\bar{\varepsilon} + Ud \ , \tag{13.73}$$

where $\bar{\varepsilon} = \bar{\varepsilon}_{\uparrow} + \bar{\varepsilon}_{\downarrow}$. Let us denote the values of q, f, $\bar{\varepsilon}$ and d for $m = 0$ by q_0, f_0, $\bar{\varepsilon}_0$ and d_0, respectively, and their changes due to m by δq, δf, $\delta\bar{\varepsilon}$ and δd. Then we have

$$\begin{aligned} q_0 &= 8d_0(1-2d_0) \ , \quad \delta q = 8(1-4d_0)\delta d - 16(\delta d)^2 \ , \\ f_0 &= 1 \ , \quad \delta f = \left[1 - \frac{1}{4}\frac{1}{(1-2d_0)^2}\right]m^2 \ , \\ \delta\bar{\varepsilon} &= \frac{1}{4N(0)}m^2 \ , \end{aligned} \tag{13.74}$$

where $N(0)$ is the density of states at the Fermi surface per lattice site and per spin. Thus the change in E_g/L is

$$\frac{\delta E_{\mathrm{g}}}{L} = q_0\delta\bar{\varepsilon} + \bar{\varepsilon}_0\delta q + q_0\bar{\varepsilon}_0\delta f + U\delta d = q_0\delta\bar{\varepsilon} + q_0\bar{\varepsilon}_0\delta f - 16\bar{\varepsilon}_0(\delta d)^2 \, . \tag{13.75}$$

Here we have used the relation (13.66) between d_0 and U. The last term of the right-hand side of (13.75) should be omitted since δd is proportional to m^2. Combining these results, we find that the change in the energy is

$$\frac{1}{L}\delta E_{\mathrm{g}} = \frac{q_0}{4N(0)}\left\{1 - p\left[1 - \frac{1}{4}\frac{1}{(1-2d_0)^2}\right]\right\}m^2 \ , \tag{13.76}$$

where p is defined by

$$p = 4|\bar{\varepsilon}_0|N(0) = \frac{1}{2}U_{\mathrm{c}}N(0) \ . \tag{13.77}$$

In terms of the susceptibility χ_{s}, $\delta E_{\mathrm{g}}/L$ can be equated to $(1/2)(\mu_{\mathrm{B}}^2/\chi_{\mathrm{s}})m^2$ so that χ_{s} is expressed as

$$\chi_{\mathrm{s}} = 2\mu_{\mathrm{B}}^2 N(0)\frac{m^*}{m}\left\{1 - p\left[1 - \frac{1}{(1+\frac{U}{U_{\mathrm{c}}})^2}\right]\right\}^{-1} \ ; \tag{13.78}$$

here we have used (13.66, 70). The susceptibility of (13.78) increases as U increases and diverges at $U = U_{\mathrm{c}}$ where the effective mass m^* diverges. On the other hand, the second factor remains finite (~ 4), since p is almost equal to 1. This result is in marked contrast to that of the Random Phase Approximation (RPA), which is described in Sect. 14.2; in the RPA, the susceptibility diverges not from the effective mass but rather from the second factor.

Anderson and *Brinkmann* [13.16] have pointed out that the Brinkman-Rice theory explains well the experimental results on the normal state of liquid ^{3}He .

14. RPA Theory of Ferromagnetism in Metals

In this chapter we describe the mean-field theory for itinerant ferromagnets, namely, the Stoner theory of ferromagnetism. Then, we discuss the elementary excitations, particularly the spin wave excitations with the use of the Random Phase Approximation. Further, we derive the dynamical susceptibility and the fluctuation-dissipation theorem for itinerant electron systems.

14.1 Stoner Theory

In the previous chapter, we discussed correlation problems in band-electron systems with short-range interaction. Here we simplify the problem by treating a conduction-electron system in one band with short-range interaction, i.e., an electron system described by the Hubbard Hamiltonian. We consider the ferromagnetism of the system again in the Hartree-Fock approximation. In the formalism of second quantization, the Hamiltonian is

$$\mathcal{H} = \sum_{k\sigma} \varepsilon_k a_{k\sigma}^\dagger a_{k\sigma} + \frac{1}{2}\frac{U}{N} \sum_{k_1 k_2 q \sigma\sigma'} a_{k_1+q\sigma}^\dagger a_{k_2-q\sigma'}^\dagger a_{k_2\sigma'} a_{k_1\sigma} \,, \tag{14.1}$$

where U is the intra-atomic Coulomb integral. The interaction term is regarded as a modified version of the Coulomb interaction of free electrons, $(1/V)4\pi e^2/q^2$ in (12.2): first, the original singularity at $q = 0$ has been removed because the screening by other electrons changes the interaction to that of the screened Coulomb type, $(1/V)4\pi e^2/(q^2 + c)$; then the q-dependence has been omitted. Even for such a simplified Hamiltonian, an accurate calculation of the electron correlation is very difficult. Examples of theories of electron correlation are the Kanamori's and Gutzwiller's theories described in the preceding section.

Here we do not go into the correlation problems, but rather treat the Hamiltonian (14.1) in both the Hartee-Fock approximation and the Random Phase Approximation (RPA), regarding U as a given parameter.

Since the $\boldsymbol{q} = 0$ part of the interaction, i.e.,the diagonal Coulomb term, should have been omitted from the beginning in (14.1), only the exchange term is to be considered in the diagonal term. With the replacement

$$\frac{1}{V}\frac{4\pi e^2}{|\boldsymbol{k}_1-\boldsymbol{k}_2|^2} \to \frac{U}{N}$$

in (12.4), the exchange term is given by

$$-\frac{1}{2}\frac{U}{N}\sum_{\boldsymbol{k}_1\boldsymbol{k}_2}(n_{\boldsymbol{k}_1\uparrow}n_{\boldsymbol{k}_2\uparrow}+n_{\boldsymbol{k}_1\downarrow}n_{\boldsymbol{k}_2\downarrow}) = -\frac{1}{2}\frac{U}{N}\left[\frac{1}{2}(N_+-N_-)^2+\frac{1}{2}N_\mathrm{e}^2\right], \quad (14.2)$$

where $N_\pm$ are the numbers of electrons with spins $\pm$. Combining (14.2) with the one-body band energy in (14.1), we represent the Hartree-Fock Hamiltonian as

$$\begin{aligned}\mathcal{H}_\mathrm{HF} &= \sum_{\boldsymbol{k}\sigma}\left[\varepsilon_{\boldsymbol{k}}-\frac{\sigma k\theta(N_+-N_-)}{N_\mathrm{e}}\right]a^\dagger_{\boldsymbol{k}\sigma}a_{\boldsymbol{k}\sigma}+\frac{1}{4}\frac{U}{N}(N_+-N_-)^2\,, \\ k\theta &= \frac{1}{2}\frac{N_\mathrm{e}}{N}U\,,\end{aligned} \quad (14.3)$$

where σ is + or − according to the direction of the spin, up or down. Equation (14.3) has the form of a one-electron Hamiltonian in a molecular field, γM ($\gamma \equiv k\theta/\mu_\mathrm{B}^2N_\mathrm{e}$), where M is the total magnetic moment. The second term in (14.3) is a correction which avoids double counting of the interaction energy. The theory of ferromagnetism based on the Hamiltonian of the simplified form (14.3) is called the *Stoner model.*

Within the model, we can rather easily calculate the temperature variation of the spontaneous magnetization, the accompanying magnetic specific heat, the Curie temperature, the susceptibility above the Curie temperature, and so on. We note here that the molecular field itself must be determined self-consistently.

The condition for ferromagnetism in the Stoner model is simpler than in the Bloch theory because the exchange interaction is independent of the wavevectors, $\boldsymbol{k}_1$ and $\boldsymbol{k}_2$. If we take the kinetic energy of free electrons, $\hbar^2k^2/2m$, as the band energy $\varepsilon_{\boldsymbol{k}}$, then the total energy is expressed by the following function of the electron numbers $N_\pm$ of up and down spins:

$$\begin{aligned}\frac{E_\mathrm{T}}{N_\mathrm{e}} &= \frac{e^4m}{2\hbar^2}\frac{3}{5}\left(\frac{9\pi}{2}\right)^{2/3}\left(\frac{a_0}{r_0}\right)^2\left[\left(\frac{N_+}{N_\mathrm{e}}\right)^{5/3}+\left(\frac{N_-}{N_\mathrm{e}}\right)^{5/3}\right] \\ &\quad -\frac{1}{2}\left(\frac{N_\mathrm{e}}{N}\right)U\left[\left(\frac{N_+}{N_\mathrm{e}}\right)^2+\left(\frac{N_-}{N_\mathrm{e}}\right)^2\right],\end{aligned} \quad (14.4)$$

where, as before, a_0 denotes the Bohr radius and r_0 is the radius of the sphere whose volume is equal to the volume per electron. Defining variables ζ and α through

$$\begin{aligned}\frac{N_+}{N_\mathrm{e}} &= \frac{1}{2}(1+\zeta)\,, & \frac{N_-}{N_\mathrm{e}} &= \frac{1}{2}(1-\zeta)\,, \\ \alpha &= \frac{1}{N}\rho(\varepsilon_\mathrm{F})U\,, & \frac{k\theta}{\varepsilon_\mathrm{F}} &= \frac{2}{3}\alpha\,,\end{aligned}$$

and using the Fermi energy in the paramagnetic state,

$$\varepsilon_{\mathrm{F}} = \frac{e^4 m}{2\hbar^2}\left(\frac{9\pi}{2}\right)^{2/3}\left(\frac{a_0}{r_0}\right)^2 ,$$

we obtain the relation

$$\frac{k\theta}{\varepsilon_{\mathrm{F}}}\zeta = \frac{1}{2}\left[(1+\zeta)^{2/3} - (1-\zeta)^{2/3}\right] , \tag{14.5}$$

as the condition that (14.4) be stationary with respect to ζ. The solution of (14.5) is obtained by plotting both sides as functions of ζ, as in Fig. 14.1; the solution corresponds to the point where the straight line representing the left-hand side crosses the curve representing the right-hand side. Thus we are led to the following results:

$$\text{paramagnetic state for } \frac{k\theta}{\varepsilon_{\mathrm{F}}} < \frac{2}{3} , \tag{14.6a}$$

$$\text{imperfect ferromagnetic state for } 2^{-1/3} > \frac{k\theta}{\varepsilon_{\mathrm{F}}} > \frac{2}{3} , \tag{14.6b}$$

$$\text{perfect ferromagnetic state for } \frac{k\theta}{\varepsilon_{\mathrm{F}}} > 2^{-1/3} . \tag{14.6c}$$

In the case (14.6c), all electrons have up (or down) spins; then a finite energy is required to reverse the spin of one electron so that the reduction of the spontaneous magnetization, $M_{T=0} - M$, varies exponentially at low temperatures. On the other hand, in the case (14.6b), any finite energy is not required to transfer an electron from the up-spin state to the down one and so the reduction of the spontaneous magnetizaton is given by a power function (T^2) of the temperature.

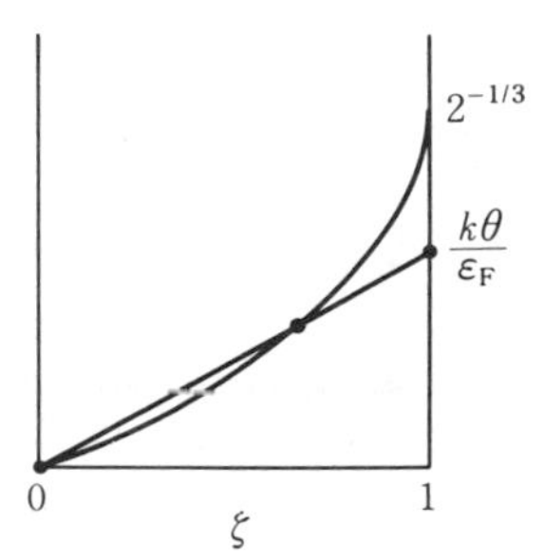

Fig. 14.1. ζ-dependence of both sides of (14.5)

14.2 Stoner Excitations

Now let us examine the energies for elementary excitations from the feromagnetic ground state satisfying the conditions (14.6b) or (14.6c). We reduce the total electron spin by unity. Suppose one makes an excited state by operating on the ground state ψ_g a suitable linear combination of the operators which transfer an electron from state $\boldsymbol{k}\uparrow$ to $\boldsymbol{k}+\boldsymbol{q}\downarrow$:

$$\sigma_{\boldsymbol{qk}-} = a^{\dagger}_{\boldsymbol{k}+\boldsymbol{q}\downarrow} a_{\boldsymbol{k}\uparrow} \; ; \tag{14.7}$$

this excited state must satisfy the Schrödinger equation:

$$\mathcal{H}\sum_{\boldsymbol{k}} f_{\boldsymbol{k}}\sigma_{\boldsymbol{qk}-}\psi_g = (E_g + \hbar\omega_{\boldsymbol{q}})\sum_{\boldsymbol{k}} f_{\boldsymbol{k}}\sigma_{\boldsymbol{qk}-}\psi_g \; , \tag{14.8}$$

where E_g is the ground-state energy, $\hbar\omega_{\boldsymbol{q}}$ is the excited-state energy measured from the ground state and the coefficients $f_{\boldsymbol{k}}$ specify the linear combination. By using the relation

$$\mathcal{H}\sigma_{\boldsymbol{qk}-} = [\mathcal{H}, \sigma_{\boldsymbol{qk}-}] + \sigma_{\boldsymbol{qk}-}\mathcal{H}$$

on the left-hand side of (14.8), we get

$$\sum_{\boldsymbol{k}} f_{\boldsymbol{k}}\{[\mathcal{H}, \sigma_{\boldsymbol{qk}-}] - \hbar\omega_{\boldsymbol{q}}\sigma_{\boldsymbol{qk}-}\}\psi_g = 0 \; . \tag{14.9}$$

The commutator between the Hamiltonian (14.1) and $\sigma_{\boldsymbol{qk}-}$ is obtained as follows by calculating the kinetic-energy part and the interaction part separately:

$$[\mathcal{H}_{\mathrm{k}}, \sigma_{\boldsymbol{qk}-}] = (\varepsilon_{\boldsymbol{k}+\boldsymbol{q}} - \varepsilon_{\boldsymbol{k}} + 2\mu_{\mathrm{B}}H)\sigma_{\boldsymbol{qk}-} \; , \tag{14.10}$$

$$[\mathcal{H}_{\mathrm{Coulomb}}, \sigma_{\boldsymbol{qk}-}] = \frac{U}{N}\sum_{\boldsymbol{k}_1\boldsymbol{q}'\sigma}\Big(a^{\dagger}_{\boldsymbol{k}+\boldsymbol{q}+\boldsymbol{q}'\downarrow}a^{\dagger}_{\boldsymbol{k}_1-\boldsymbol{q}'\sigma}a_{\boldsymbol{k}_1\sigma}a_{\boldsymbol{k}\uparrow} - a^{\dagger}_{\boldsymbol{k}+\boldsymbol{q}\downarrow}a^{\dagger}_{\boldsymbol{k}_1-\boldsymbol{q}'\sigma}a_{\boldsymbol{k}_1\sigma}a_{\boldsymbol{k}-\boldsymbol{q}'\uparrow}\Big) \; . \tag{14.11}$$

In (14.10) a term due to an external magnetic field is added. Equation (14.11) contains products of four a and $a^{\dagger}$ operators. In order to reduce them to products of two operators we keep for $\sigma=\uparrow$ in the first term on the right-hand side, for example, only the diagonal element of $a^{\dagger}_{\boldsymbol{k}_1-\boldsymbol{q}'\uparrow}a_{\boldsymbol{k}\uparrow}$ with respect to ψ_g by taking $\boldsymbol{k}_1 = \boldsymbol{k}+\boldsymbol{q}'$. Nondiagonal elements with $\boldsymbol{k}_1 \neq \boldsymbol{k}+\boldsymbol{q}'$ are neglected. Such an approximation is called the *Random Phase Approximation* or *RPA* for short; it can be regarded as an extension of the Hartree-Fock approximation. By operating a similar procedure to terms with $\sigma=\downarrow$ we can finally approximate (14.11) as

$$[\mathcal{H}_{\mathrm{Coulomb}}, \sigma_{\boldsymbol{qk}-}]_{\mathrm{RPA}} = \frac{U}{N}\sum_{\boldsymbol{k}_1}(n_{\boldsymbol{k}_1\uparrow} - n_{\boldsymbol{k}_1+\boldsymbol{q}\downarrow})\sigma_{\boldsymbol{qk}-} - \frac{U}{N}(n_{\boldsymbol{k}\uparrow} - n_{\boldsymbol{k}+\boldsymbol{q}\downarrow})\sum_{\boldsymbol{k}_1}\sigma_{\boldsymbol{qk}_1-} \; . \tag{14.12}$$

Substituting (14.10, 12) into (14.9), we obtain

$$\sum_{k} f_k \Big[\big(\varepsilon_{k+q} - \varepsilon_k + 2\mu_{\rm B} H + \frac{N_{\rm e}}{N} U\zeta - \hbar\omega_q \big) \sigma_{qk-} - \frac{U}{N} (n_{k\uparrow} - n_{k+q\downarrow}) \sum_{k_1} \sigma_{qk_1-} \Big] \psi_{\rm g} = 0 \,. \tag{14.13}$$

On taking the inner product of this equation with $\sigma_{qk-}\psi_{\rm g}$, we are led to

$$(\varepsilon_{k+q} - \varepsilon_k + 2\mu_{\rm B} H + \frac{N_{\rm e}}{N} U\zeta - \hbar\omega_q) f_k = \frac{U}{N} \sum_{k} f_k (n_{k\uparrow} - n_{k+q\downarrow}) \,. \tag{14.14}$$

At this point we compare (14.13) with the initial equation (14.8). If the $k\uparrow$ state is unoccupied in the ground state $\psi_{\rm g}$, then (14.8) always vanishes for the annihilation operator $a_{k\uparrow}$. While in (14.13), $n_{k+q\downarrow}$ appears singly in the second term and (14.13) would not vanish even if $k\uparrow$ is unoccupied by electrons. The discrepancy arises from using the relation $\mathcal{H}\psi_{\rm g} = E_{\rm g}\psi_{\rm g}$ on going from (14.8) to (14.9); in fact, the relation holds only approximately because $\psi_{\rm g}$ is the ground-state for $\mathcal{H}_{\rm HF}$ with energy $E_{\rm g}$. So instead of using this relation, we directly calculate

$$\Big\langle \sigma_{qk-}\psi_{\rm g} \Big| \sum_{k'} f_{k'} \sigma_{qk'-} \mathcal{H} \Big| \psi_{\rm g} \Big\rangle \,;$$

then another term

$$-\frac{U}{N} n_{k\uparrow} (1 - n_{k+q\downarrow}) \sum_{k'} f_{k'} n_{k'+q\downarrow} (1 - n_{k'\uparrow})$$

appears in addition to $f_k n_{k\uparrow}(1 - n_{k+q\downarrow}) E_{\rm g}$. Addition of this term to (14.14) reduces the right-hand side to

$$\frac{U}{N} \sum_{k} f_k n_{k\uparrow} (1 - n_{k+q\downarrow}) \,. \tag{14.15}$$

From (14.14), as modified by (14.15), we obtain for f_k

$$f_k = \frac{U}{N} \frac{\sum f_k n_{k\uparrow}(1 - n_{k+q\downarrow})}{\varepsilon_{k+q} - \varepsilon_k + 2\mu_{\rm B} H + \frac{N_{\rm e}}{N} U\zeta - \hbar\omega_q} \,. \tag{14.16}$$

Multiplying both sides by $n_{k\uparrow}(1 - n_{k+q\downarrow})$ and summing over k, we are led to the eigenvalue equation determining the excitation energy $\hbar\omega_q$:

$$1 = \frac{U}{N} \sum_{k} \frac{n_{k\uparrow}(1 - n_{k+q\downarrow})}{\varepsilon_{k+q} - \varepsilon_k + 2\mu_{\rm B} H + \frac{N_{\rm e}}{N} U\zeta - \hbar\omega_q} \,. \tag{14.17}$$

The eigenvalues $\hbar\omega_q$ are obtained by setting the denominator to zero under the condition that the numerator does not vanish (i.e., the state $k\uparrow$ is occupied while the state $k+q\downarrow$ is empty):

$$\hbar\omega_{q,k} = \varepsilon_{k+q} - \varepsilon_k + 2\mu_{\rm B} H + \frac{N_{\rm e}}{N} U\zeta \,. \tag{14.18}$$

The energy thus obtained depends also on $\boldsymbol{k}$ and so it forms a continuous spectrum in the region specified by the above condition. These excitations are electron-hole excitations of one-body type and are called *Stoner excitations.*

The Stoner excitation spectrum starts at $\boldsymbol{q} = 0$ with the difference in the $\pm$ spin band shifts (equal to the exchange splitting Δ_{exch} for $H = 0$),

$$\Delta = 2\mu_{\text{B}}H + \frac{N_{\text{e}}}{N}U\zeta\ , \tag{14.19}$$

and with increasing q it forms an energy band whose width is caused by the magnitude and the direction of $\boldsymbol{k}$. The shaded area in Fig. 14.2 represents the region in k-space satisfying the condition $n_{\boldsymbol{k}\uparrow}(1-n_{\boldsymbol{k}+\boldsymbol{q}\downarrow}) = 1$ for given $\boldsymbol{q}$. The upper and lower bounds of the energy band formed by the Stoner excitations are given from (14.18) as

$$\frac{\hbar^2}{2m}(\pm 2k_{\text{F}\uparrow}q + q^2) + \Delta\ . \tag{14.20}$$

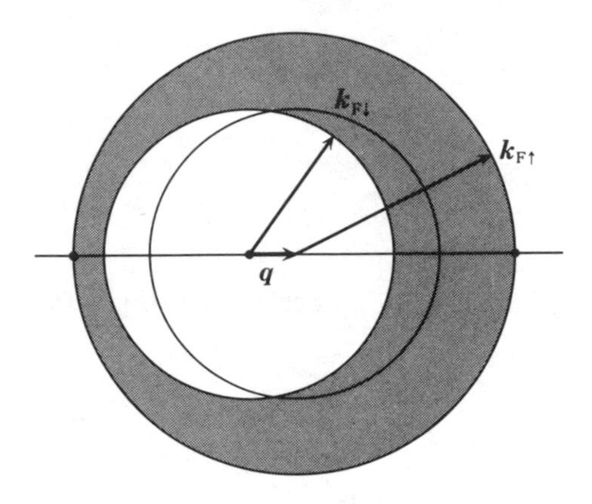

Fig. 14.2. Region where Stoner excitations are possible for given $\boldsymbol{q}$. $k_{\text{F}\uparrow,\downarrow}$ are the radii of the Fermi spheres for $\pm$ spin. The filled dots represent the upper and lower bounds of the excitation energies

It is evident from this relation that the upper bound increases with q while the lower bound first decreases and then, in the case of perfect ferromagnetism, increases after passing through a minimum. In the case of imperfect ferromagnetism, the lower bound coincides with the ground-state energy for a certain range of q; this range is determined by setting the expression (14.20) with $-$ sign to zero:

$$q^2 - 2k_{\text{F}\uparrow}q + \frac{2m}{\hbar^2}\Delta = 0\ .$$

In terms of the two roots

$$q_0 = k_{F\uparrow} \pm \sqrt{k_{F\uparrow}^2 - \frac{2m}{\hbar^2}\Delta} = k_{F\uparrow} \pm k_{F\downarrow}$$

of the equation, the range is given as

$$k_{F\uparrow} - k_{F\downarrow} \leq q \leq k_{F\uparrow} + k_{F\downarrow} \ . \tag{14.21}$$

The condition $k_{F\uparrow} \pm k_{F\downarrow} = q_0$ is the condition that the two Fermi spheres in Fig. 14.2 touch each other. When the two Fermi spheres intersect, the lower bound of the excitation energy is always zero; the wavevectors $\boldsymbol{k}$ of the lowest excitations are given by the points on the circumference of the intersection of two spheres. The contribution to the reduction of the spontaneous magnetization which is proportional to T^2 comes from the Stoner excitations around this region. When q exceeds $k_{F\uparrow} + k_{F\downarrow}$, the lower bound becomes positive as shown in Fig. 14.3.

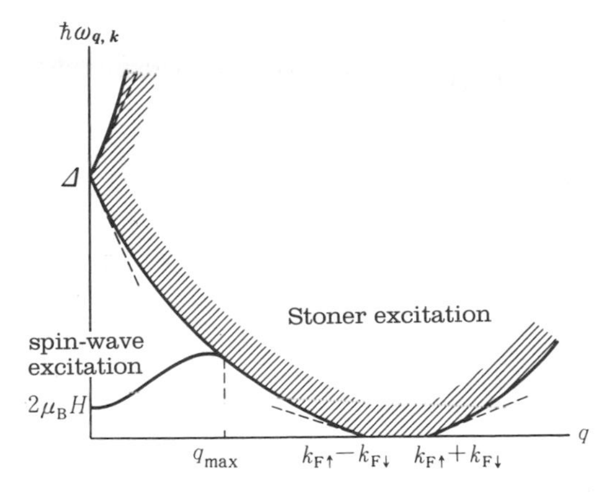

Fig. 14.3. Stoner excitations and spin-wave excitations for imperfect ferromagnetism

14.3 Spin-Wave Excitations

Equation (14.17) determining the excitation energies has solutions corresponding to collective spin-wave excitations in addition to the Stoner excitations above mentioned; the spin-wave excitations lie isolated below the continuous band of the Stoner excitations. The fact that spin-wave excitations can exist in the itinerant electron model as well as in the Heisenberg model was pointed out by *Herring* and *Kittel* [14.1]. In particular, *Herring* calculated the spin-wave excitation energy for an electron system with bare Coulomb interactions [14.2].

The energy of spin-wave excitations starts with $2\mu_B H$ at $q = 0$ and increases proportionally to q^2; it merges into the continuous band in contact with it at $q = q_{max}$. For small q, $\hbar\omega_{\boldsymbol{q}} - 2\mu_B H$ is much smaller than the exchange splitting Δ_{exch}, and we may expand (14.17) with respect to $\hbar\omega_q - 2\mu_B H$; to first order we obtain

$$\hbar\omega_q - 2\mu_B H = \left[1 - \frac{U}{N}\sum_k \frac{n_{k\uparrow}(1-n_{k+q\downarrow})}{\varepsilon_{k+q} - \varepsilon_k + \Delta_{\text{exch}}}\right] \times \left[\frac{U}{N}\sum_k \frac{n_{k\uparrow}(1-n_{k+q\downarrow})}{(\varepsilon_{k+q} - \varepsilon_k + \Delta_{\text{exch}})^2}\right]^{-1} . \quad (14.22)$$

In the denominator and the numerator on the right-hand side, we further make expansion with respect to $\varepsilon_{k+q} - \varepsilon_k$; to order q^2, the excitation energy is

$$\hbar\omega_q = 2\mu_B H + \frac{\hbar^2 q^2}{2m}\frac{1}{\zeta}\left\{1 - \frac{1}{5}\frac{\varepsilon_F}{k\theta}\frac{[(1+\zeta)^{5/3} - (1-\zeta)^{5/3}]}{\zeta}\right\} , \quad (14.23)$$

where ε_F is the Fermi energy in the paramagnetic state and $k\theta$ is $(1/2)(N_e/N)$ $\times U$ as defined before. If $k\theta/\varepsilon_F$ is larger than $2^{-1/3}$ and $\zeta = 1$, then $\hbar\omega_q$ is given by

$$\hbar\omega_q = 2\mu_B H + \frac{\hbar^2 q^2}{2m}\left(1 - \frac{2}{5}\frac{2^{2/3}\varepsilon_F}{k\theta}\right) . \quad (14.24)$$

In the case that ζ is much smaller than unity, we expand (14.23) with respect to ζ to get

$$\hbar\omega_q = 2\mu_B H + \frac{\hbar^2 q^2}{2m}\frac{1}{\zeta}\left[1 - \frac{1}{\alpha}\left(1 - \frac{1}{27}\zeta^2\right)\right] . \quad (14.25)$$

If one puts $H = 0$ here, the condition for $\hbar\omega_q > 0$ coincides with that for the appearance of ferromagnetism, (14.6a).

14.4 Paramagnetic Dynamical Susceptibility

Now we calculate the paramagnetic dynamical susceptibility in an external magnetic field which varies in space and time, also in the RPA.

Suppose that a field rotating around the z-axis, with wavevector $\boldsymbol{Q}$ and angular frequency ω, is externally applied to the system; that is,

$$\begin{aligned} H_x &= H(\boldsymbol{Q},\omega)\cos(\boldsymbol{Q}\cdot\boldsymbol{r} - \omega t), \\ H_y &= H(\boldsymbol{Q},\omega)\sin(\boldsymbol{Q}\cdot\boldsymbol{r} - \omega t), \end{aligned} \quad (14.26)$$

The Zeeman energy corresponding to the magnetic field is

$$\mathcal{H}_{\text{Zeeman}} = -\frac{1}{2}g\mu_B H(\boldsymbol{Q},\omega)\sum_i \left(s_{i-}e^{i(\boldsymbol{Q}\cdot\boldsymbol{r}_i - \omega t)} + s_{i+}e^{-i(\boldsymbol{Q}\cdot\boldsymbol{r}_i - \omega t)}\right) ; \quad (14.27)$$

here $s_{i\pm}$ is defined in terms of the x- and y-components of the ith electron spin operators s_{ix}, s_{iy} by

$$s_{i\pm} = s_{ix} \pm \mathrm{i} s_{iy} \ . \qquad (14.28)$$

Equation (14.27) can be rewritten in the second-quantized form by using the Fermi operators $a^\dagger_{\boldsymbol{k}\sigma}$, $a_{\boldsymbol{k}\sigma}$ for electrons:

$$\mathcal{H}_{\mathrm{Zeeman}} = -\frac{1}{2} g\mu_{\mathrm{B}} H(\boldsymbol{Q},\omega)(\sigma^+_{\boldsymbol{Q}+}\mathrm{e}^{-\mathrm{i}\omega t} + \sigma_{\boldsymbol{Q}+}\mathrm{e}^{\mathrm{i}\omega t}) \ . \qquad (14.29)$$

In (14.29), the operators $\sigma_{\boldsymbol{Q}+}$ and $\sigma^+_{\boldsymbol{Q}+}$ are related to $\sigma_{\boldsymbol{Q}x}$ and $\sigma_{\boldsymbol{Q}y}$, the tranverse components of the Fourier transform of the spin density; the latter are defined by

$$\sigma(\boldsymbol{r}) = \sum_i s_i \delta(\boldsymbol{r} - \boldsymbol{r}_i) = \frac{1}{V} \sum_{\boldsymbol{q}} \sigma_{\boldsymbol{q}} \mathrm{e}^{\mathrm{i}\boldsymbol{q}\cdot\boldsymbol{r}} \ . \qquad (14.30)$$

The following relations hold between these operators:

$$\begin{aligned} \sigma_{\boldsymbol{Q}+} &= \sigma_{\boldsymbol{Q}x} + \mathrm{i}\sigma_{\boldsymbol{Q}y} = \sum_{\boldsymbol{k}} a^\dagger_{\boldsymbol{k}-\boldsymbol{Q}\uparrow} a_{\boldsymbol{k}\downarrow} \ , \\ \sigma^+_{\boldsymbol{Q}+} &= \sigma_{-\boldsymbol{Q}-} = \sigma_{-\boldsymbol{Q}x} - \mathrm{i}\sigma_{-\boldsymbol{Q}y} = \sum_{\boldsymbol{k}} a^\dagger_{\boldsymbol{k}+\boldsymbol{Q}\downarrow} a_{\boldsymbol{k}\uparrow} \ . \end{aligned} \qquad (14.31)$$

Because of the Zeeman energy given by (14.29), a spin-density wave with wavevector $\boldsymbol{Q}$ is induced in the electron system. The spin-density wave causes an internal field through the Coulomb interaction of (14.1): In the interaction part of (14.1) one may interchange $a^\dagger_{\boldsymbol{k}_2-\boldsymbol{q}\sigma'}$ and $a_{\boldsymbol{k}_2\sigma'}$ and keep only the parts with $\sigma \neq \sigma'$ in the summation over $\boldsymbol{k}_1$, $\boldsymbol{k}_2$, $\boldsymbol{q}$, σ and σ'. Moreover, if one replaces $\boldsymbol{k}_1 - \boldsymbol{k}_2 + \boldsymbol{q}$ by $\boldsymbol{q}$, one obtains

$$-\frac{U}{N} \sum_{\boldsymbol{k}_1\boldsymbol{k}_2\boldsymbol{q}} a^\dagger_{\boldsymbol{k}_1+\boldsymbol{q}\downarrow} a_{\boldsymbol{k}_1\uparrow} a^\dagger_{\boldsymbol{k}_2-\boldsymbol{q}\uparrow} a_{\boldsymbol{k}_2\downarrow} \ . \qquad (14.32)$$

Here we pick up terms with $\boldsymbol{q} = \boldsymbol{Q}$ and replace $\sum_{\boldsymbol{k}_1} a^\dagger_{\boldsymbol{k}_1+\boldsymbol{Q}\downarrow} a_{\boldsymbol{k}_1\uparrow}$ and $\sum_{\boldsymbol{k}_2} a^\dagger_{\boldsymbol{k}_2-\boldsymbol{Q}\uparrow} a_{\boldsymbol{k}_2\downarrow}$ with the average value estimated by the RPA:

$$\begin{aligned} \sum_{\boldsymbol{k}_1} a^\dagger_{\boldsymbol{k}_1+\boldsymbol{Q}\downarrow} a_{\boldsymbol{k}_1\uparrow} &= \sigma^+_{\boldsymbol{Q}+} \to \langle \sigma^+_{\boldsymbol{Q}+} \rangle = \Delta^*_{\boldsymbol{Q}} \mathrm{e}^{\mathrm{i}\omega t} \ , \\ \sum_{\boldsymbol{k}_1} a^\dagger_{\boldsymbol{k}_1-\boldsymbol{Q}\uparrow} a_{\boldsymbol{k}_1\downarrow} &= \sigma_{\boldsymbol{Q}+} \to \langle \sigma_{\boldsymbol{Q}+} \rangle = \Delta_{\boldsymbol{Q}} \mathrm{e}^{-\mathrm{i}\omega t} \ . \end{aligned} \qquad (14.33)$$

The Zeeman energy including this molecular field becomes

$$\mathcal{H}_{\mathrm{Zeeman}} = -\Big(\frac{1}{2} g\mu_{\mathrm{B}} H(\boldsymbol{Q},\omega) + \frac{U}{N}\Delta_{\boldsymbol{Q}}\Big)\sigma^+_{\boldsymbol{Q}+}\mathrm{e}^{-\mathrm{i}\omega t} + \mathrm{h.c.} \ . \qquad (14.34)$$

By treating the Zeeman energy thus obtained as the perturbation $\mathcal{H}'$, we calculate $\langle\sigma_{Q+}\rangle$ in first-order perturbation theory. Then the dynamical paramagnetic susceptibility is given by the ratio

$$\chi(\boldsymbol{Q},\omega)=\frac{g\mu_{\mathrm{B}}\Delta_{Q}}{H(\boldsymbol{Q},\omega)}\,. \tag{14.35}$$

To calculate $\langle\sigma_{Q+}^{+}\rangle$ and $\langle\sigma_{Q+}\rangle$ we may follow the usual method of time-dependent perturbation theory. In the Schrödinger equation

$$\mathcal{H}\psi(t)=-\frac{\hbar}{\mathrm{i}}\frac{d}{dt}\psi(t) \tag{14.36}$$

we represent the wave function $\psi(t)$ as a linear combination of the eigenfunction ψ_n of the unperturbed system:

$$\psi(t)=\sum_n \mathrm{e}^{-\mathrm{i}E_nt/\hbar}a_n(t)\psi_n\,, \tag{14.37}$$

where E_n is the energy eigenvalue of ψ_n; we take $n=0$ for the ground state. The perturbation $\mathcal{H}'$ is assumed to set in adiabatically from $t=-\infty$, and thus it is multiplied by the factor, $\mathrm{e}^{\eta t}$, where η is a positive infinitesimal. The initial condition is

$$a_n(t\to-\infty)=\delta_{n0}\,. \tag{14.38}$$

Substitution of (14.37) into (14.36) leads to a set of equations for the expansion coefficients $a_n(t)$:

$$-\frac{\hbar}{\mathrm{i}}\frac{d}{dt}a_n(t)=\sum_{n'}\langle n|\mathcal{H}'|n'\rangle\mathrm{e}^{\mathrm{i}(E_n-E_{n'})t/\hbar}a_{n'}(t)\,. \tag{14.39}$$

In the lowest-order approximation, we use the initial condition (14.38) for $a_{n'}(t)$ on the right-hand side and neglect terms other than $n'=0$. Using (14.34) for $\mathcal{H}'$, we obtain

$$\begin{aligned}
a_n(t)&=\left(\frac{1}{2}g\mu_{\mathrm{B}}H(\boldsymbol{Q},\omega)+\frac{U}{N}\Delta_Q^*\right)\langle\sigma_{Q+}\rangle_{n0}\frac{\mathrm{e}^{\mathrm{i}(\omega+\omega_{n0})t+\eta t}}{\hbar\omega+\hbar\omega_{n0}-\mathrm{i}\eta}\,,\\
a_{n'}(t)&=\left(\frac{1}{2}g\mu_{\mathrm{B}}H(\boldsymbol{Q},\omega)+\frac{U}{N}\Delta_Q\right)\langle\sigma_{Q+}^{+}\rangle_{n'0}\frac{\mathrm{e}^{-\mathrm{i}(\omega-\omega_{n'0})t+\eta t}}{-\hbar\omega+\hbar\omega_{n'0}-\mathrm{i}\eta}\,,
\end{aligned} \tag{14.40}$$

where $\langle\sigma_{Q+}^{+}\rangle_{n0}$ is the matrix element of σ_{Q+}^{+} between ψ_n and ψ_0, while $\omega_{n0}=(E_n-E_0)/\hbar$. We have taken into account the fact that $\langle\sigma_{Q+}^{+}\rangle_{n0}=0$ for ψ_n which gives $\langle\sigma_{Q+}\rangle_{n0}\neq0$.

Once the expansion coefficients $a_n(t)$ for $\psi(t)$ are found, then $\langle\sigma_{Q+}\rangle$ is obtained to first order in $\mathcal{H}'$ as follows:

$$\langle\sigma_{Q+}\rangle = \Delta_Q e^{-i\omega t} = \sum_n e^{-i\omega_{n0}t} a_n(t)\langle 0|\sigma_{Q+}|n\rangle + \sum_n e^{i\omega_{n0}t} a_n^*(t)\langle n|\sigma_{Q+}|0\rangle \ . \tag{14.41}$$

On using (14.40) for $a_n(t)$, we have finally

$$\Delta_Q = \left(\frac{1}{2}g\mu_B H(\boldsymbol{Q},\omega) + \frac{U}{N}\Delta_Q\right)\sum_n \left(\frac{|\langle\sigma_{Q+}^+\rangle_{n0}|^2}{-\hbar\omega+\hbar\omega_{n0}-i\eta} + \frac{|\langle\sigma_{Q+}\rangle_{n0}|^2}{\hbar\omega+\hbar\omega_{n0}+i\eta}\right) . \tag{14.42}$$

Solving (14.42) for Δ_Q, we obtain, through (14.35), an expression for the susceptibility $\chi(\boldsymbol{Q},\omega)$ in the RPA:

$$\chi_{\mathrm{RPA}}(\boldsymbol{q},\omega) = \frac{1}{2}\chi_{\mathrm{Pauli}} f'(\boldsymbol{q},\omega) \Big/ \left(1 - \frac{1}{2}\alpha f'(\boldsymbol{q},\omega)\right) , \quad \chi_{\mathrm{Pauli}} = \frac{1}{2}g^2\mu_B^2\rho(\varepsilon_F) , \tag{14.43}$$

$$f'(\boldsymbol{q},\omega) = \frac{2}{\rho(\varepsilon_F)}\sum_n \left(\frac{|\langle\sigma_{q+}^+\rangle_{n0}|^2}{-\hbar\omega + \hbar\omega_{n0} - i\eta} + \frac{|\langle\sigma_{q+}\rangle_{n0}|^2}{\hbar\omega + \hbar\omega_{n0} + i\eta}\right) , \tag{14.44}$$

where we have changed the notation for the wavevector from $\boldsymbol{Q}$ to general $\boldsymbol{q}$. Using (14.31) for σ_{Q+} and σ_{Q+}^+, we have

$$|\langle\sigma_{q+}\rangle_{n0}|^2 = n_{k\downarrow}(1 - n_{k-q\uparrow}) , \quad \hbar\omega_{n0} = \varepsilon_{k-q\uparrow} - \varepsilon_{k\downarrow} ,$$
$$|\langle\sigma_{q+}^+\rangle_{n0}|^2 = n_{k\uparrow}(1 - n_{k+q\downarrow}) , \quad \hbar\omega_{n0} = \varepsilon_{k+q\downarrow} - \varepsilon_{k\uparrow} ;$$

thus it follows

$$f'(\boldsymbol{q},\omega) = \frac{2}{\rho(\varepsilon_F)}\sum_k \left[\frac{n_{k\uparrow}(1 - n_{k+q\downarrow})}{\varepsilon_{k+q\downarrow} - \varepsilon_{k\uparrow} - \hbar\omega - i\eta} + \frac{n_{k\downarrow}(1 - n_{k-q\uparrow})}{\varepsilon_{k-q\uparrow} - \varepsilon_{k\downarrow} + \hbar\omega + i\eta}\right] .$$

By making the replacement $\boldsymbol{k} \to \boldsymbol{k}+\boldsymbol{q}$ or $\boldsymbol{k}-\boldsymbol{q} \to \boldsymbol{k}$ in the second term, we may rewrite $f'(\boldsymbol{q},\omega)$ as

$$f'(\boldsymbol{q},\omega) = \frac{2}{\rho(\varepsilon_F)}\sum_k \frac{n_{k\uparrow} - n_{k+q\downarrow}}{\varepsilon_{k+q\downarrow} - \varepsilon_{k\uparrow} - \hbar\omega - i\eta} . \tag{14.45}$$

In the paramagnetic state and in the absence of static magnetic field along the z-axis, we may omit the spin dependences of n_k and ε_k in (14.45). Then $f'(\boldsymbol{q},\omega)$ can be represented as

$$f'(\boldsymbol{q},\omega) = \frac{2}{\rho(\varepsilon_F)}\sum_k n_k \left(\frac{1}{\varepsilon_{k+q} - \varepsilon_k - \hbar\omega - i\eta} + \frac{1}{\varepsilon_{k+q} - \varepsilon_k + \hbar\omega + i\eta}\right) . \tag{14.46}$$

The susceptibility in the noninteracting case ($\alpha = 0$) is given by $(1/2)\chi_{\text{Pauli}} f'(\boldsymbol{q}, \omega)$; in the limit of $\omega \to 0$ this coincides with the previous expression (11.25) for the static susceptibility $\chi(\boldsymbol{q})$.

By using the relation

$$\frac{1}{x + \mathrm{i}\eta} = \mathrm{P}\frac{1}{x} - \mathrm{i}\pi\delta(x) \tag{14.47}$$

in (14.46), we obtain the real and imaginary parts of $f'(\boldsymbol{q}, \omega)$, respectively, as

$$f'_{\mathrm{R}}(\boldsymbol{q}, \omega) = \frac{2}{\rho(\varepsilon_{\mathrm{F}})} \sum_{\boldsymbol{k}} n_{\boldsymbol{k}} \left(\mathrm{P}\frac{1}{\varepsilon_{\boldsymbol{k}+\boldsymbol{q}} - \varepsilon_{\boldsymbol{k}} - \hbar\omega} + \mathrm{P}\frac{1}{\varepsilon_{\boldsymbol{k}+\boldsymbol{q}} - \varepsilon_{\boldsymbol{k}} + \hbar\omega} \right), \tag{14.48}$$

and

$$f'_{\mathrm{I}}(\boldsymbol{q}, \omega) = \frac{2\pi}{\rho(\varepsilon_{\mathrm{F}})} \sum_{\boldsymbol{k}} n_{\boldsymbol{k}} \left[\delta(\varepsilon_{\boldsymbol{k}+\boldsymbol{q}} - \varepsilon_{\boldsymbol{k}} - \hbar\omega) - \delta(\varepsilon_{\boldsymbol{k}+\boldsymbol{q}} - \varepsilon_{\boldsymbol{k}} + \hbar\omega) \right], \tag{14.49}$$

where P means the principal value. From these results one finds that f'_{R} is an even function of ω, while f'_{I} is odd. For free electron systems one can carry out the integration over $\boldsymbol{k}$ inside the Fermi sphere in (14.48, 49). (For finite temperatures, $n_{\boldsymbol{k}}$ should be replaced by the Fermi distribution function $f_{\boldsymbol{k}} = 1/[\mathrm{e}^{\beta(\varepsilon_{\boldsymbol{k}} - \varepsilon_{\mathrm{F}})} + 1]$.) The results of the calculation are summarized in the following:

$$\begin{aligned} f'_{\mathrm{R}}(\boldsymbol{q}, \omega) &= \frac{1}{2}\left(1 - \frac{\omega}{v_{\mathrm{F}} q}\frac{2k_{\mathrm{F}}}{q}\right) f\left(\frac{q}{2k_{\mathrm{F}}} - \frac{\omega}{v_{\mathrm{F}} q}\right) \\ &\quad + \frac{1}{2}\left(1 + \frac{\omega}{v_{\mathrm{F}} q}\frac{2k_{\mathrm{F}}}{q}\right) f\left(\frac{q}{2k_{\mathrm{F}}} + \frac{\omega}{v_{\mathrm{F}} q}\right), \end{aligned} \tag{14.50}$$

$$\begin{aligned} f'_{\mathrm{I}}(\boldsymbol{q}, \omega) &= 0\,, \qquad \text{for} \quad \frac{\omega}{2 v_{\mathrm{F}} k_{\mathrm{F}}} > \frac{q}{2k_{\mathrm{F}}} + \left(\frac{q}{2k_{\mathrm{F}}}\right)^2 \\ &= \frac{\pi}{4}\frac{2k_{\mathrm{F}}}{q}\left[1 - \left(\frac{q}{2k_{\mathrm{F}}} - \frac{\omega}{v_{\mathrm{F}} q}\right)^2\right], \\ &\qquad\qquad \text{for} \quad \frac{q}{2k_{\mathrm{F}}} - \left(\frac{q}{2k_{\mathrm{F}}}\right)^2 < \frac{\omega}{2 v_{\mathrm{F}} k_{\mathrm{F}}} < \frac{q}{2k_{\mathrm{F}}} + \left(\frac{q}{2k_{\mathrm{F}}}\right)^2 \\ &= \pi\frac{\omega}{v_{\mathrm{F}} q}\,, \quad \text{for} \quad \frac{\omega}{2 v_{\mathrm{F}} k_{\mathrm{F}}} < \frac{q}{2k_{\mathrm{F}}} - \left(\frac{q}{2k_{\mathrm{F}}}\right)^2 \end{aligned} \tag{14.51}$$

where $f(x)$ is again the function defined by (11.26).

In the limit $\omega \to 0$, we have

$$\chi_{\mathrm{RPA}}(\boldsymbol{q}, 0) = \frac{\frac{1}{2}\chi_{\mathrm{Pauli}} f\left(\frac{q}{2k_{\mathrm{F}}}\right)}{1 - \frac{1}{2}\alpha f\left(\frac{q}{2k_{\mathrm{F}}}\right)}\,. \tag{14.52}$$

Since $f(q/2k_{\rm F})$ decreases monotonically from its value $f(0) = 2$ at $q = 0$, when q increases, the minimum of the denominator in (14.52) occurs at $q = 0$. Therefore, when α increases from 0, the susceptibility at $q = 0$ diverges at $\alpha = 1$. This is nothing but the condition for the occurrence of ferromagnetism in the Hartree-Fock approximation. However, for general band structures, it may happen that the susceptibility $\chi_0(\boldsymbol{q}) = (1/2)\chi_{\rm Pauli} f'_{\rm R}(\boldsymbol{q}, 0)$ for the noninteracting system has a maximum at some $\boldsymbol{q}$ not equal to zero. For such a band structure, when α is increased the susceptibility diverges first at $\boldsymbol{Q}$ where a maximum of $\chi_0(\boldsymbol{q})$ occurs. It implies the emergence of spiral structures or spin-density wave structures with wavevector $\boldsymbol{Q}$. Actually the spin-density waves in chromium metal can be understood from this point of view.

In calculating the susceptibility in the RPA, we have treated the Zeeman energy of the total magnetic field (i.e., the sum of the external field and the internal field induced by the interaction) as a perturbation to the free-elecron state.

This method of calculation still works even if we take the total Hamiltonian including the electron-electron interaction as the unperturbed one and take the Zeeman energy due to the external magnetic field only as the perturbation. In this case the susceptibility is represented by following (14.42) as

$$\chi(\boldsymbol{q}, \omega) = \sum_n \left(\frac{|\langle \sigma_{q+}^+ \rangle_{n0}|^2}{\hbar\omega_{n0} - \hbar\omega - \mathrm{i}\eta} + \frac{|\langle \sigma_{q+} \rangle_{n0}|^2}{\hbar\omega_{n0} + \hbar\omega + \mathrm{i}\eta} \right) . \tag{14.53}$$

We have omitted here the factor $(1/2)g^2\mu_{\rm B}^2$. Note that the matrix elements are taken with respect to ψ_0 and ψ_n, the ground state and excited one,respectively, of the Hamiltonian including the interaction. Thus (14.53) is an exact expression. Its poles are at $\hbar\omega = \pm\hbar\omega_{n0} - \mathrm{i}\eta$. Since η is a positive infinitesimal, $\chi(\boldsymbol{q}, \omega)$ is an analytic function in the upper half-plane (including the real axis) in the complex ω-plane. Therefore the Fourier transform of $\chi(\boldsymbol{q}, \omega)$ with respect to time

$$\chi(\boldsymbol{q}, t) = \frac{1}{2\pi} \int_{-\infty}^{\infty} \chi(\boldsymbol{q}, \omega) \mathrm{e}^{-\mathrm{i}\omega t} d\omega \tag{14.54}$$

vanishes for $t < 0$, because now we can close the integration path by adding the semicircle at infinity in the upper complex plane. The fact that $\chi(\boldsymbol{q}, t) = 0$ for $t < 0$ is a natural result since the function $\chi(\boldsymbol{q}, t)$ represents the response to an impulse of delta-function type applied at $t = 0$.

Next we examine the integral

$$\oint_\Gamma \frac{\chi(\boldsymbol{q}, \omega')}{\omega - \omega'} d\omega' = 0 , \tag{14.55}$$

where the path of the integration is given in Fig. 14.4; the integral vanishes because $\chi(\boldsymbol{q}, \omega)$ is analytic in the upper half of the complex plane. Since

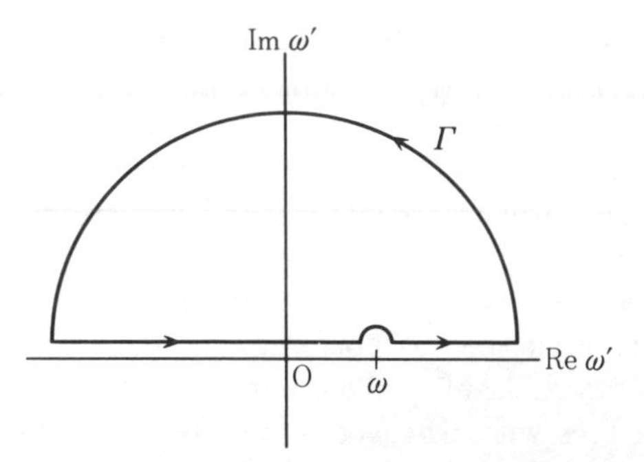

Fig. 14.4. Integration path for (14.55)

$\chi(\boldsymbol{q},\omega')$ vanishes as $1/\omega'^2$ on the semicircle at infinity, the integral is the sum of the integral along the real axis and the integral along the small semicircle centered at ω. Therefore (14.55) leads to

$$\int_{-\infty}^{\infty} \chi(\boldsymbol{q},\omega')\mathrm{P}\Big(\frac{1}{\omega-\omega'}\Big)d\omega' + \pi\mathrm{i}\chi(\boldsymbol{q},\omega) = 0\ ; \tag{14.56}$$

here P means to take principal value. If one separates the real and imaginary parts of the equation, one can obtain two relations between the real part χ' and the imaginary part χ'' of $\chi(\boldsymbol{q},\omega)$:

$$\chi'(\boldsymbol{q},\omega) = -\frac{1}{\pi}\int_{-\infty}^{\infty} \chi''(\boldsymbol{q},\omega')\mathrm{P}\Big(\frac{1}{\omega-\omega'}\Big)d\omega' \tag{14.57}$$

and

$$\chi''(\boldsymbol{q},\omega) = \frac{1}{\pi}\int_{-\infty}^{\infty} \chi'(\boldsymbol{q},\omega')\mathrm{P}\Big(\frac{1}{\omega-\omega'}\Big)d\omega'. \tag{14.58}$$

These relations are known as the Kramers-Kronig relations and naturally hold at finite temperatures as well.

The susceptibilitiy at finite temperatures is given by generalizing (14.53) as

$$\chi(\boldsymbol{q},\omega) = Z^{-1}\sum_{mn} \mathrm{e}^{-\beta E_m}\Big(\frac{|\langle\sigma_{q+}^{+}\rangle_{nm}|^2}{-\hbar\omega+\hbar\omega_{nm}-\mathrm{i}\eta} + \frac{|\langle\sigma_{q+}\rangle_{nm}|^2}{\hbar\omega+\hbar\omega_{nm}+\mathrm{i}\eta}\Big), \tag{14.59}$$

where β is $1/kT$ and Z denotes the partition function:

$$Z = \sum_m \mathrm{e}^{-\beta E_m}\ . \tag{14.60}$$

We define here transition probabilities related to σ_{q+}^{+} and σ_{q+} as

$$S^+(\boldsymbol{q},\omega) = Z^{-1}\sum_{mn} \mathrm{e}^{-\beta E_m} |\langle \sigma^+_{q+}\rangle_{nm}|^2 \delta(\omega - \omega_{nm}) \ , \tag{14.61}$$

$$S(\boldsymbol{q},\omega) = Z^{-1}\sum_{mn} \mathrm{e}^{-\beta E_m} |\langle \sigma_{q+}\rangle_{nm}|^2 \delta(\omega - \omega_{nm}) \ ; \tag{14.62}$$

these are identical to the Fourier components of the spin correlation functions introduced in Sect. 7.1. Then the imaginary part of $\chi(\boldsymbol{q},\omega)$ is represented as

$$\chi''(\boldsymbol{q},\omega) = \frac{\pi}{\hbar}\Big[S^+(\boldsymbol{q},\omega) - S(\boldsymbol{q},-\omega)\Big] \ . \tag{14.63}$$

By interchanging the suffices m and n in (14.62), and using the relation $\langle \sigma_{q+}\rangle_{nm} = \langle \sigma^+_{q+}\rangle^*_{mn}$, we obtain

$$S(\boldsymbol{q},\omega) = Z^{-1}\sum_{mn} \mathrm{e}^{-\beta E_n} |\langle \sigma^+_{q+}\rangle_{nm}|^2 \delta(\omega + \omega_{nm}) = \mathrm{e}^{\beta\hbar\omega} S^+(\boldsymbol{q},-\omega) \ . \tag{14.64}$$

Therefore (14.63) can be written as

$$\chi''(\boldsymbol{q},\omega) = \frac{\pi}{\hbar}(1 - \mathrm{e}^{-\beta\hbar\omega}) S^+(\boldsymbol{q},\omega) \ . \tag{14.65}$$

The last relation is known as the fluctuation-dissipation theorem [14.3–5] mentioned before in (7.9) of Part II.

The interaction part of the Hamiltonian given by (14.1),

$$\mathcal{H}_{\text{int}} = \frac{1}{2}\frac{U}{N}\sum_{k_1 k_2 q \sigma\sigma'} a^\dagger_{k_1+q\sigma} a^\dagger_{k_2-q\sigma'} a_{k_2\sigma'} a_{k_1\sigma} \ , \tag{14.66}$$

which is described in terms of the Fermi operators $a_{k\sigma}$ for the Bloch orbitals, can be written in terms of the operator $a_{n\sigma}$ for the Wannier localized orbitals as

$$\mathcal{H}_{\text{int}} = \frac{1}{2}U\sum_{n\sigma\sigma'} a^\dagger_{n\sigma} a^\dagger_{n\sigma'} a_{n\sigma'} a_{n\sigma} \ ,$$

if we use the relation (3.5),

$$a_{k\sigma} = \frac{1}{\sqrt{N}}\sum_{n} \mathrm{e}^{\mathrm{i}\boldsymbol{k}\cdot\boldsymbol{n}} a_{n\sigma}$$

between $a_{k\sigma}$ and $a_{n\sigma}$. This relation holds for the single band Bloch orbitals and the Wannier orbitals without orbital degeneracy. In the above expression for $\mathcal{H}_{\text{int}}$, $\sigma = \sigma'$ terms identically vanish because of the anti-commutation relations for the fermi operators. Therefore, by dropping the terms of $\sigma = \sigma'$, (14.66) can be written as

$$\mathcal{H}_{\text{int}} = -\frac{1}{2}\frac{U}{N}\sum_{k_1 k_2 q} (a^\dagger_{k_1+q\uparrow} a_{k_2\downarrow} a^\dagger_{k_2-q\downarrow} a_{k_1\uparrow} + a^\dagger_{k_1+q\downarrow} a_{k_2\uparrow} a^\dagger_{k_2-q\uparrow} a_{k_1\downarrow}) \ . \tag{14.67}$$

This expression can be used for the Hubbard Hamiltonian for the localized orbitals without degeneracy. In the cases where the conduction electrons are described by plane waves and also the localized orbitals have orbital degeneracy there is no reason for dropping $\sigma = \sigma'$ terms.

If one takes the diagonal terms in (14.66), namely, $\boldsymbol{q} = 0$ terms, we obtain as an expectation value

$$\langle \mathcal{H}_{\text{int}} \rangle_{\text{exp}} = \frac{U}{N} \sum_{\boldsymbol{k}_1 \boldsymbol{k}_2} n_{\boldsymbol{k}_1 \uparrow} n_{\boldsymbol{k}_2 \downarrow} ,$$

where the repulsive energy U/N works only between electrons with antiparallel spins. The exchange energy given by (14.2) corresponds to the energy measured from

$$\frac{1}{2}\frac{U}{N} \sum_{\boldsymbol{k}_1} (n_{\boldsymbol{k}_1 \uparrow} + n_{\boldsymbol{k}_1 \downarrow}) \sum_{\boldsymbol{k}_2} (n_{\boldsymbol{k}_2 \uparrow} + n_{\boldsymbol{k}_2 \downarrow}) = \frac{1}{2}\frac{U}{N} N_{\text{e}}^2 .$$

Now, we come back to (14.67). If we put $\boldsymbol{k}_1 + \boldsymbol{q} = \boldsymbol{k}_2 + \boldsymbol{q}'$, (14.67) can be expressed by the sum over $\boldsymbol{q}'$ of $\sigma^+_{\boldsymbol{q}'+}\sigma_{\boldsymbol{q}'+}$ and $\sigma_{\boldsymbol{q}'+}\sigma^+_{\boldsymbol{q}'+}$. Therefore, the thermal average can be written as

$$\langle \mathcal{H}_{\text{int}} \rangle = -\frac{1}{2}\frac{U}{N} Z^{-1} \sum_{mn\boldsymbol{q}} \text{e}^{-\beta E_m} \left(|(\sigma^+_{\boldsymbol{q}+})_{nm}|^2 + |(\sigma_{\boldsymbol{q}+})_{nm}|^2 \right) . \tag{14.68}$$

With the use of (14.61, 62), this can be expressed as the sum of the ω-integral of $S^+(\boldsymbol{q},\omega)$ and $S(\boldsymbol{q},\omega)$. Since $S(\boldsymbol{q},\omega)$ is equal to $\text{e}^{\beta\hbar\omega} S^+(\boldsymbol{q},-\omega)$ by (14.63), $\langle \mathcal{H}_{\text{int}} \rangle$ is given by

$$\langle \mathcal{H}_{\text{int}} \rangle = -\frac{1}{2}\frac{U}{N} \sum_{\boldsymbol{q}} \int_{-\infty}^{\infty} (1 + \text{e}^{-\beta\hbar\omega}) S^+(\boldsymbol{q},\omega) d\omega ,$$

and using (14.65) we obtain the final expression for $\langle \mathcal{H}_{\text{int}} \rangle$ as

$$\langle \mathcal{H}_{\text{int}} \rangle = -\alpha \frac{\hbar}{\rho(\varepsilon_{\text{F}})} \sum_{\boldsymbol{q}} \int_{-\infty}^{\infty} \coth \frac{\beta\hbar\omega}{2} \chi''(\boldsymbol{q},\omega) \frac{d\omega}{2\pi} , \tag{14.69}$$

where $\rho(\varepsilon_{\text{F}})U/N$ is put as α.

Now we invoke the following relation which is often used

$$g\frac{\partial E_n(g)}{\partial g} = \langle \psi_n(g) | g\mathcal{H}' | \psi_n(g) \rangle , \tag{14.70}$$

where the interaction $\mathcal{H}'$ is multiplied by g and ψ_n and E_n are an eigenfunction and an eigenvalue for the total Hamiltonian $\mathcal{H}_0 + g\mathcal{H}'$. This relation is easily derived by differentiating

$$E_n(g) = \langle \psi_n(g) | \mathcal{H}_0 + g\mathcal{H}' | \psi_n(g) \rangle$$

with respect to g and noting that $\psi_n(g)$ is stationary to the change of g. Using the relation (14.70) for (14.69), we obtain as the part of the free energy which is brought by the interaction,

$$\Delta F = -\frac{\hbar}{\rho(\varepsilon_{\mathrm{F}})} \int_0^{\alpha} d\alpha \sum_{\boldsymbol{q}} \int_{-\infty}^{\infty} \frac{d\omega}{2\pi} \operatorname{Im} \chi(\boldsymbol{q}, \omega; \alpha) \coth \frac{\beta\hbar\omega}{2} . \tag{14.71}$$

As shown by (14.65), the imaginary part of the dynamical susceptibility is given by the Fourier component of the spin correlation function in the same way as for the localized spin system. Therefore it is closely related to the scattering cross-section of the neutron beam [14.6], the relaxation time of the nuclear spin [14.7] and furthermore the part of the free-energy arising from the interaction [14.8, 9]. Particularly, *Doniach* and *Engelsberg* [14.8] and also *Berk* and *Schrieffer* [14.9] derived T^2-term and $T^4 \log T$-term in the free-energy due to the spin-fluctuation by using the susceptibility obtained in the RPA (14.43) for χ'' in (14.71). This T^2-term gives rise to the mass-enhancement for the T-linear electronic specific heat. This mass-enhancement is proportonal to $\log(1-\alpha)$ and shows a singular behavior, namely logarithmic divergence at $\alpha = 1$.

However, the RPA is not regarded as a good approximation for conduction-electron sytems with strong correlation. As mentioned previously, the $T = 0$ properties of the ferromagnetic metals Fe, Co and Ni, etc. are roughly grasped by the Stoner model and by its RPA extension, when the band structure is taken into account. On the other hand, if we notice, for instance, the susceptibility above the Curie temperature where the ferromagnetism disappears, the RPA cannot explain the Curie-Weiss law observed above T_{c}. *Murata* and *Doniach* [14.10] and also *Moriya* and *Kawabata* [14.11] have derived the Curie-Weiss law for the susceptibility above T_{c} by taking further steps beyond the RPA to include mode–mode couplings between spin fluctuations. Further, Moriya has developed more generally a theory of ferromagnetic and antiferromagnetic metals from the same point of view and has succeeded in explaining several properties of various metals, i.e., the susceptibility above the transition temperature, the electrical resistivity, the nuclear-spin relaxation time and the lattice expansion coefficient. A specialized review article in detail on these topics has been published by *Moriya* [14.12].

Part IV

Magnetism of Dilute Alloys

15. *s-d* Model and Anderson Hamiltonian

In general, when an impurity atom is introduced into a metal, the conduction electron states are disturbed through the local potential change induced by the impurity. The electronic states of the impurity atom also change at the same time. Such a change of the electronic states should show up also in the magnetism of alloys; this causes characteristic magnetism in the alloys containing transition metal atoms, such as the CuMn alloys discussed later. In the following, we discuss the change of electronic states induced by the interaction between an impurity and the conduction electrons.

15.1 Basis of Scattering Theory

Let us consider first the problem of scattering due to an impurity potential. In the presence of the local potential $V(\boldsymbol{r})$, the Schrödinger equation for the free electron is written as

$$-\frac{\hbar^2}{2m}\nabla^2\psi(\boldsymbol{r}) + V(\boldsymbol{r})\psi(\boldsymbol{r}) = E\psi(\boldsymbol{r}) \, , \tag{15.1}$$

where $\psi(\boldsymbol{r})$ and E are the wave function and the energy of the electron, respectively. With the replacements of E by $\hbar^2k^2/2m$ and $2mV/\hbar^2$ by V, (15.1) is rewritten as

$$\nabla^2\psi(\boldsymbol{r}) + [k^2 - V(\boldsymbol{r})]\psi(\boldsymbol{r}) = 0 \, . \tag{15.2}$$

The integrated form of this differential equation, which can be easily checked by a direct substitution, is written as

$$\psi(\boldsymbol{r}) = \mathrm{e}^{\mathrm{i}\boldsymbol{k}\cdot\boldsymbol{r}} - \frac{1}{4\pi}\int \frac{\mathrm{e}^{\mathrm{i}k|\boldsymbol{r}-\boldsymbol{\rho}|}}{|\boldsymbol{r}-\boldsymbol{\rho}|} V(\boldsymbol{\rho})\psi(\boldsymbol{\rho})d\boldsymbol{\rho} \, . \tag{15.3}$$

Now let us solve the Schrödinger equation (15.2) under such boundary condition that in the region far from the scattering center $\psi(\boldsymbol{r})$ is represented by a superposition of the incoming wave $\mathrm{e}^{\mathrm{i}\boldsymbol{k}\cdot\boldsymbol{r}}$ and the scattered wave as

$$\psi(\boldsymbol{r}) \to \mathrm{e}^{\mathrm{i}\boldsymbol{k}\cdot\boldsymbol{r}} + f(\theta_{\boldsymbol{kr}}, \phi)\frac{\mathrm{e}^{\mathrm{i}kr}}{r} \, , \tag{15.4}$$

where $\theta_{\boldsymbol{k}\boldsymbol{r}}$ is the angle between $\boldsymbol{r}$ and $\boldsymbol{k}$. Here we assume that $V(\boldsymbol{r})$ is spherically symmetric and we fix the z-axis in the direction of $\boldsymbol{k}$. By expanding $\psi(\boldsymbol{r})$ in terms of partial waves as

$$\psi(\boldsymbol{r}) = \sum_l c_l u_l(r) \mathrm{P}_l(\cos\theta) \ , \tag{15.5}$$

we obtain the following equation for the radial wave function $u_l(r)$:

$$-\frac{1}{r^2}\frac{d}{dr}\left(r^2\frac{du_l}{dr}\right) + \left[\frac{l(l+1)}{r^2} + V(r)\right]u_l = k^2 u_l \ . \tag{15.6}$$

The potential $V(r)$ vanishes at $r = \infty$; in the region where $V(r) = 0$ the differential equation (15.6) reduces to the Bessel equation, which possesses two independent solutions, $j_l(kr)$ and $n_l(kr)$. Here j_l is the spherical Bessel function which is finite at $r = 0$, and n_l is the spherical Neumann function which diverges at $r = 0$. Therefore, in the region of large r, the solution of (15.6) is given by the linear combination of the two functions

$$u_l(r) = \cos\delta_l j_l(kr) - \sin\delta_l n_l(kr) \ . \tag{15.7}$$

The functions $j_l(\rho)$ and $n_l(\rho)$ are given for $\rho \to \infty$ by the asymptotic forms

$$j_l(\rho) \to \frac{1}{\rho}\cos\left[\rho - \frac{1}{2}(l+1)\pi\right] = \frac{1}{\rho}\sin\left(\rho - \frac{1}{2}l\pi\right) , \tag{15.8}$$

$$n_l(\rho) \to \frac{1}{\rho}\sin\left[\rho - \frac{1}{2}(l+1)\pi\right] = -\frac{1}{\rho}\cos\left(\rho - \frac{1}{2}l\pi\right) . \tag{15.9}$$

By substituting these asymptotic forms into (15.7), we obtain the asymptotic form of $u_l(r)$, which is given by

$$u_l(r) \to \frac{1}{r}\sin\left(kr - \frac{1}{2}l\pi + \delta_l\right) . \tag{15.10}$$

On the other hand, in the absence of the potential $V(r)$ we must choose $j_l(kr)$, since (15.7) should hold also at the origin $r = 0$. The asymptotic form of the solution in this case is obtained by putting $\delta_l = 0$ in (15.10). The parameter δ_l, which is introduced as the coefficient of the linear combination of (15.7), is the phase shift caused by the potential $V(r)$ in the asymptotic form of $u_l(r)$, and is an important physical quantity in the scattering problem.

By substitution of (15.10) into (15.5), the boundary condition (15.4) is rewritten as

$$\sum_{l=0}^{\infty} c_l r^{-1}\sin\left(kr - \frac{1}{2}l\pi + \delta_l\right)\mathrm{P}_l(\cos\theta) = \mathrm{e}^{\mathrm{i}\boldsymbol{k}\cdot\boldsymbol{r}} + f(\theta)\frac{\mathrm{e}^{\mathrm{i}kr}}{r} . \tag{15.11}$$

By using the expansion formula for $\mathrm{e}^{\mathrm{i}\boldsymbol{k}\cdot\boldsymbol{r}}$

$$e^{i\boldsymbol{k}\cdot\boldsymbol{r}} = \sum_{l=0}^{\infty} i^l (2l+1) j_l(kr) P_l(\cos\theta) \tag{15.12}$$

and equating the coefficients of $e^{\pm ikr}$ on both sides of (15.11), we obtain the relations

$$2ikf(\theta) + \sum_{l=0}^{\infty}(2l+1)i^l e^{-i(l/2)\pi} P_l(\cos\theta) = \sum_{l=0}^{\infty} kc_l e^{i[\delta_l-(l/2)\pi]} P_l(\cos\theta), \tag{15.13}$$

$$\sum_{l=0}^{\infty}(2l+1)i^l e^{i(l/2)\pi} P_l(\cos\theta) = \sum_{l=0}^{\infty} kc_l e^{-i[\delta_l-(l/2)\pi]} P_l(\cos\theta)\ . \tag{15.14}$$

From (15.14), we obtain

$$kc_l = (2l+1)i^l e^{i\delta_l}\ . \tag{15.15}$$

Substituting this result into (15.13), we obtain the following expression for $f(\theta)$:

$$f(\theta) = (2ik)^{-1}\sum_{l=0}^{\infty}(2l+1)(e^{2i\delta_l}-1)P_l(\cos\theta)\ . \tag{15.16}$$

Therefore the scattering probability of the incident wave in the direction of θ, namely the differential scattering cross section, is given by

$$\sigma(\theta) = |f(\theta)|^2 = \frac{1}{k^2}\left|\sum_{l=0}^{\infty}(2l+1)e^{i\delta_l}\sin\delta_l P_l(\cos\theta)\right|^2\ . \tag{15.17}$$

The total cross section is obtained as

$$\sigma = 2\pi\int_0^{\pi}\sigma(\theta)\sin\theta d\theta = \frac{4\pi}{k^2}\sum_{l=0}^{\infty}(2l+1)\sin^2\delta_l\ . \tag{15.18}$$

Since $P_l(\cos\theta)_{\theta=0} = 1$, we obtain for the forward scattering amplitude

$$f(0) = k^{-1}\sum_{l=0}^{\infty}(2l+1)e^{i\delta_l}\sin\delta_l\ . \tag{15.19}$$

Comparing (15.18) with (15.19), we can derive the relation

$$\sigma = \frac{4\pi}{k}\,\mathrm{Im}\{f(0)\} = \frac{2\pi}{ik}[f(0)-f^*(0)]\ . \tag{15.20}$$

This relation, which is called the *optical theorem*, holds generally for general potentials $V(\boldsymbol{r})$ and also when inelastic scattering and absorption are included in σ.

In the k representation (15.3) which is equivalent to the Schrödinger equation is written as

$$|\boldsymbol{k}\rangle^{+} = |\boldsymbol{k}\rangle + \frac{1}{\varepsilon_{\boldsymbol{k}} - \mathcal{H}_0 + \mathrm{i}\eta} V |\boldsymbol{k}\rangle^{+} \quad , \tag{15.21}$$

where η is an infinitesimal positive number. The superscript + of $|\boldsymbol{k}\rangle^{+}$ means the scattered outgoing wave. $\psi_{\boldsymbol{k}}(\boldsymbol{r})$ in (15.3) is the transformation function between the $\boldsymbol{k}$ and $\boldsymbol{r}$ representations:

$$(2\pi)^{-3/2}\psi_{\boldsymbol{k}}(\boldsymbol{r}) = \langle \boldsymbol{r}|\boldsymbol{k}\rangle^{+} \quad ; \tag{15.22}$$

the incident wave is given by

$$(2\pi)^{-3/2}\mathrm{e}^{\mathrm{i}\boldsymbol{k}\cdot\boldsymbol{r}} = \langle \boldsymbol{r}|\boldsymbol{k}\rangle \ . \tag{15.23}$$

Here the normalization of plane waves in infinite space is assumed so as to satisfy the relations

$$\langle \boldsymbol{k}'|\boldsymbol{k}\rangle = \delta(\boldsymbol{k} - \boldsymbol{k}') \quad , \qquad \langle \boldsymbol{r}|\boldsymbol{r}'\rangle = \delta(\boldsymbol{r} - \boldsymbol{r}') \ .$$

Now we introduce a matrix T defined by

$$T(\varepsilon_{\boldsymbol{k}} + \mathrm{i}\eta)|\boldsymbol{k}\rangle = V|\boldsymbol{k}\rangle^{+} \ . \tag{15.24}$$

This matrix is called *T-matrix*. Substituting (15.24) into (15.21), we obtain the relation

$$\begin{aligned} \frac{\mathrm{e}^{ikr}}{r} f(\theta_{\boldsymbol{k}\boldsymbol{r}}) &= (2\pi)^{3/2} \lim_{r\to\infty} \left(\langle \boldsymbol{r}|\boldsymbol{k}\rangle^{+} - \langle \boldsymbol{r}|\boldsymbol{k}\rangle \right) \\ &= (2\pi)^{3/2} \lim_{r\to\infty} \left\langle \boldsymbol{r} \middle| \frac{1}{\varepsilon_{\boldsymbol{k}} - \mathcal{H}_0 + \mathrm{i}\eta} T(\varepsilon_{\boldsymbol{k}} + \mathrm{i}\eta) \middle| \boldsymbol{k} \right\rangle \ . \end{aligned} \tag{15.25}$$

The right-hand side is rewritten with use of (15.23) as

$$-\frac{1}{2}(2\pi)^{1/2}\frac{\mathrm{e}^{ikr}}{r} \int \mathrm{e}^{-\mathrm{i}\boldsymbol{k}'\cdot\boldsymbol{\rho}} \langle \boldsymbol{\rho}|T(\varepsilon_{\boldsymbol{k}} + \mathrm{i}\eta)|\boldsymbol{k}\rangle d\boldsymbol{\rho} \ ,$$

where $\boldsymbol{k}' = k\boldsymbol{r}/r$. Thus, the scattering amplitude $f(\theta)$ is written as

$$f(\theta_{\boldsymbol{k}\boldsymbol{r}}) = f(\theta_{\boldsymbol{k}\boldsymbol{k}'}) = -2\pi^2 \langle \boldsymbol{k}'|T(\varepsilon_{\boldsymbol{k}} + \mathrm{i}\eta)|\boldsymbol{k}\rangle|_{|\boldsymbol{k}'|=|\boldsymbol{k}|} \tag{15.26}$$

and is proportional to the matrix element of the T-matrix between the two points, $\boldsymbol{k}$ and $\boldsymbol{k}'$ on the sphere $|\boldsymbol{k}'| = |\boldsymbol{k}|$. On the other hand, by expanding the $\boldsymbol{k}'\boldsymbol{k}$ element of the T-matrix in partial waves as

$$\langle \boldsymbol{k}'|T(\varepsilon_{\boldsymbol{k}} + \mathrm{i}\eta)|\boldsymbol{k}\rangle|_{|\boldsymbol{k}'|=|\boldsymbol{k}|} = \sum_{l} (2l+1) T_l(\varepsilon_{\boldsymbol{k}} + \mathrm{i}\eta) \mathrm{P}_l(\cos\theta_{\boldsymbol{k}\boldsymbol{k}'}) \tag{15.27}$$

and comparing (15.26, 27) with (15.16), we can derive the relation

$$-2\pi^2 k T_l(\varepsilon_{\boldsymbol{k}} + \mathrm{i}\eta) = \frac{1}{2\mathrm{i}}(\mathrm{e}^{2\mathrm{i}\delta_l} - 1) = \mathrm{e}^{\mathrm{i}\delta_l} \sin\delta_l \ . \tag{15.28}$$

This relation can be generalized with use of the density of states,

$$\rho(\varepsilon) = \frac{4\pi k^2}{d\varepsilon/dk} = 2\pi k \tag{15.29}$$

defined by

$$\int d\Omega\, k^2 dk = \int \rho(\varepsilon) d\varepsilon \ .$$

The generalized form is

$$\pi\rho(\varepsilon)T_l(\varepsilon_l + \mathrm{i}\eta) = -\mathrm{e}^{\mathrm{i}\delta_l}\sin\delta_l \ . \tag{15.30}$$

From this equation the relation corresponding to the optical theorem is derived immediately as

$$\mathrm{Im}\, T_l = -\pi\rho(\varepsilon)|T_l|^2 \ ; \tag{15.31}$$

the phase of T_l is δ_l.

Now we define S_l, the l wave component of the *S-matrix* , by

$$S_l(\varepsilon) = \mathrm{e}^{2\mathrm{i}\delta_l(\varepsilon)} \ . \tag{15.32}$$

From (15.30, 32), the relation between S_l and T_l is given by

$$S_l(\varepsilon) = 1 - 2\pi\mathrm{i}\rho(\varepsilon)T_l(\varepsilon) \ . \tag{15.33}$$

The optical theorem (15.31) is derived from the unitarity relation

$$S_l S_l^* = 1 \ . \tag{15.34}$$

The relation (15.33) is rewritten in the k-representation as

$$S_{\boldsymbol{k}\boldsymbol{k}'} = \delta(\boldsymbol{k} - \boldsymbol{k}') - 2\pi\mathrm{i}\delta(\varepsilon_{\boldsymbol{k}} - \varepsilon_{\boldsymbol{k}'})T_{\boldsymbol{k}\boldsymbol{k}'}(\varepsilon_{\boldsymbol{k}} + \mathrm{i}\eta) \ . \tag{15.35}$$

Now we discuss the Friedel sum rule. With use of the wave function of (15.23), the scattering wave is written by the substitution of c_l given by (15.15) into (15.5) as

$$\psi(\boldsymbol{r}) = \sum_l (2l+1)(2\pi)^{-3/2}\mathrm{i}^l \mathrm{e}^{\mathrm{i}\delta_l} k^{-1} \mathrm{P}_l(\cos\theta) u_l(r) \ , \tag{15.36}$$

where θ is the angle between $\boldsymbol{k}$ (the wavevector of the incident wave) and $\boldsymbol{r}$. By taking the square of the absolute value of (15.36) and integrating over volume of radius R and over $\boldsymbol{k}$ inside the Fermi sphere, we obtain the number of electrons inside the sphere of radius R. The number (including the spin sum) is given by

$$\int_0^R \rho(r) 4\pi r^2 dr = \int_0^{k_F} dk \left[\frac{4}{\pi} \sum_l (2l+1) \int_0^R \psi_l^2(r) dr \right] , \tag{15.37}$$

where $\psi_l(r)$ is defined by

$$\psi_l(r) = r u_l(r) . \tag{15.38}$$

From (15.6), the differential equation for $\psi_l(r)$ is given by

$$\frac{d^2\psi_l}{dr^2} + \left[k^2 - V(r) - \frac{l(l+1)}{r^2} \right] \psi_l = 0 . \tag{15.39}$$

Combining this equation and that for k', we obtain

$$(k'^2 - k^2) \int_0^R \psi_l \psi_l' dr = \int_0^R \left(\psi_l' \frac{d^2\psi_l}{dr^2} - \psi_l \frac{d^2\psi_l'}{dr^2} \right) dr = \left[\psi_l' \frac{d\psi_l}{dr} - \psi_l \frac{d\psi_l'}{dr} \right]_0^R .$$

Taking the limit $k' \to k$ in the above equation, we obtain

$$2k \int_0^R \psi_l^2 dr = \left[\frac{\partial \psi_l}{\partial r} \frac{\partial \psi_l}{\partial k} - \psi_l \frac{\partial^2 \psi_l}{\partial k \partial r} \right]_0^R . \tag{15.40}$$

Here the contribution from the lower limit of the integral vanishes. Assuming sufficiently large R and using the asymptotic form of $\psi_l(R)$ at the upper limit, namely

$$\psi_l(R) = \sin\left(kR + \delta_l(k) - \frac{l}{2}\pi \right) , \tag{15.41}$$

we obtain

$$\int_0^R \psi_l^2(r) dr = \frac{1}{2k} \left[k \left(R + \frac{d\delta_l}{dk} \right) - \frac{1}{2} \sin 2 \left(kR + \delta_l - \frac{l\pi}{2} \right) \right] . \tag{15.42}$$

Using this relation in (15.37), and taking the difference between the two cases with and without impurity potential, we obtain the change of electronic charge induced by the impurity as

$$\int_0^R [\rho(r) - \rho_0(r)] 4\pi r^2 dr = \frac{2}{\pi} \int_0^{k_F} dk \sum_l (2l+1) \times \left[\frac{d\delta_l}{dk} - \frac{1}{k} \sin\delta_l \cos(2kR + \delta_l - l\pi) \right] . \tag{15.43}$$

We can treat the second term, $(\sin\delta_l)/k$, as a slowly varying function of k compared with the cosine function and obtain the asymptotic form by the partial integration of cosine term.

Thus, from (15.43) one obtains after the integration over k

$$\frac{2}{\pi}\sum_l (2l+1)\left[\delta_l - \frac{1}{2kR}\sin\delta_l \sin(2kR+\delta_l - l\pi)\right]_0^{k_F} .$$

In the case where the potential is not so strong to create a bound state below the conduction band, the phase shift at $k = 0$ is 0. Therefore both the first and the second terms give no contributions at the lower limit and (15.43) becomes

$$\int_0^R [\rho(r)-\rho_0(r)]4\pi r^2 dr = \frac{2}{\pi}\sum_l (2l+1)\left[\delta_l(\varepsilon_F) - \frac{1}{2k_F R}\sin\delta_l(\varepsilon_F)\sin(2k_F R + \delta_l(\varepsilon_F) - l\pi)\right] . \quad (15.44)$$

In the case where the potential is large enough to create a bound state, for example an s-wave bound state with $l = 0, \delta_0(k)$ starts at π at $k = 0$. In this case the second term still vanishes at $k = 0$, but the first term at the lower limit gives the contribution $-(2/\pi)\delta_0(k = 0) = -2$. Since the bound state should be occupied by two electrons with up and down spins, the contribution of the lower limit is canceled out by adding the contribution of two occupied electrons. Therefore, whether the bound states exist or not, (15.44) holds without any modification. Though the phase shift at $k = 0$ changes discontinuously from zero to π at the appearance of the bound state, the change of the phase shift at the sufficiently large Fermi energy is generally smooth. This fact is directly related to the theorem of *Kohn* and *Majumdar* [15.1] which says that the physical quantities of the total electron system can be written in terms of analytical functions of the potential value when the intensity of the local potential changes. However, the case in which strong interaction exists among the electrons should be excluded.

Now it is noted that the second term of (15.44) vanishes in the limit $R \to \infty$. Therefore the change of the total charge is given by the first term. The charge distribution decreases as R^{-3}, oscillating around the impurity at $R = 0$. The local change of total charge, namely the screening charge, is equal to the extra charge Ze of the impurity ion in absolute value, due to charge neutrality in metals, and should completely cancel it out. The number Z is also the difference in valence numbers between host metal atom and solute atom. Therefore the relation

$$\frac{2}{\pi}\sum_l (2l+1)\delta_l(\varepsilon_F) = Z \quad (15.45)$$

holds. This relation was derived by *Friedel* [15.2] and is called the *Friedel sum rule*. This relation can be considered as the self-consistent condition to be satisfied by the impurity potential; conversely, this can be used to estimate the phase shift of electrons at the Fermi surface from the value of Z.

15.2 s-d Model

When iron-group elements are introduced into nonmagnetic metals, there occur two cases; in one case the solute atoms maintain the spins existing in the free-atom state and in the other case they lose their spins. Examples of the former are Fe and Mn doped in Cu and an example of the latter is Mn doped in Al. Whether the iron-group elements possess the spins or not is judged from whether the temperature dependence of the susceptibility due to the impurity obeys the Curie law or not. When a spin $\boldsymbol{S}$ exists on the impurity atom, an interaction appears between this localized spin and the spins of the conduction electrons. This interaction can be written in the form

$$\mathcal{H}_{\text{exch}} = -2Jv \sum_i \delta(\boldsymbol{r}_i)(\boldsymbol{s}_i \cdot \boldsymbol{S}) = -\frac{2J}{N} \sum_{\boldsymbol{q}i} \mathrm{e}^{-\mathrm{i}\boldsymbol{q}\cdot\boldsymbol{r}_i}(\boldsymbol{s}_i \cdot \boldsymbol{S}) , \qquad (15.46)$$

where $\boldsymbol{s}_i$ and $\boldsymbol{r}_i$ represent the spin and the position vector of the conduction electron, respectively. When conduction electrons arrive at the position of the impurity atom (where we put the origin of the position vector), the exchange interaction works between conduction electrons and the localized spin of the impurity atom. The δ function of (15.46) is introduced to show that the interaction is local. J is the exchange integral between conduction electrons and the electron localized at the impurity, and v is the volume per lattice point.

The spin $\boldsymbol{S}$ localized at the impurity induces a polarization of the conduction electrons by the s-d exchange interaction (15.46). If the z-component of the localized spin is assumed to be S_z, the exchange interaction (15.46) is equal to the Zeeman energy due to the external field, the Fourier component of which is given by

$$H_{\boldsymbol{q}} = \frac{2J}{Ng\mu_{\mathrm{B}}} S_z . \qquad (15.47)$$

Therefore the magnetic-moment density of conduction electrons induced by this magnetic field is given, within the linear approximation with respect to J, by (11.20)

$$\sigma(\boldsymbol{r}) = \frac{1}{V} \sum_{\boldsymbol{q}} \chi_{\boldsymbol{q}} H_{\boldsymbol{q}} \mathrm{e}^{-\mathrm{i}\boldsymbol{q}\cdot\boldsymbol{r}} .$$

We substitute (15.47) into $H_{\boldsymbol{q}}$ and (11.25) calculated for the free electron into $\chi_{\boldsymbol{q}}$ and carry out the integration over $\boldsymbol{q}$. Thus we obtain

$$\sigma(r) = \frac{12\pi}{V} \frac{N_e}{N} \frac{J}{g\mu_{\mathrm{B}}} S_z \chi_{\text{Pauli}} F(2k_{\mathrm{F}} r) , \qquad (15.48)$$

$$F(x) = \frac{-x\cos x + \sin x}{x^4} . \qquad (15.49)$$

As shown by these equations, the spin polarization of the conduction electrons decays as $1/r^3$ at large distances, oscillating with period $1/2k_{\mathrm{F}}$, the reciprocal of the diameter of the Fermi sphere. By integrating (15.48) over a sphere of radius R, we obtain

$$\int_0^R \sigma(r) 4\pi r^2 dr = \frac{2J}{Ng\mu_{\mathrm{B}}} S_z \chi_{\mathrm{Pauli}} \left(1 - \frac{\sin 2k_{\mathrm{F}}R}{2k_{\mathrm{F}}R}\right) . \tag{15.50}$$

The first and second terms correspond to the first and second ($l = 0$) terms of (15.44), respectively. The first term is just equal to the total magnetic moment of conduction electrons induced by the uniform magnetic field which is equal to $2JS_z/Ng\mu_{\mathrm{B}}$, that is, the contribution of the $\boldsymbol{q} = 0$ component of the Fourier transform of $\sigma(r)$.

The *s-d* exchange interaction was studied by *Fröhlich* and *Nabarro* [15.3] long time ago as the origin of the ferromagnetism of nuclear spins in metals. After that, it was also considered by *Zener* [15.4] as the origin of the ferromagnetism of iron-group metals. In these theories, only the $\boldsymbol{q} = 0$ component of the spin polarization of conduction electrons was taken into account.

Now we consider the case in which there exist two localized spins at lattice points $\boldsymbol{R}_n$ and $\boldsymbol{R}_m$. By the interaction between spin S_{mz} localized at $\boldsymbol{R}_m$ and the spin density of conduction electrons polarized by spin S_{nz} localized at $\boldsymbol{R}_n$, the following interaction between $\boldsymbol{S}_n$ and $\boldsymbol{S}_m$ is found as

$$\begin{aligned} & -2J\frac{V}{N} \int \delta(\boldsymbol{r} - \boldsymbol{R}_m) \frac{1}{g\mu_{\mathrm{B}}} \sigma(\boldsymbol{r} - \boldsymbol{R}_n) S_{mz} d\boldsymbol{r} \\ & = -9\pi \frac{J^2}{\varepsilon_{\mathrm{F}}} \left(\frac{N_e}{N}\right)^2 F(2k_{\mathrm{F}}|\boldsymbol{R}_n - \boldsymbol{R}_m|) S_{mz} S_{nz} . \end{aligned} \tag{15.51}$$

When we include the transverse components of spins, $S_{nz}S_{mz}$ in (15.51) is replaced by $\boldsymbol{S}_n \cdot \boldsymbol{S}_m$. This interaction was derived directly by *Ruderman* and *Kittel* [15.5] for the nuclear spins $\boldsymbol{I}_n$ in metals as the second-order term in the perturbation expansion with respect to (15.46). In this case, J is replaced by the hyperfine coupling between the conduction electrons and the nuclear spins, and $\boldsymbol{S}_n$ is replaced by the nuclear spin $\boldsymbol{I}_n$. Using this interaction, *Kasuya* [15.6] also discussed the ferromagnetism of the rare-earth metals. On the other hand, *Yosida* [15.7] discussed on the basis of the *s-d* interaction the magnetism and the electrical resistivity of dilute CuMn alloys and studied the spin polarization of conduction electrons around the localized spin. The interaction (15.51) between localized spins in metals, which is mediated by the conduction electrons, is now called the *RKKY interaction.*

Let us turn to (15.48) and discuss further the spin polarization. $F(x)$ diverges as $1/3x$ in the limit of small x. This divergence originates from the approximation which treats the *s-d* exchange interaction as a δ function. In the real case the interaction should extend over the atomic radius and this extension of the interaction leads to a finite value at $x = 0$. In another way, by assuming a finite band closed at the top instead of the open band for free electrons we can also obtain a finite value of $F(x)$ at $x = 0$. Anyway, this divergence gives no problem since the integrated value converges.

Since the second term of (15.50) is an oscillatory damped function, the spin polarization of the conduction electrons given by the first term is localized around the localized spin. Therefore, by including this polarization in the localized spin, we can consider that the value of the localized spin is reduced ($J < 0$) or enhanced ($J > 0$). The total value of the reduction or enhancement is determined by $\chi_{\rm pauli}$, namely, the density of states at the Fermi surface, $\rho(\varepsilon_{\rm F})$. However, the spatial distribution of the polarization around the impurity spin depends on the dispersion of the energy band of conduction electrons, $\varepsilon_{\boldsymbol{k}}$. For general energy bands, the asymptotic form for large R can be obtained analytically and is essentially equivalent to that given by the free-electron approximation [15.8] in showing a damped oscillation as $1/R^3$. However, for small R, the real calculation is rather complicated, since the asymptotic approximation is not applicable.

The systems in which the RKKY interaction plays an important role are dilute magnetic alloys like CuMn, and rare-earth metals and their intermetallic compounds. In case of CuMn, Mn atoms are distributed dilutely at lattice points; by the localized spins of the magnetic impurities the polarization of conduction electrons is induced around the localized spins. It gives the internal magnetic field at the positions of Cu nuclei through the hyperfine interaction. The value of the internal field changes from site to site depending on the position of Cu relative to Mn and distributes with a finite width. Due to this effect, NMR signals of Cu nuclei are spread over a rather wide width, though the peak position does not shift compared with pure Cu metals [15.9]. This is one of the characteristics of CuMn.

In dilute alloys of CuMn, the RKKY interaction acts among Mn impurity spins. Since Mn spins occupy randomly the points of the Cu lattice, the internal magnetic field acting on the spin of one Mn atom differs from that acting on other spins because of the different distribution of surrounding Mn atoms. This internal magnetic field differs from others not only in its magnitude but also in its direction. At sufficiently low temperatures, the Mn spins are frozen in the direction of the randomly oriented internal field. This state of spins is called a *spin glass*. In CuMn alloys, the susceptibility shows a broad peak with decreasing temperature, which is not so sharp as that observed in antiferromagnets under ordinary external magnetic fields [15.10], but it shows a cusp under weak external fields. At low temperatures the characteristic features of the spin-glass are observed in other physical properties such as the specific heat which is linear in T. To study the spin-glass state theoretically is difficult because of its statistical problem and is the subject for which the active study is still continued today as an important field of solid-state physics.

The magnetism of the rare-earth metals originates from the magnetic moment of the incomplete $4f$-shell. As the $4f$-shell exists inside the ion core, the direct exchange interaction between $4f$-electrons is very small even at the nearest neighboring atoms. Therefore, the interaction between the $4f$-electrons in the rare-earth metals arises from the RKKY interaction via conduction electrons. However, in the rare-earth atoms, the LS coupling is so strong that the total angular momentum $\boldsymbol{J} = \boldsymbol{L}+\boldsymbol{S}$ becomes a good quantum number in the $4f$-shell. Therefore, the s-f exchange interaction is obtained by projection of $\boldsymbol{S}$ on the direction of $\boldsymbol{J}$ in (15.46), which represents the s-d exchange interaction, as

$$\mathcal{H}_{\text{exch}} = -\frac{2J}{N}(g_J - 1)\sum_{qi} \mathrm{e}^{-\mathrm{i}q(\boldsymbol{r}_i - \boldsymbol{R}_n)}(\boldsymbol{s}_i \cdot \boldsymbol{J}_n) \,. \tag{15.52}$$

Thus, the RKKY interaction

$$\mathcal{H} = -\frac{9\pi}{2}\left(\frac{N_e}{N}\right)^2 \frac{J^2}{\varepsilon_{\mathrm{F}}}(g_J - 1)^2 \sum_{n \neq m} F(2k_{\mathrm{F}}|\boldsymbol{R}_n - \boldsymbol{R}_m|)(\boldsymbol{S}_n \cdot \boldsymbol{S}_m) \,, \tag{15.53}$$

works among the $4f$-spins of rare-earth metals. Here, $\boldsymbol{S}$ should be written in principle as $\boldsymbol{J}$, but for convenience it is written as $\boldsymbol{S}$ and called the localized spin. It is considered that the ordering of the $4f$-spins in the rare-earth metals such as Gd, Tb, Dy, ..., originates from the RKKY interaction.

15.3 Anderson Hamiltonian

In systems where the localized spin is maintained in nonmagnetic metals, such as Mn atoms in Cu metal, the s-d interaction of the exchange type works between spins of the impurity and conduction electrons. These systems can be well described by the so called s-d model. However, the same Mn atom introduced in Al metal loses its localized spin. Therefore, in order to answer what state the $3d$-electron of an iron-group impurity atom takes in metals, we have to consider it on the basis of a microscopic standpoint. *Anderson* [15.11] presented the following model for the impurity atom in metals, which takes explicitly d-orbitals of iron-group atoms into account:

$$\mathcal{H} = \sum_{k\sigma} \varepsilon_k a^\dagger_{k\sigma} a_{k\sigma} + \sum_{\sigma} E_{\mathrm{d}} a^\dagger_{\mathrm{d}\sigma} a_{\mathrm{d}\sigma} + \sum_{k\sigma}(V_{k\mathrm{d}} a^\dagger_{k\sigma} a_{\mathrm{d}\sigma} + V_{\mathrm{d}k} a^\dagger_{\mathrm{d}\sigma} a_{k\sigma})$$
$$+ U a^\dagger_{\mathrm{d}\uparrow} a_{\mathrm{d}\uparrow} a^\dagger_{\mathrm{d}\downarrow} a_{\mathrm{d}\downarrow} \,. \tag{15.54}$$

Here $a^\dagger_{k\sigma}$ and $a_{k\sigma}$ are the Fermi operators for the conduction electrons with wave vector $\boldsymbol{k}$ and spin σ, and ε_k is the band energy. In addition to the orbitals of the conduction electrons, the $3d$-orbital localized at the impurity atom is introduced and Fermi operators $a^\dagger_{\mathrm{d}\sigma}$ and $a_{\mathrm{d}\sigma}$ are defined for it. σ denotes the spin state $\uparrow$ or $\downarrow$. E_{d} is the energy level of d-orbital and is assumed to be lower than the Fermi energy ε_{F}. Here we neglect the 5-fold degeneracy of d-orbitals for simplicity. The third term in the above equation denotes the electron transfer between the localized d-orbital and conduction electron orbitals and is called the s-d mixing term. $V_{k\mathrm{d}}$ is the matrix element of the impurity potential between the above two orbitals. The fourth term represents the energy due to the Coulomb repulsion between two electrons with antiparallel spins occupying the localized d-orbital, and U is the Coulomb integral between two d-orbitals. Today this Hamiltonian is called Anderson Hamiltonian and is the most fundamental model to describe the impurity atom in metals.

It is the last term describing the repulsive interaction between d-electrons that is the most important in this Anderson Hamiltonian. If this term can be ignored, our problem is reduced to a one-body problem; conduction electrons and the localized d-orbital mix with each other by the mixing term to make new one-body eigenfunctions. The ground state is given by the state in which electrons with up and down spins occupy the new eigenfunctions in the order of energy. Therefore the same number of electrons with up and down spins occupy the d-orbitals of the impurity atom and the spins of the localized d-orbital cancel out. That is, the localized d-electrons have no spins. Mn in Al metal is considered to be this state. It is due to the repulsive potential between d-electrons that d-electrons at the impurity atom possess the degree of spin freedom. Now we consider the limit of small mixing term and put $V_{\mathrm{sd}} = 0$. In this case one electron with either up or down spin occupies the localized d-orbital, because E_{d} is lower than the Fermi energy ε_{F}. When another electron with opposite spin occupies the d-orbital, the the energy of the second electron becomes $E_{\mathrm{d}} + U$ owing to the Coulomb repulsion by the occupied d-electron. If this energy is higher than ε_{F}, the second electron is rejected from the d-orbital. The result is that only one electron can occupy the d-orbital. Thus the d-electron has a spin degree of freedom.

If the mixing matrix V_{sd} is not 0, the Hamiltonian (15.54) describes a many body problem. Therefore, it is very difficult to obtain the eigenvalue and eigenstate. The usual procedure in this case is to use the mean-field approximation, namely the Hartree-Fock approximation. In the fourth term denoting the interaction between d-electrons,

$$\mathcal{H}_c = U n_{\mathrm{d}\uparrow} n_{\mathrm{d}\downarrow} , \tag{15.55}$$

we write the d-electron number $n_{\mathrm{d}\uparrow,\downarrow}$ of $\pm$ spin as the average value $\langle n_{\mathrm{d}\uparrow,\downarrow} \rangle$ and the deviation from it:

$$n_{\mathrm{d}\uparrow,\downarrow} = \langle n_{\mathrm{d}\uparrow,\downarrow} \rangle + (n_{\mathrm{d}\uparrow,\downarrow} - \langle n_{\mathrm{d}\uparrow,\downarrow} \rangle) ; \tag{15.56}$$

we neglect the square of the deviation to replace $U n_{\mathrm{d}\uparrow} n_{\mathrm{d}\downarrow}$ as

$$\mathcal{H}_c \cong -U \langle n_{\mathrm{d}\uparrow} \rangle \langle n_{\mathrm{d}\downarrow} \rangle + U(\langle n_{\mathrm{d}\uparrow} \rangle n_{\mathrm{d}\downarrow} + \langle n_{\mathrm{d}\downarrow} \rangle n_{\mathrm{d}\uparrow}) . \tag{15.57}$$

By this approximation the Anderson Hamiltonian is reduced to the one-body Hamiltonian written as

$$\mathcal{H}_{\mathrm{HF}} = \sum_{k\sigma} \varepsilon_k a_{k\sigma}^\dagger a_{k\sigma} + \sum_{\sigma} (E_{\mathrm{d}} + U\langle n_{\mathrm{d}-\sigma}\rangle) a_{\mathrm{d}\sigma}^\dagger a_{\mathrm{d}\sigma} + \sum_{k\sigma} (V_{k\mathrm{d}} a_{k\sigma}^\dagger a_{\mathrm{d}\sigma} + V_{\mathrm{d}k} a_{\mathrm{d}\sigma}^\dagger a_{k\sigma}) \ . \tag{15.58}$$

It is easy to obtain the eigenvalue of energy ε_n and eigenfunction ψ_n of (15.58). Fermi operators $a_{n\sigma}^\dagger, a_{n\sigma}$ diagonalizing $\mathcal{H}_{\mathrm{HF}}$ are written by a linear combination of $a_{k\sigma}^\dagger$ and $a_{\mathrm{d}\sigma}^\dagger$ as

$$a_{n\sigma}^\dagger = \sum_{k} \langle n|k\rangle_\sigma a_{k\sigma}^\dagger + \langle n|\mathrm{d}\rangle_\sigma a_{\mathrm{d}\sigma}^\dagger \ , \tag{15.59}$$

where $\langle n|k\rangle_\sigma$ and $\langle n|\mathrm{d}\rangle_\sigma$ are the coefficients of $\psi_{n\sigma}$ expanded in $\psi_{k\sigma}$ and $\psi_{\mathrm{d}\sigma}$, respectively. Since the d-orbital is included in the eigenfunction $\psi_{n\sigma}$ with weight $|\langle n|\mathrm{d}\rangle_\sigma|^2$, the energy level isolated at $E_{\mathrm{d}} + U\langle n_{\mathrm{d}-\sigma}\rangle$ is spread over some energy width by the s-d mixing. This kind of d-state was called the virtual bound state by Friedel. The density of states $\rho_{\mathrm{d}\sigma}(\varepsilon)$ for this state is given by

$$\rho_{\mathrm{d}\sigma}(\varepsilon) = \sum_{n} \delta(\varepsilon - \varepsilon_n) |\langle n|\mathrm{d}\rangle_\sigma|^2 \ . \tag{15.60}$$

Now we introduce the Green function

$$G(\varepsilon + \mathrm{i}\eta) = \frac{1}{\varepsilon + \mathrm{i}\eta - \mathcal{H}} \ , \tag{15.61}$$

where η is an infinitesimal positive number and $\mathcal{H}$ is $\mathcal{H}_{\mathrm{HF}}$ in (15.58). Since the imaginary part of $G(\varepsilon+\mathrm{i}\eta)$ is given by $-\pi\delta(\varepsilon-\mathcal{H})$, (15.60) can be written with use of this Green function as

$$\rho_{\mathrm{d}\sigma}(\varepsilon) = -\frac{1}{\pi} \operatorname{Im} G_{\mathrm{dd}}^\sigma(\varepsilon) \ . \tag{15.62}$$

Here, G_{dd}^σ is the diagonal element of the Green function with respect to the d-state and the argument ε in Green function is assumed to include $\mathrm{i}\eta$. Similarly, the total density of states, including the $\boldsymbol{k}$-states, is represented as

$$\rho_\sigma(\varepsilon) = -\frac{1}{\pi} \operatorname{Im}\{\operatorname{Tr} G^\sigma(\varepsilon)\} \ ; \tag{15.63}$$

here Tr means the diagonal sum with respect to $\boldsymbol{k}$ and d states. Therefore, we have only to calculate the Green function to obtain $\rho_{\mathrm{d}\sigma}(\varepsilon)$ and $\rho_\sigma(\varepsilon)$. For this purpose, we substitute (15.58) into the equation

$$\sum_{\nu} (\varepsilon - \mathcal{H})_{\mu\nu} G_{\nu\kappa} = \delta_{\mu\kappa} \ , \tag{15.64}$$

which is satisfied by (15.61), and take $\boldsymbol{k}$ and d as μ and κ in (15.64). Then we can derive the four relations among the matrix elements of the Green function:

$$(\varepsilon - E_\sigma)G^\sigma_{\mathrm{dd}} - \sum_{\boldsymbol{k}} V_{\mathrm{d}\boldsymbol{k}} G^\sigma_{\boldsymbol{k}\mathrm{d}} = 1 , \tag{15.65a}$$

$$(\varepsilon - \varepsilon_{\boldsymbol{k}})G^\sigma_{\boldsymbol{k}\mathrm{d}} - V_{\boldsymbol{k}\mathrm{d}} G^\sigma_{\mathrm{dd}} = 0 , \tag{15.65b}$$

$$(\varepsilon - E_\sigma)G^\sigma_{\mathrm{d}\boldsymbol{k}} - \sum_{\boldsymbol{k}'} V_{\mathrm{d}\boldsymbol{k}'} G^\sigma_{\boldsymbol{k}'\boldsymbol{k}} = 0 , \tag{15.65c}$$

$$(\varepsilon - \varepsilon_{\boldsymbol{k}'})G^\sigma_{\boldsymbol{k}'\boldsymbol{k}} - V_{\boldsymbol{k}'\mathrm{d}} G^\sigma_{\mathrm{d}\boldsymbol{k}} = \delta_{\boldsymbol{k}\boldsymbol{k}'} . \tag{15.65d}$$

Here, for simplicity we have put

$$E_\sigma = E_{\mathrm{d}} + U\langle n_{\mathrm{d}-\sigma}\rangle . \tag{15.66}$$

Substituting $G^\sigma_{\boldsymbol{k}\mathrm{d}}$ obtained from (15.65b) into (15.65a), we obtain G^σ_{dd} as

$$G^\sigma_{\mathrm{dd}} = \left(\varepsilon - E_\sigma - \sum_{\boldsymbol{k}} \frac{|V_{\boldsymbol{k}\mathrm{d}}|^2}{\varepsilon - \varepsilon_{\boldsymbol{k}}}\right)^{-1} . \tag{15.67}$$

The third term of the denominator is given by

$$\sum_{\boldsymbol{k}} \frac{|V_{\boldsymbol{k}\mathrm{d}}|^2}{\varepsilon - \varepsilon_{\boldsymbol{k}}} = \Delta E_{\mathrm{d}} - \mathrm{i}\Delta , \tag{15.68}$$

$$\Delta E_{\mathrm{d}} = \mathrm{P} \sum_{\boldsymbol{k}} \frac{|V_{\boldsymbol{k}\mathrm{d}}|^2}{\varepsilon - \varepsilon_{\boldsymbol{k}}} , \tag{15.69}$$

$$\Delta = \pi \langle |V_{\boldsymbol{k}\mathrm{d}}|^2 \rangle \rho(\varepsilon) . \tag{15.70}$$

Here ΔE_{d} is the term representing the energy shift of the *d*-orbital and P means the principal part. Δ is the term giving the broadening of the d level and $\langle \quad \rangle$ means the average over $\boldsymbol{k}$ states with $\varepsilon_{\boldsymbol{k}} = \varepsilon$. $\rho(\varepsilon)$ is the density of states of conduction electrons defined by

$$\rho(\varepsilon) = \sum_{\boldsymbol{k}} \delta(\varepsilon - \varepsilon_{\boldsymbol{k}}) . \tag{15.71}$$

Ignoring the energy shift ΔE_{d}, replacing $\rho(\varepsilon)$ with the value at the Fermi surface, ρ and neglecting the energy dependence of Δ, we obtain from (15.67, 68)

$$G^\sigma_{\mathrm{dd}} = \frac{1}{\varepsilon - E_\sigma + \mathrm{i}\Delta} . \tag{15.72}$$

Thus we obtain the density of states for the *d*-orbital, which is given by a Lorentzian:

$$\rho_{\mathrm{d}\sigma}(\varepsilon) = \frac{1}{\pi} \frac{\Delta}{(\varepsilon - E_\sigma)^2 + \Delta^2} . \tag{15.73}$$

By a similar calculation, the $\boldsymbol{k}'\boldsymbol{k}$ component of Green function is obtained as

$$G^\sigma_{\boldsymbol{k}'\boldsymbol{k}}(\varepsilon) = \frac{\delta_{\boldsymbol{k}'\boldsymbol{k}}}{\varepsilon - \varepsilon_{\boldsymbol{k}}} + \frac{V_{\boldsymbol{k}'\mathrm{d}} V_{\mathrm{d}\boldsymbol{k}}}{(\varepsilon - \varepsilon_{\boldsymbol{k}'})(\varepsilon - E_\sigma - \Delta E_{\mathrm{d}} + \mathrm{i}\Delta)(\varepsilon - \varepsilon_{\boldsymbol{k}})} . \tag{15.74}$$

The number of d-electrons with $\pm$ spin can be calculated, with use of the density of states for the d orbital $\rho_{\mathrm{d}}^{\sigma}(\varepsilon)$ obtained in (15.73), as

$$\langle n_{\mathrm{d}\sigma}\rangle = \frac{1}{\pi}\int_{-\infty}^{\varepsilon_{\mathrm{F}}} \frac{\Delta d\varepsilon}{(\varepsilon - E_\sigma)^2 + \Delta^2} = \frac{1}{\pi}\cot^{-1}\left(\frac{E_\sigma - \varepsilon_{\mathrm{F}}}{\Delta}\right) . \tag{15.75}$$

This equation is nothing but the relation which determines $\langle n_{\mathrm{d}\sigma}\rangle$ self-consistently, since E_σ contains $\langle n_{\mathrm{d}-\sigma}\rangle$ as seen in (15.66). By using (15.66), (15.75) gives the equations

$$\cot \pi\langle n_{\mathrm{d}\uparrow}\rangle = \frac{1}{\Delta}(E_{\mathrm{d}} - \varepsilon_{\mathrm{F}} + U\langle n_{\mathrm{d}\downarrow}\rangle) , \tag{15.76a}$$

$$\cot \pi\langle n_{\mathrm{d}\downarrow}\rangle = \frac{1}{\Delta}(E_{\mathrm{d}} - \varepsilon_{\mathrm{F}} + U\langle n_{\mathrm{d}\uparrow}\rangle) . \tag{15.76b}$$

These equations always have a solution

$$\langle n_{\mathrm{d}\uparrow}\rangle = \langle n_{\mathrm{d}\downarrow}\rangle = \langle n_{\mathrm{d}}\rangle . \tag{15.77}$$

In this solution, d-electron numbers with up and down spin are equal and the impurity atom has no magnetic moment [Fig. 15.1(a)].

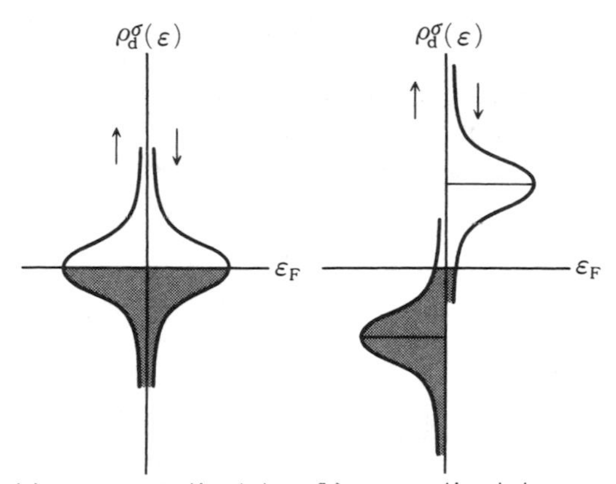

Fig. 15.1. Density of states for d-orbital, $\rho_{\mathrm{d}}(\varepsilon)$

However, as U/Δ increases, this solution becomes unstable and the solution with $\langle n_{\mathrm{d}\uparrow}\rangle \neq \langle n_{\mathrm{d}\downarrow}\rangle$ appears. The condition for this solution is given by

$$\frac{U}{\Delta} \geq \frac{\pi}{\sin^2 \pi\langle n_{\mathrm{d}}\rangle} . \tag{15.78}$$

This condition can be rewritten with use of $\rho_{\mathrm{d}}^{\sigma}(\varepsilon)$ as

$$\rho_{\mathrm{d}}(\varepsilon_{\mathrm{F}})U \geq 1 . \tag{15.79}$$

$\rho_{\rm d}(\varepsilon_{\rm F})$ is the density of states for d-electron at the Fermi surface in case of $\langle n_{{\rm d}\uparrow}\rangle - \langle n_{{\rm d}\downarrow}\rangle$. This form is the same as the condition for the appearance of ferromagnetism in the Hartree-Fock approximation. The electron distribution in the case of $\langle n_{{\rm d}\uparrow}\rangle \neq \langle n_{{\rm d}\downarrow}\rangle$ is shown in Fig. 15.1(b); d-electron numbers with up and down spins are out of balance and the impurity atom possesses the magnetic moment.

When an external magnetic field is applied to the nonmagnetic state, the Zeeman energy $\mp(g/2)\mu_{\rm B}H$ is added to E_σ and E_σ suffers a change given by

$$\delta E_{\uparrow,\downarrow} = \mp\frac{g}{2}\mu_{\rm B}H + U\langle\delta n_{{\rm d}\downarrow,\uparrow}\rangle \,. \tag{15.80}$$

The change of $\langle n_{{\rm d}\uparrow,\downarrow}\rangle$ induced by this change is determined by

$$-\left(\frac{\pi}{\sin^2\pi\langle n_{\rm d}\rangle}\right)\langle\delta n_{{\rm d}\uparrow,\downarrow}\rangle = \frac{\delta E_{\uparrow,\downarrow}}{\Delta}\,, \tag{15.81}$$

which can be derived from (15.76a, b). By solving this equation, we obtain the following expression for the susceptibility due to the d-electron:

$$\begin{aligned}\chi_{\rm d} &= \frac{1}{H}\frac{g\mu_{\rm B}}{2}(\delta n_\uparrow - \delta n_\downarrow) = \frac{1}{2}g^2\mu_{\rm B}^2\Big/\left(\frac{\pi\Delta}{\sin^2\pi\langle n_{\rm d}\rangle} - U\right)\\ &= \frac{1}{2}g^2\mu_{\rm B}^2\rho_{\rm d}(\varepsilon_{\rm F})/[1-\rho_{\rm d}(\varepsilon_{\rm F})U]\,.\end{aligned} \tag{15.82}$$

This susceptibility increases with U and diverges at $\rho_{\rm d}(\varepsilon_{\rm F})U = 1$; for U larger than this value, the impurity atom has a magnetic moment.

Next, we consider the density of states for the conduction electrons $\rho^\sigma_{\rm cond}(\varepsilon)$. Since this is given by

$$\rho^\sigma_{\rm cond}(\varepsilon) = -\frac{1}{\pi}\,{\rm Im}\sum_{\boldsymbol{k}} G^\sigma_{\boldsymbol{kk}}(\varepsilon)\,, \tag{15.83}$$

we use (15.74) for $G^\sigma_{\boldsymbol{kk}}(\varepsilon)$ and obtain

$$\rho^\sigma_{\rm cond}(\varepsilon) = -\frac{1}{\pi}\,{\rm Im}\int_{-\infty}^{\infty} d\varepsilon_{\boldsymbol{k}}\frac{\rho(\varepsilon_{\boldsymbol{k}})}{\varepsilon-\varepsilon_{\boldsymbol{k}}}\left(1+\frac{|V_{{\rm d}\boldsymbol{k}}|^2(\varepsilon-E_\sigma-{\rm i}\Delta)}{(\varepsilon-\varepsilon_{\boldsymbol{k}})[(\varepsilon-E_\sigma)^2+\Delta^2]}\right)\,. \tag{15.84}$$

The second term in this expression is a modification due to the s-d mixing. This part is calculated by expanding $\rho(\varepsilon_{\boldsymbol{k}})$ as

$$\rho(\varepsilon_{\boldsymbol{k}}) = \rho(\varepsilon) - \left.\frac{d\rho}{d\varepsilon_{\boldsymbol{k}}}\right|_\varepsilon(\varepsilon-\varepsilon_{\boldsymbol{k}}) + \ldots\,,$$

noting that ε includes iη and integrating over the complex plane as

$$\Delta\rho^\sigma_{\rm cond}(\varepsilon) = -\frac{d\rho}{d\varepsilon}\frac{|V_{{\rm d}\boldsymbol{k}}|^2(\varepsilon-E_\sigma)}{(\varepsilon-E_\sigma)^2+\Delta^2}\,. \tag{15.85}$$

From this result, we can see that when $d\rho/d\varepsilon$ is small, the density of states for conduction electrons hardly changes.

Since the s-d mixing brings only the electron transfer between d-orbital and conduction electron orbital without any change of spin, any spin change in total electron system including the d-electron does not occur by this mixing. However, when the localized spin is up, conduction electrons with down spin can transfer to the d-orbital with down spin to reduce the self-energy, but conduction electrons with up spin cannot gain this energy reduction because of the Pauli principle. Therefore, the number of conduction electrons with down spin increases. Thus, as a whole, the spin polarization occurs in the opposite direction to the localized d spin. Since the number of conduction electrons with $\pm$ spin hardly changes, as seen from (15.85), the polarization of opposite spin is brought by the d electrons. In real calculation $\langle n_{\mathrm{d}\uparrow}\rangle$ decreases a little from unity and $\langle n_{\mathrm{d}\downarrow}\rangle$ increases by the same amount in case of charge neutrality ($\varepsilon_{\mathrm{F}} - E_{\mathrm{d}} = U/2$). Since the spin of localized d-electron in the s-d model is always fixed to 1/2, this contraction is brought by the conduction electrons.

Let us calculate the total electron number n_σ including d-electron. This is obtained from (15.63) as

$$n_\sigma = -\frac{1}{\pi}\,\mathrm{Im}\int_{-\infty}^{\varepsilon_{\mathrm{F}}}\left[G^\sigma_{\mathrm{dd}}(\varepsilon) + \sum_{\boldsymbol{k}} G^\sigma_{\boldsymbol{kk}}(\varepsilon)\right] d\varepsilon\,. \tag{15.86}$$

With use of (15.74), the sum of $G^\sigma_{\boldsymbol{kk}}(\varepsilon)$ over $\boldsymbol{k}$ is rewritten as

$$\begin{aligned}\sum_{\boldsymbol{k}} G^\sigma_{\boldsymbol{kk}}(\varepsilon) &= \sum_{\boldsymbol{k}}\frac{1}{\varepsilon-\varepsilon_{\boldsymbol{k}}} + \sum_{\boldsymbol{k}}\frac{|V_{\mathrm{d}\boldsymbol{k}}|^2}{(\varepsilon-\varepsilon_{\boldsymbol{k}})^2}\,\frac{1}{\varepsilon - E_\sigma - \sum_{\boldsymbol{k}}\frac{|V_{\boldsymbol{k}\mathrm{d}}|^2}{\varepsilon-\varepsilon_{\boldsymbol{k}}}} \\ &= \sum_{\boldsymbol{k}}\frac{1}{\varepsilon-\varepsilon_{\boldsymbol{k}}} + \frac{d}{d\varepsilon}\log\left(\varepsilon - E_\sigma - \sum_{\boldsymbol{k}}\frac{|V_{\boldsymbol{k}\mathrm{d}}|^2}{\varepsilon-\varepsilon_{\boldsymbol{k}}}\right) \\ &\quad - \frac{1}{\varepsilon - E_\sigma - \sum_{\boldsymbol{k}}\frac{|V_{\boldsymbol{k}\mathrm{d}}|^2}{\varepsilon-\varepsilon_{\boldsymbol{k}}}}\,.\end{aligned} \tag{15.87}$$

Thus, the integrand of (15.86) is given by

$$G^\sigma_{\mathrm{dd}}(\varepsilon) + \sum_{\boldsymbol{k}} G^\sigma_{\boldsymbol{kk}}(\varepsilon) = \sum_{\boldsymbol{k}}\frac{1}{\varepsilon-\varepsilon_{\boldsymbol{k}}} + \frac{d}{d\varepsilon}\left[\log\left(\varepsilon - E_\sigma - \sum_{\boldsymbol{k}}\frac{|V_{\boldsymbol{k}\mathrm{d}}|^2}{\varepsilon-\varepsilon_{\boldsymbol{k}}}\right)\right]. \tag{15.88}$$

The change of n_σ brought about by the s-d mixing is given by the second term. On the other hand, from (15.74), the T-matrix giving the transition from $\boldsymbol{k}$ to $\boldsymbol{k}'$ is given by

$$T_{\boldsymbol{k}'\boldsymbol{k}} = \frac{V_{\boldsymbol{k}'\mathrm{d}}V_{\mathrm{d}\boldsymbol{k}}}{\varepsilon - E_\sigma - \sum_{\boldsymbol{k}}\frac{|V_{\boldsymbol{k}\mathrm{d}}|^2}{\varepsilon-\varepsilon_{\boldsymbol{k}}}}\,. \tag{15.89}$$

Since the phase of T-matrix is the phase shift $\delta_\sigma(\varepsilon), \delta n_\sigma$ is given in term of the phase shift as

$$\delta n_\sigma = \frac{1}{\pi}\int_{-\infty}^{\varepsilon_{\rm F}} \frac{d}{d\varepsilon}\delta_\sigma(\varepsilon)d\varepsilon = \frac{1}{\pi}\delta_\sigma(\varepsilon_{\rm F}) . \qquad (15.90)$$

This is nothing but the Friedel sum rule ($l = 0$) derived before. Here, it should be noted that in the phase shift of conduction electrons due to the impurity scattering, the localized d-electron is also included. In particular, since the conduction electron number hardly changes by the scattering, $\delta_\sigma(\varepsilon_{\rm F})/\pi$ of (15.90) is considered to express the number of the localized d-electrons.

In contrast with the Hartree-Fock approximation (which replaces the interaction term with the mean field and reduces the many-body problem to the one-body one), there exist other methods which treat the problem by perturbational methods. For this purpose, there are two possibilities: the first method treats the interaction U as a perturbation under the assumption of small Coulomb repulsion and the second treats the s-d mixing as perturbation by including the interaction U in the unperturbed term under the assumption of large U. In real iron-group ions, the Coulomb repulsion is considered to be large, and we adopt here the second treatment. In the limit of large U and $|E_{\rm d}|$, as mentioned before, when

$$E_{\rm d} < \varepsilon_{\rm F}, \qquad U + E_{\rm d} > \varepsilon_{\rm F} \qquad (15.91)$$

are satisfied, the d-orbital is occupied by one electron of up or down spin, and the impurity atom possesses the localized spin. Therefore the unperturbed states are doubly degenerate in the spin direction of electron. The two wave functions, denoted as $\psi_\uparrow$ and $\psi_\downarrow$, are written in terms of the wave function ψ_v (of conduction electron states occupied up to the Fermi surface) as

$$\psi_\uparrow = a_{{\rm d}\uparrow}^+\psi_v, \qquad \psi_\downarrow = a_{{\rm d}\downarrow}^+\psi_v . \qquad (15.92)$$

In the second-order perturbation processes for $\psi_\uparrow$, the following four cases are possible.

$$\begin{array}{lll}
\text{(i)} & \boldsymbol{k}\downarrow\to {\rm d}\downarrow\to \boldsymbol{k}'\downarrow ; & \dfrac{V_{\boldsymbol{k}'{\rm d}}V_{{\rm d}\boldsymbol{k}}}{\varepsilon_{\boldsymbol{k}} - U - E_{\rm d}} a_{\boldsymbol{k}'\downarrow}^\dagger a_{{\rm d}\downarrow} a_{{\rm d}\downarrow}^\dagger a_{\boldsymbol{k}\downarrow} \\
\text{(ii)} & \boldsymbol{k}\downarrow\to {\rm d}\downarrow & \\
 & \quad {\rm d}\uparrow\to \boldsymbol{k}'\uparrow ; & \dfrac{V_{\boldsymbol{k}'{\rm d}}V_{{\rm d}\boldsymbol{k}}}{\varepsilon_{\boldsymbol{k}} - U - E_{\rm d}} a_{\boldsymbol{k}'\uparrow}^\dagger a_{{\rm d}\uparrow} a_{{\rm d}\downarrow}^\dagger a_{\boldsymbol{k}\downarrow} \\
\text{(iii)} & {\rm d}\uparrow\to \boldsymbol{k}'\uparrow & \\
 & \quad \boldsymbol{k}\uparrow\to {\rm d}\uparrow ; & \dfrac{V_{{\rm d}\boldsymbol{k}}V_{\boldsymbol{k}'{\rm d}}}{E_{\rm d} - \varepsilon_{\boldsymbol{k}'}} a_{{\rm d}\uparrow}^\dagger a_{\boldsymbol{k}\uparrow} a_{\boldsymbol{k}'\uparrow}^\dagger a_{{\rm d}\uparrow} \\
\text{(iv)} & {\rm d}\uparrow\to \boldsymbol{k}'\uparrow & \\
 & \quad \boldsymbol{k}\downarrow\to {\rm d}\downarrow ; & \dfrac{V_{{\rm d}\boldsymbol{k}}V_{\boldsymbol{k}'{\rm d}}}{E_{\rm d} - \varepsilon_{\boldsymbol{k}'}} a_{{\rm d}\downarrow}^\dagger a_{\boldsymbol{k}\downarrow} a_{\boldsymbol{k}'\uparrow}^\dagger a_{{\rm d}\uparrow}
\end{array} \qquad (15.93)$$

In (i), an s-electron with $\downarrow$ spin and wave vector $\boldsymbol{k}$ transfers to the d-orbital, and then this electron transfers to $\boldsymbol{k}'$ state. In (ii), an s-electron with $\downarrow$

spin and wave vector $\boldsymbol{k}$ transfers to the d-orbital and then a d-electron with $\uparrow$ spin transfers to $\boldsymbol{k}'$ orbital. In (iii) and (iv), firstly, a d-electron with $\uparrow$ spin transfers and then a conduction electron with $\uparrow$ or $\downarrow$ spin and wave vector $\boldsymbol{k}$ transfers to the d-orbital. The second-order perturbation energy for each process is shown to the right of the process. Summing up (i)~(iv) and adding the contributions from the four processes for $\psi_\downarrow$ (which is obtained by replacing $\pm$ spin with $\mp$ spin), we obtain

$$\sum_{\boldsymbol{k}} \frac{|V_{\boldsymbol{k}\mathrm{d}}|^2}{E_\mathrm{d} - \varepsilon_{\boldsymbol{k}}} (a^\dagger_{\mathrm{d}\uparrow} a_{\mathrm{d}\uparrow} + a^\dagger_{\mathrm{d}\downarrow} a_{\mathrm{d}\downarrow}) \tag{15.94}$$

$$+ \sum_{\boldsymbol{k}\boldsymbol{k}'} \frac{V_{\boldsymbol{k}'\mathrm{d}} V_{\mathrm{d}\boldsymbol{k}}}{\varepsilon_{\boldsymbol{k}} - U - E_\mathrm{d}} (a^\dagger_{\boldsymbol{k}'\downarrow} a_{\boldsymbol{k}\downarrow} + a^\dagger_{\boldsymbol{k}'\uparrow} a_{\boldsymbol{k}\uparrow}) \tag{15.95}$$

$$- \sum_{\boldsymbol{k}\boldsymbol{k}'} V_{\boldsymbol{k}'\mathrm{d}} V_{\mathrm{d}\boldsymbol{k}} \left(\frac{1}{\varepsilon_{\boldsymbol{k}} - U - E_\mathrm{d}} + \frac{1}{E_\mathrm{d} - \varepsilon_{\boldsymbol{k}'}} \right) (a^\dagger_{\boldsymbol{k}'\uparrow} a_{\boldsymbol{k}\uparrow} a^\dagger_{\mathrm{d}\uparrow} a_{\mathrm{d}\uparrow}$$
$$+ a^\dagger_{\boldsymbol{k}'\downarrow} a_{\boldsymbol{k}\downarrow} a^\dagger_{\mathrm{d}\downarrow} a_{\mathrm{d}\downarrow} + a^\dagger_{\boldsymbol{k}'\uparrow} a_{\boldsymbol{k}\downarrow} a^\dagger_{\mathrm{d}\downarrow} a_{\mathrm{d}\uparrow} + a^\dagger_{\boldsymbol{k}'\downarrow} a_{\boldsymbol{k}\uparrow} a^\dagger_{\mathrm{d}\uparrow} a_{\mathrm{d}\downarrow}) \,. \tag{15.96}$$

Now we rewrite the sum of the first and the second term in the last bracket in (15.96) as

$$\frac{1}{2}(a^\dagger_{\boldsymbol{k}'\uparrow} a_{\boldsymbol{k}\uparrow} - a^\dagger_{\boldsymbol{k}'\downarrow} a_{\boldsymbol{k}\downarrow})(a^\dagger_{\mathrm{d}\uparrow} a_{\mathrm{d}\uparrow} - a^\dagger_{\mathrm{d}\downarrow} a_{\mathrm{d}\downarrow})$$
$$+ \frac{1}{2}(a^\dagger_{\boldsymbol{k}'\uparrow} a_{\boldsymbol{k}\uparrow} + a^\dagger_{\boldsymbol{k}'\downarrow} a_{\boldsymbol{k}\downarrow})(a^\dagger_{\mathrm{d}\uparrow} a_{\mathrm{d}\uparrow} + a^\dagger_{\mathrm{d}\downarrow} a_{\mathrm{d}\downarrow}) \,.$$

On the other hand, denoting the spin of the d-electron as $\boldsymbol{S}$ and using the relations

$$\frac{1}{2}(a^\dagger_{\mathrm{d}\uparrow} a_{\mathrm{d}\uparrow} - a^\dagger_{\mathrm{d}\downarrow} a_{\mathrm{d}\downarrow}) = S_z \,,$$
$$a^\dagger_{\mathrm{d}\uparrow} a_{\mathrm{d}\downarrow} = S_+ \,, \qquad a^\dagger_{\mathrm{d}\downarrow} a_{\mathrm{d}\uparrow} = S_- \,, \tag{15.97}$$

we can write the sum of (15.95, 96) as

$$\mathcal{H}_\mathrm{eff} = \mathcal{H}_\mathrm{imp} + \mathcal{H}_\mathrm{exch} \,, \tag{15.98}$$

$$\mathcal{H}_\mathrm{imp} = \sum_{\boldsymbol{k}\boldsymbol{k}'} V_{\boldsymbol{k}'\mathrm{d}} V_{\mathrm{d}\boldsymbol{k}} \left[\frac{1}{\varepsilon_{\boldsymbol{k}} - U - E_\mathrm{d}} - \frac{1}{2} n_\mathrm{d} \left(\frac{1}{\varepsilon_{\boldsymbol{k}} - U - E_\mathrm{d}} + \frac{1}{E_\mathrm{d} - \varepsilon_{\boldsymbol{k}'}} \right) \right]$$
$$\times (a^\dagger_{\boldsymbol{k}'\uparrow} a_{\boldsymbol{k}\uparrow} + a^\dagger_{\boldsymbol{k}'\downarrow} a_{\boldsymbol{k}\downarrow}) \,, \tag{15.99}$$

$$\mathcal{H}_\mathrm{exch} = - \sum_{\boldsymbol{k}\boldsymbol{k}'} V_{\boldsymbol{k}'\mathrm{d}} V_{\mathrm{d}\boldsymbol{k}} \left(\frac{1}{\varepsilon_{\boldsymbol{k}} - U - E_\mathrm{d}} + \frac{1}{E_\mathrm{d} - \varepsilon_{\boldsymbol{k}'}} \right)$$
$$\times \left[(a^\dagger_{\boldsymbol{k}'\uparrow} a_{\boldsymbol{k}\uparrow} - a^\dagger_{\boldsymbol{k}'\downarrow} a_{\boldsymbol{k}\downarrow}) S_z + a^\dagger_{\boldsymbol{k}'\uparrow} a_{\boldsymbol{k}\downarrow} S_- + a^\dagger_{\boldsymbol{k}'\downarrow} a_{\boldsymbol{k}\uparrow} S_+ \right] \,. \tag{15.100}$$

Here, n_d is the number of d-electrons (unity in our case). Thus the effective Hamiltonian derived by second-order perturbation theory with respect to the s-d mixing is composed of two parts, (15.99) representing the impurity scattering and (15.100) representing the exchange interaction between conduction electrons and localized spin $\boldsymbol{S}$, except the term of (15.94) giving the self-energy of the d-electron [15.12].

In (15.99), we put $n_\mathrm{d} = 1$ and neglect the $\boldsymbol{k}$-dependence of the transfer matrix $V_{\boldsymbol{k}\mathrm{d}}$; we also fix the origin of energy at the Fermi energy ε_F and ignore $\varepsilon_{\boldsymbol{k}}$ in the denominator compared with U and $|E_\mathrm{d}|$. Then we obtain the coefficient giving the strength of the impurity scattering as

$$\frac{1}{2}|V|^2\left(-\frac{1}{E_\mathrm{d}} - \frac{1}{U+E_\mathrm{d}}\right) . \tag{15.101}$$

For the case $|E_\mathrm{d}| < U + E_\mathrm{d}$ this gives a repulsive potential, while for $|E_\mathrm{d}| > U + E_\mathrm{d}$ it gives an attractive potential; for the former conduction electrons avoid the impurity atom and for the latter they gather around the impurity atom. In the case where $|E_\mathrm{d}|$ is just equal to $U + E_\mathrm{d}$ and two d-orbitals with $\pm$ spin are symmetric with respect to ε_F, there is no impurity potential and charge neutrality is maintained even in the presence of the s-d mixing.

With the same approximation for (15.100) we obtain

$$\mathcal{H}_\mathrm{exch} = -\frac{J_\mathrm{eff}}{N}\sum_{\boldsymbol{k}\boldsymbol{k}'}\left[(a^\dagger_{\boldsymbol{k}'\uparrow}a_{\boldsymbol{k}\uparrow} - a^\dagger_{\boldsymbol{k}'\downarrow}a_{\boldsymbol{k}\downarrow})S_z + a^\dagger_{\boldsymbol{k}'\uparrow}a_{\boldsymbol{k}\downarrow}S_- + a^\dagger_{\boldsymbol{k}'\downarrow}a_{\boldsymbol{k}\uparrow}S_+\right] , \tag{15.102}$$

$$J_\mathrm{eff} = -N|V|^2\left(\frac{1}{U+E_\mathrm{d}} + \frac{1}{-E_\mathrm{d}}\right) < 0 . \tag{15.103}$$

Equation (15.102) is the same as the Hamiltonian in which the following exchange interaction works between the localized spin and the conduction electrons,

$$\mathcal{H}_\mathrm{exch} = -2J_\mathrm{eff}v\sum_i \delta(\boldsymbol{r}_i)(\boldsymbol{s}_i\cdot\boldsymbol{S}) = -2\frac{J_\mathrm{eff}}{N}\sum_{\boldsymbol{q}}\sum_i \mathrm{e}^{-\mathrm{i}\boldsymbol{q}\cdot\boldsymbol{r}_i}(\boldsymbol{s}_i\cdot\boldsymbol{S}) , \tag{15.104}$$

where v is the atomic volume V/N. Note that the sign of J_eff corresponding to the exchange integral is negative. For the case in which E_d and $U + E_\mathrm{d}$ are symmetric with respect to ε_F, $J_\mathrm{eff} = -4N|V|^2/U$ holds. As we have seen, we arrive at the conclusion that the Anderson Hamiltonian is equivalent to the s-d exchange interaction (15.102) in the limit of large U and $|E_\mathrm{d}|$.

16. Kondo Effect

The essential properties of dilute magnetic alloys such as CuMn, which contain iron-group atoms as impurities, have been understood on the basis of the *s-d* model for the interaction between conduction electrons and a localized spin of magnetic impurity. However, the phenomenon of the resistance minimum, observed in the temperature dependence of the electrical resistivity in magnetic alloys with low concentration (less than 0.1%), has still remained to be a difficult unsolved problem. At high temperatures, the resistance of dilute alloys arises from the scattering of conduction electrons by the random deviation of the periodic potential due to lattice vibrations. This resistance is proportional to T at high temperatures ; at lower temperatures it decreases more rapidly, approaching zero proportionally to T^5. In addition to the resistance due to lattice vibrations, the scattering by impurities contributes to the resistance in dilute alloys. Usually, the electrical resistance due to impurity scattering is independent of temperature. Therefore the electrical resistance in dilute alloys is proportional to T at high temperatures and decreases rapidly with decreasing temperature to reach a finite value at $T = 0$, which is called the residual resistance. However, in dilute magnetic alloys, though the resistance decreases with decreasing temperature, it reaches a minimum value around 10K, below which it increases with decreasing temperature. The phenomenon of the resistance minimum was discovered in the early 1930's and research on the phenomenon continued for many years since then. In particular, after the second world war, research on the resistance minimum was undertaken at the Kamerlingh-Onnes Laboratory at Leiden, the University of California at Berkeley and the Université de Paris *etc.* In 1964, the key to understand the phenomenon of the resistance minimum was found by *Kondo* [16.1]. In this and the following chapters the developments of the theory related to the localized spins of impurity atoms in metals, which started with the resistance minimum (now called Kondo effect), will be described in due order. We begin with discussion on the electrical resistivity due to the *s-d* exchange interaction.

16.1 Electrical Resistivity in the *s*-*d* Model

Conduction electrons in metals are scattered by localized spins on impurity atoms. We consider the electrical resistivity arising from this scattering. The current densities due to electrons with up and down spin, respectively, which are induced by the external field E (taken parallel to the x-axis), are obtained from

$$j_{\uparrow\downarrow} = -\frac{e}{V}\sum_{\boldsymbol{k}} \frac{\hbar k_x}{m}[f_{\uparrow\downarrow}(\varepsilon_{\boldsymbol{k}}) - f_0(\varepsilon_{\boldsymbol{k}})] , \tag{16.1}$$

where $f_{\uparrow\downarrow}(\varepsilon_{\boldsymbol{k}})$ is the distribution function in k-space in the presence of the electric current under the external field, and $f_0(\varepsilon_{\boldsymbol{k}})$ is that in the absence of electric field, namely the Fermi distribution function

$$f_0(\varepsilon) = \frac{1}{e^{(\varepsilon-\varepsilon_F)/kT} + 1} . \tag{16.2}$$

Here, we use the Fermi energy ε_F as the chemical potential. We intoduce the average lifetime, $\tau_l(\varepsilon_{\boldsymbol{k}})$, for the scattering of electrons with wave vector $\boldsymbol{k}$ by impurity atoms. Then the difference of the distribution function appearing in (16.1) is given by

$$\frac{\hbar k_x}{m} eE\tau_{l\uparrow\downarrow}(\varepsilon_{\boldsymbol{k}})\frac{\partial f_0}{\partial \varepsilon_{\boldsymbol{k}}} .$$

Substitution of this expression into (16.1) gives the electrical conductivity σ as

$$\sigma = -\frac{e^2}{V}\frac{4}{3m}\int \rho(\varepsilon_{\boldsymbol{k}})\varepsilon_{\boldsymbol{k}}\tau_l(\varepsilon_{\boldsymbol{k}})\frac{\partial f_0}{\partial \varepsilon_{\boldsymbol{k}}} d\varepsilon_{\boldsymbol{k}} . \tag{16.3}$$

Here $\rho(\varepsilon_{\boldsymbol{k}})$ is the density of states for conduction electrons and the mean-free times of conduction electrons with up and down spins are put equal to each other. The lifetime $\tau_l(\varepsilon_{\boldsymbol{k}})$ is obtained by calculating the probability that an electron with wave vector $\boldsymbol{k}$ and up spin makes a transition to $\boldsymbol{k}'\uparrow$ or $\boldsymbol{k}'\downarrow$ due to the *s*-*d* exchange interaction. The transition probability is given by the golden rule as

$$W(\boldsymbol{k}\uparrow\rightarrow \boldsymbol{k}'\uparrow) = \frac{2\pi}{\hbar}\sum_M w_M |T(\boldsymbol{k}'\uparrow M|\boldsymbol{k}\uparrow M)|^2 \delta(\varepsilon_{\boldsymbol{k}} - \varepsilon_{\boldsymbol{k}'}) , \tag{16.4a}$$

$$W(\boldsymbol{k}\uparrow\rightarrow \boldsymbol{k}'\downarrow) = \frac{2\pi}{\hbar}\sum_M w_M |T(\boldsymbol{k}'\downarrow M+1|\boldsymbol{k}\uparrow M)|^2 \delta(\varepsilon_{\boldsymbol{k}} - \varepsilon_{\boldsymbol{k}'}) , \tag{16.4b}$$

where T is the transition matrix. The probability w_M depends on M, the z component of the localized spin and we assume zero magnetic field. Summing (16.4) over $\boldsymbol{k}'$ and adding expressions (16.4a, b), we find that the inverse of the mean-free time (per impurity atom) is

$$\frac{1}{\tau_l(\varepsilon_{\boldsymbol{k}})} = \frac{2\pi}{\hbar}\rho(\varepsilon_{\boldsymbol{k}})\Big(\sum_M w_M|T(\varepsilon_{\boldsymbol{k}'\uparrow}M|\varepsilon_{\boldsymbol{k}\uparrow}M)\,|^2_{\varepsilon_{\boldsymbol{k}'}=\varepsilon_{\boldsymbol{k}}} + \sum_M w_M\,|\,T(\varepsilon_{\boldsymbol{k}'\downarrow}M+1\,|\,\varepsilon_{\boldsymbol{k}\uparrow}M)|^2_{\varepsilon_{\boldsymbol{k}'}=\varepsilon_{\boldsymbol{k}}}\Big)\,, \tag{16.5}$$

where we have assumed the s-wave scattering. In the general case, the transition matrix T depends also on the angle between $\boldsymbol{k}$ and $\boldsymbol{k}'$; for this case, we expand T in partial waves as follows:

$$T(\boldsymbol{k}'\uparrow|\boldsymbol{k}\uparrow) = \sum_l (2l+1)T_l(\varepsilon_{\boldsymbol{k}})\mathrm{P}_l(\cos\theta_{\boldsymbol{k}\boldsymbol{k}'})\,, \tag{16.6}$$

where each term in the brackets in (16.5) is replaced by

$$\sum_M w_M \int\frac{d\Omega}{4\pi}\left|\sum_l(2l+1)T_l(\varepsilon_{\boldsymbol{k}}M)\mathrm{P}_l(\cos\theta_{\boldsymbol{k}\boldsymbol{k}'})\right|^2_{\varepsilon_{\boldsymbol{k}}=\varepsilon_{\boldsymbol{k}'}}(1-\cos\theta_{\boldsymbol{k}\boldsymbol{k}'})\,. \tag{16.7}$$

Here we have inserted the factor $(1-\cos\theta_{\boldsymbol{k}\boldsymbol{k}'})$ to take into account the fact that the forward scattering gives no contribution to the resistivity. This factor, however, is not necessary for s-wave ($l=0$) scattering, since the angular integral of $\cos\theta$ vanishes.

For the s-d exchange interaction described by (15.46) or (15.102), in which only the s-wave part is taken into account, (that is, we neglected $\boldsymbol{k}, \boldsymbol{k}'$ dependence of J_{eff} there and treated it as a constant value), T-matrix for one electron scattering can be written as

$$T(\varepsilon+\mathrm{i}\eta) = t(\varepsilon+\mathrm{i}\eta) + \tau(\varepsilon+\mathrm{i}\eta)(\boldsymbol{\sigma}\cdot\boldsymbol{S})\,, \tag{16.8}$$

by considering the existence of localized spin. Here, $t(\varepsilon)$ and $\tau(\varepsilon)$ are the parts from spin conserving or spin non-flip and spin-flip processes, respectively; σ is the Pauli spin matrix. For simplicity, we write $\varepsilon+\mathrm{i}\eta$ as ε hereafter. On substituting (16.8) for T into (16.5), we obtain

$$\begin{aligned} T_{\uparrow\uparrow}(\varepsilon_{\boldsymbol{k}}) &= t(\varepsilon_{\boldsymbol{k}}) + \tau(\varepsilon_{\boldsymbol{k}})M\,,\\ T_{\downarrow\uparrow}(\varepsilon_{\boldsymbol{k}}) &= \tau(\varepsilon_{\boldsymbol{k}})\sqrt{S(S+1)-M(M+1)}\,. \end{aligned} \tag{16.9}$$

Thus, the inverse of the mean-free time τ_l of conduction electrons (per impurity atom) becomes

$$\frac{1}{\tau_l} = \frac{2}{\hbar}\pi\rho(\varepsilon_{\boldsymbol{k}})\left[|t(\varepsilon_{\boldsymbol{k}})|^2 + |\tau(\varepsilon_{\boldsymbol{k}})|^2S(S+1)\right]\,. \tag{16.10}$$

On the other hand, from the unitarity of the scattering S-matrix, or the optical theorem for the case with two channels the relation

$$-\operatorname{Im}t_{\boldsymbol{k}\boldsymbol{k}} = \pi\rho(\varepsilon_{\boldsymbol{k}})\left[|t(\varepsilon_{\boldsymbol{k}})|^2 + |\tau(\varepsilon_{\boldsymbol{k}})|^2S(S+1)+\ldots\right] \tag{16.11}$$

holds. With use of this relation the mean-free time is given by

$$\frac{1}{\tau_l} = -\frac{2}{\hbar} \operatorname{Im} t_{\boldsymbol{kk}} \; . \tag{16.12}$$

The transition matrix as given by the first-order approximation (the Born approximation) with respect to the *s-d* exchange interaction (15.46) or (15.102), is

$$t^{(1)}(\varepsilon_{\boldsymbol{k}}) = 0 \; , \qquad \tau^{(1)}(\varepsilon_{\boldsymbol{k}}) = -\frac{J}{2N} \; . \tag{16.13}$$

Henceforth, we use J instead of $2J$ as the coefficient of exchange interaction J. On substituting (16.13) into (16.10), and substituting this mean-free time τ_l into (16.3), we obtain the resistivity (the inverse of σ) per impurity atom in the Born approximation, R_{B}, as

$$R_{\mathrm{B}} = \frac{3}{2}\frac{m\pi}{e^2\hbar}\frac{V}{\varepsilon_{\mathrm{F}}}\frac{J^2}{4N^2}S(S+1) \; . \tag{16.14}$$

Thus the *s-d* exchange interaction adds a new term to the residual resistivity due to the impurity scattering. As seen clearly from the above calculation, $1/3$ of $S(S+1)$ in (16.14) is the contribution from the z-component of the *s-d* exchange interaction and $2/3$ from the transverse component. In the cases where an external magnetic field is applied or an internal field exists owing to the mutual interaction between localized spins, the contribution of the transverse component is suppressed and that of the z-component is increased. At sufficiently low temperatures and in the limit of large magnetic field, the contribution of the z-component is given by S^2. Therefore, when the localized spin can rotate freely (at high temperatures and in weak magnetic field), the contribution is given by $S(S+1)$, while for the localized spin fixed to the z-direction (at low temperatures and in strong magnetic field), the contribution is S^2. For the latter case, the resistivity is small. Therefore the Born approximation cannot explain the resistance minimum.

16.2 Second-Order Scattering Matrix

In 1964 *Kondo* calculated the scattering matrix for conduction electrons to order J^2, the next order term beyond the Born approximation in order to explain the phenomenon of the resistance minimum. He found that there exists a singular term proportional to $\log kT$.

The scattering matrix of conduction electrons in the *s-d* model consists of the spin-conserving part $t(\varepsilon)$ and the spin flip part $\tau(\varepsilon)$. To obtain these two parts, we need only calculate one component of the T-matrix, $T_{\uparrow\uparrow}(\varepsilon)$. For the second-order processes where electron in state $\boldsymbol{k}\uparrow$ is scattered to the state $\boldsymbol{k}'\uparrow$, there are the two types shown in Fig. 16.1.

In (a), the electron in state $\boldsymbol{k}\uparrow$ is scattered by the localized spin to the state $\boldsymbol{k}''\uparrow$ or $\boldsymbol{k}''\downarrow$ and the z-component of the localized spin is unchanged

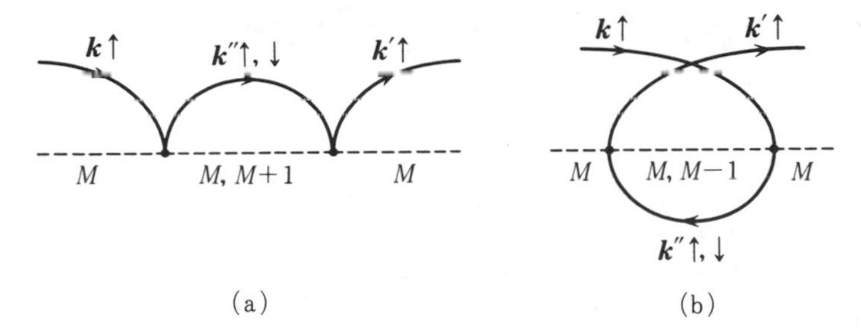

Fig. 16.1. Two processes giving J^2 terms of T-matrix

or changed from M to $M+1$. After this process, the electron $\boldsymbol{k}''\uparrow$ or $\boldsymbol{k}''\downarrow$ is scattered again by the localized spin to the $\boldsymbol{k}'\uparrow$ state and at the same time the z-component of the localized spin returns to the original value M. In process (b), the electron at first in $\boldsymbol{k}''\uparrow$ or $\boldsymbol{k}''\downarrow$ is excited to $\boldsymbol{k}'\uparrow$ and then the electron $\boldsymbol{k}\uparrow$ combines with the hole $\boldsymbol{k}''\uparrow$ or $\boldsymbol{k}''\downarrow$ to leave an electron in $\boldsymbol{k}'\uparrow$. In these processes, the energy is invariant between the initial and final states and $\varepsilon_{\boldsymbol{k}} = \varepsilon_{\boldsymbol{k}'} = \varepsilon$. The T-matrix for these two processes is

$$T_a(\boldsymbol{k}'\uparrow,\boldsymbol{k}\uparrow) = \left(-\frac{J}{2N}\right)^2 \left(S_z^2 \sum_{\boldsymbol{k}''} \frac{\langle a^\dagger_{\boldsymbol{k}'\uparrow}\psi_v | a^\dagger_{\boldsymbol{k}'\uparrow} a_{\boldsymbol{k}''\uparrow} a^\dagger_{\boldsymbol{k}''\uparrow} a_{\boldsymbol{k}\uparrow} | a^\dagger_{\boldsymbol{k}\uparrow}\psi_v\rangle}{\varepsilon - \varepsilon_{\boldsymbol{k}''} + \mathrm{i}\eta} + S_- S_+ \sum_{\boldsymbol{k}''} \frac{\langle a^\dagger_{\boldsymbol{k}'\uparrow}\psi_v | a^\dagger_{\boldsymbol{k}'\uparrow} a_{\boldsymbol{k}''\downarrow} a^\dagger_{\boldsymbol{k}''\downarrow} a_{\boldsymbol{k}\uparrow} | a^\dagger_{\boldsymbol{k}\uparrow}\psi_v\rangle}{\varepsilon - \varepsilon_{\boldsymbol{k}''} + \mathrm{i}\eta}\right), \tag{16.15}$$

$$T_b(\boldsymbol{k}'\uparrow,\boldsymbol{k}\uparrow) = \left(-\frac{J}{2N}\right)^2 \left(S_z^2 \sum_{\boldsymbol{k}''} \frac{\langle a^\dagger_{\boldsymbol{k}'\uparrow}\psi_v | a^\dagger_{\boldsymbol{k}''\uparrow} a_{\boldsymbol{k}\uparrow} a^\dagger_{\boldsymbol{k}'\uparrow} a_{\boldsymbol{k}''\uparrow} | a^\dagger_{\boldsymbol{k}\uparrow}\psi_v\rangle}{\varepsilon_{\boldsymbol{k}''} - \varepsilon - \mathrm{i}\eta} + S_+ S_- \sum_{\boldsymbol{k}''} \frac{\langle a^\dagger_{\boldsymbol{k}'\uparrow}\psi_v | a^\dagger_{\boldsymbol{k}''\downarrow} a_{\boldsymbol{k}\uparrow} a^\dagger_{\boldsymbol{k}'\uparrow} a_{\boldsymbol{k}''\downarrow} | a^\dagger_{\boldsymbol{k}\uparrow}\psi_v\rangle}{\varepsilon_{\boldsymbol{k}''} - \varepsilon - \mathrm{i}\eta}\right), \tag{16.16}$$

where ψ_v represents the Fermi sphere state. T_a and T_b are calculated with the use of distribution function $f(\varepsilon_{\boldsymbol{k}''})$, respectively, as

$$T_a(\boldsymbol{k}'\uparrow,\boldsymbol{k}\uparrow) = \left(\frac{J}{2N}\right)^2 [S(S+1) - S_z] \int_{-\infty}^{\infty} \rho(\varepsilon') \frac{1 - f(\varepsilon')}{\varepsilon - \varepsilon' + \mathrm{i}\eta} d\varepsilon' , \tag{16.17}$$

$$T_b(\boldsymbol{k}'\uparrow,\boldsymbol{k}\uparrow) = \left(\frac{J}{2N}\right)^2 [S(S+1) + S_z] \int_{-\infty}^{\infty} \rho(\varepsilon') \frac{f(\varepsilon')}{\varepsilon - \varepsilon' + \mathrm{i}\eta} d\varepsilon' . \tag{16.18}$$

The sum of (16.17, 18) gives the second-order term of the T-matrix as

$$T(\boldsymbol{k}'\uparrow M, \boldsymbol{k}\uparrow M) = \frac{J^2}{4N^2}S(S+1)\int_{-\infty}^{\infty}\rho(\varepsilon')\frac{1}{\varepsilon-\varepsilon'+\mathrm{i}\eta}d\varepsilon'$$
$$-\frac{J^2}{4N^2}M\int_{-\infty}^{\infty}\rho(\varepsilon')\frac{1-2f(\varepsilon')}{\varepsilon-\varepsilon'+\mathrm{i}\eta}d\varepsilon' . \tag{16.19}$$

The first term is the normal part which can be treated as one-body problem in a similar way to impurity scattering, since it doesn't contain the Fermi distribution function in the integrand, and it is also the lowest-order term of the spin non-flip part t, as it does not depend on M. The imaginary part

$$\mathrm{Im}\, t^{(2)}(\varepsilon) = -\pi\rho(\varepsilon)\frac{J^2}{4N^2}S(S+1) \tag{16.20}$$

satisfies the optical theorem combined with the Born term of the spin-flip part $\tau(\varepsilon)$.

The real part of the second term is the singular term discovered by Kondo. The term is singular because, as shown below, the integral gives $\log|\varepsilon|$ or $\log T$ which diverges logarithmically as ε or T tends to zero. Also, the real part of the second term itself originates from the fact that S_+ and S_- are quantum variables and do not commute; note that the result depends on the distribution of the other electrons, which means the many-body effect. The second term of (16.19) is proportional to M and gives the J^2 term of the spin-flip part τ. That is,

$$\tau^{(2)}(\varepsilon) = -\frac{J^2}{4N^2}\int_{-\infty}^{\infty}\rho(\varepsilon')\mathrm{P}\frac{1-2f(\varepsilon')}{\varepsilon-\varepsilon'}d\varepsilon' + \mathrm{i}\pi\frac{J^2}{4N^2}\rho(\varepsilon)[1-2f(\varepsilon)] . \tag{16.21}$$

Taking the zero of energy at ε_{F} and assuming a flat density of states $\rho(\varepsilon)$ given by

$$\rho(\varepsilon) = \begin{cases} \text{constant} & (-D<\varepsilon<D) , \\ 0 & (|\varepsilon|>D) , \end{cases} \tag{16.22}$$

we calculate the real part by partial integration, obtaining

$$\mathrm{Re}\,\tau^{(2)}(\varepsilon) = -\frac{J^2}{4N^2}\rho\Bigg(-\log|D-\varepsilon| - \log|D+\varepsilon|$$
$$-2\int_{-D}^{D}\log|\varepsilon-\varepsilon'|\frac{df'}{d\varepsilon'}d\varepsilon'\Bigg) .$$

In the integral of the third term, we can approximate the factor $df'/d\varepsilon'$ as finite only in a range of width kT around $\varepsilon'=0$; the third term then gives $2\log|\varepsilon|$ for the case $|\varepsilon|>kT$. On the other hand, the third term is calculated for the case $|\varepsilon|<kT$ as

$$2\log kT + \int_0^{\infty}\log x\ \mathrm{sech}^2\frac{x}{2}dx\left(=-2\log\frac{2e^{\gamma}}{\pi}\right) ,$$

where γ is the Euler number $\simeq 0.577$. Thus, for ε sufficiently small compared with D, we obtain

$$\mathrm{Re}\,\tau^{(2)}(\varepsilon) = \begin{cases} -\dfrac{J}{2N}\dfrac{J\rho}{N}\log\dfrac{|\varepsilon|}{D} & (|\varepsilon| > kT)\,, \\ -\dfrac{J}{2N}\dfrac{J\rho}{N}\log\dfrac{kT}{D} & (|\varepsilon| < kT)\,. \end{cases} \tag{16.23}$$

Even if $|J\rho/N|$ is small, the second-order term cannot be neglected compared with the Born term for sufficiently small ε and kT. Thus, including the Born term, we obtain $\tau(\varepsilon)$ as

$$\mathrm{Re}\,\tau(\varepsilon) = -\frac{J}{2N}\left(1 + \frac{J\rho}{N}\log\frac{|\varepsilon|}{D} + \ldots\right)\,. \tag{16.24}$$

If we use this result in place of the Born term in (16.13), we obtain the electrical resistivity as

$$R = R_{\mathrm{B}}\left(1 + 2\frac{J\rho}{N}\log\frac{kT}{D}\right)\,. \tag{16.25}$$

For the case $J < 0$, the factor multiplying the Born term increases with decreasing temperature from a value near unity at high temperatures. Combining this term with the resistance due to lattice vibrations ($\propto T^5$), we obtain the resistance minimum. This is the explanation of the resistance minimum due to *Kondo* [16.1]. It is important that J be negative for the resistance minimum to be realized. This fact provides strong evidence that the *s-d* exchange interaction in the dilute iron-group alloys originates from the mechanism of *s-d* mixing.

16.3 Resistivity, Susceptibility and Specific Heat in Higher-Order Perturbation Theory

The singular term $(J\rho/N)\log(kT/D)$ obtained in the second Born approximation for $J < 0$ increases with decreasing temperature and diverges as $T \to 0$. What does this divergence mean? At low temperatures where $J\log(T/D)$ becomes of order unity, terms of yet higher order than the second Born term also give important contributions. Therefore, in order to clarify the divergence in the singular terms obtained by Kondo, we must study higher order terms. *Abrikosov* [16.2] studied the scattering T-matrix in higher order and showed that the terms most divergent in the limit of $\varepsilon \to 0$ are given by $-(J/2N)[(J\rho/N)\log(|\varepsilon|/D)]^{n-1}$ in the nth order perturbation, with the coefficient of unity.

For example, in third order there are such scattering processes as shown in Fig. 16.2. In these figures, the diagrams from (a) to (f) can be reduced to

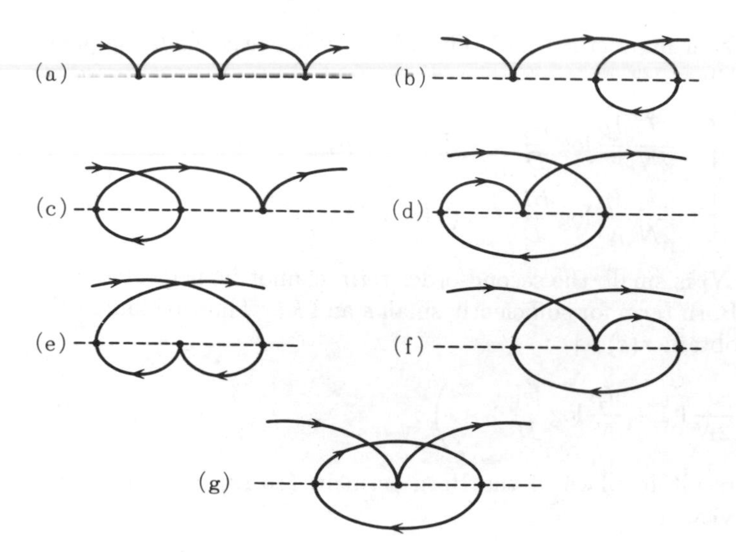

Fig. 16.2. Third-order processes of the scattering T-matrix

diagram (a) or (b) of Fig. 16.1, by joining a pair of neighboring points into a point. The diagram (a) or (b) thus obtained can be reduced to the simple diagram with one interaction point corresponding to the Born approximation by the same procedure. Such diagrams are called parquet diagrams: the processes corresponding to these diagrams give the most divergent terms. On the other hand, diagram (g) cannot be reduced to the simple diagram by the procedure mentioned above. Even if divergent terms appear from the processes corresponding to such diagrams, the order of divergence is low. If only the most divergent terms are retained, the spin-flip part $\tau(\varepsilon)$ of the scattering T-matrix is obtained by taking the sum of the geometric series as

$$\tau(\varepsilon) = -\frac{J}{2N} \frac{1}{1 - \dfrac{J\rho}{N} \log \dfrac{|\varepsilon|}{D}} . \tag{16.26}$$

The divergence of the spin-conserving part $t(\varepsilon)$ is weaker.

In this way, by summing all orders of the most divergent terms, we can obtain results which converge to zero in the limits $|\varepsilon| \to 0$ or $T \to 0$, for the case of $J > 0$. However, for the case with $J < 0$ (as in CuMn) the divergence still appears at the energy ε_{K} given by

$$kT_{\mathrm{K}} = \varepsilon_{\mathrm{K}} = D \exp\left(\frac{N}{J\rho}\right) . \tag{16.27}$$

The temperature T_{K} corresponding to this energy is called the *Kondo temperature*. The electrical resistivity calculated from (16.26) increases gradually with decreasing temperature from a value given by the Born approximation and then diverges at T_{K}. Therefore the results are incorrect at least in the temperature range below T_{K}.

In order to avoid the divergence brought by the perturbation calculation, attempts have been made to obtain a closed-form solution with the use of some approximate theories. The approaches along this line are represented by *Suhl* 's method [16.3] on the basis of the scattering theory and *Nagaoka*'s method [16.4] which pursues the equation of motion for the two-time Green's function and decouples higher-order Green's functions. By using these methods we can avoid the divergence at finite temperatures, but a singularity still remains near $T = 0$.

What does the divergence of electrical resistivity at $T = T_{\mathrm{K}}$ mean? To clarify this, we have to inspect physical quantities other than the electrical resistivity. We inspect first the susceptibility of the localized spin. Denoting kT as β^{-1}, we can express the free energy of the system as

$$F = -\frac{1}{\beta}\log Z\ , \qquad Z = \mathrm{Tr}\{\mathrm{e}^{-\beta(\mathcal{H}+\mathcal{H}_Z)}\}\ . \tag{16.28}$$

Here Z is the partition function and Tr means to take the trace; $\mathcal{H}_Z$ is the Zeeman energy given by

$$\mathcal{H}_Z = -g\mu_{\mathrm{B}}HS_Z\ , \tag{16.29}$$

and $\mathcal{H}$ is the Hamiltonian of the *s*-*d* system in the absence of the magnetic field. Assuming that $\mathcal{H}_Z$ is a small perturbation, we expand the exponential function as

$$\begin{aligned}\mathrm{e}^{-\beta(\mathcal{H}+\mathcal{H}_Z)} = \mathrm{e}^{-\beta\mathcal{H}}\bigg(1 &- \int_0^\beta \mathrm{e}^{\lambda\mathcal{H}}\mathcal{H}_Z\mathrm{e}^{-\lambda\mathcal{H}}d\lambda \\ &+ \int_0^\beta d\lambda_1 \int_0^{\lambda_1} d\lambda_2 \mathrm{e}^{\lambda_1\mathcal{H}}\mathcal{H}_Z\mathrm{e}^{-(\lambda_1-\lambda_2)\mathcal{H}}\mathcal{H}_Z\mathrm{e}^{-\lambda_2\mathcal{H}} + \dots\bigg)\ .\end{aligned} \tag{16.30}$$

On substituting this expansion into (16.28), we obtain the free energy up to second-order terms as

$$F = -\frac{1}{\beta}\bigg(\log Z_0 + (g\mu_{\mathrm{B}}H)^2\int_0^\beta d\lambda_1\int_0^{\lambda_1} d\lambda_2\langle \mathrm{e}^{(\lambda_1-\lambda_2)\mathcal{H}}S_Z\mathrm{e}^{-(\lambda_1-\lambda_2)\mathcal{H}}S_Z\rangle\bigg)\ ; \tag{16.31}$$

the first term is the free energy in the absence of magnetic field. Writing the second term as $-(1/2)\chi H^2$, we obtain the susceptibility from

$$\chi = (g\mu_{\mathrm{B}})^2\int_0^\beta d\lambda\langle \mathrm{e}^{\lambda\mathcal{H}}S_Z\mathrm{e}^{-\lambda\mathcal{H}}S_Z\rangle\ , \tag{16.32}$$

where $\langle\ \ \rangle$ represents the thermal average. Assuming the Hamiltonian in the absence of magnetic field to be

$$\mathcal{H} = \mathcal{H}_0 + \mathcal{H}_{\rm sd} \ , \qquad \mathcal{H}_0 = \sum_{k\sigma} \varepsilon_{k} a^\dagger_{k\sigma} a_{k\sigma} \ ,$$

we expand $\exp(\lambda\mathcal{H})$ in powers of $\mathcal{H}_{\rm sd}$, and obtain the susceptibility as the following series [16.5] in J:

$$\chi = \chi_0 \left[1 - \left(\frac{J}{N}\right)^2 \sum_{kk'} \frac{f_k(1-f_{k'})}{(\varepsilon_k - \varepsilon_{k'})^2} \left(1 - \frac{2}{\beta}\frac{1}{(\varepsilon_{k'} - \varepsilon_k)}\right) + \dots \right] . \tag{16.33}$$

Here $\chi_0 = g^2\mu_{\rm B}^2 S(S+1)/(3kT)$ is the susceptibility of an isolated localized spin.

In general, the transverse component of the *s-d* exchange interaction changes the direction of the localized spin. By this effect, the magnitude of the localized spin is contracted on average. If the contraction factor is a, the second term in the square brackets in (16.33) corresponds to $2a$. Now, we note that the ground state of the unperturbed Hamiltonian is $(2S+1)$-fold degenerate due to the direction of the localized spin at $T = 0$. As an example, we consider that one of the $(2S+1)$ states ψ_{M0} changes by the *s-d* interaction to the state ψ_M:

$$\psi_M = \left(1 + \frac{1-P}{E - \mathcal{H}_0}\mathcal{H}_{\rm sd} + \dots\right)\psi_{M0} \ , \tag{16.34}$$

where P is the projection operator on ψ_{M0} and E is the ground state energy. Since the expectation value of S_Z in the state ψ_M is given by

$$\langle S_Z \rangle = \frac{\langle \psi_M | S_Z | \psi_M \rangle}{\langle \psi_M | \psi_M \rangle} \ , \tag{16.35}$$

(16.34) for ψ_M gives

$$\langle S_Z \rangle = M \left[1 - \frac{1}{2}\left(\frac{J}{N}\right)^2 \sum_{kk'} \frac{f_k(1-f_{k'})}{(\varepsilon_k - \varepsilon_{k'})^2} + \dots \right] . \tag{16.36}$$

At $T = 0$, for which f_k is replaced by a step function, the integral in the second term of (16.36) diverges logarithmically as $N_e \to \infty$,

$$-\frac{1}{2}\left(\frac{J\rho}{N}\right)^2 \int_{0+}^{D} d\varepsilon' \int_{-D}^{0-} d\varepsilon \frac{1}{(\varepsilon' - \varepsilon)^2} \cong -\frac{1}{2}\left(\frac{J\rho}{N}\right)^2 \log N_e \ . \tag{16.37}$$

Here N_e is the number of conduction electrons with *s*-wave symmetry, and the separation between the energy levels at Fermi surface $(0_+ - 0_-)$ is D/N_e. As seen clearly from (16.35, 36), this divergence is related to that of the integral for the normalization of ψ_M,

$$\langle \psi_M | \psi_M \rangle = 1 + \frac{1}{2} S(S+1) \left(\frac{J}{N}\right)^2 \sum_{kk'} \frac{f_k(1-f_{k'})}{(\varepsilon_{k'} - \varepsilon_k)^2} + \dots \ . \tag{16.38}$$

At finite temperatures, in the limit $N_e \to \infty$, the divergence is replaced by

$$\log N_e \to -\log \frac{kT}{D} .$$

In fact, the integral of the second term in (16.33) gives $(J\rho/N)^2 \log(kT/D)$. (It should be noted that the second term in (16.33) vanishes in the limit $\varepsilon_{\boldsymbol{k}'} \to \varepsilon_{\boldsymbol{k}}$.) Then, we can consider that the $\log T$ term appearing in the susceptibility originates from the divergence of $\langle \psi_M | \psi_M \rangle$. We shall discuss this point again, later.

The most divergent terms in the higher-order terms of (16.36) [16.6], form also geometric series, and by collecting these most divergent terms, the expectation value of $\boldsymbol{S}$ can be written in the form:

$$\langle \boldsymbol{S} \rangle = \boldsymbol{S} \left[1 + \frac{1}{2} \left(\frac{J\rho}{N} \right)^2 \log \frac{kT}{D} \left(1 - \frac{J\rho}{N} \log \frac{kT}{D} \right)^{-1} \right] . \tag{16.39}$$

On the other hand, the total polarization of the conduction-electron spins is given by

$$\langle \boldsymbol{\sigma} \rangle = \frac{J\rho}{2N} \langle \boldsymbol{S} \rangle . \tag{16.40}$$

By considering the sum of (16.39, 40) as the average value of $\boldsymbol{S}$ and multiplying χ_0 by its square, we obtain [16.6], for the susceptibility due to the localized spin in the *s-d* system,

$$\chi = \chi_0 \left[1 + \frac{J\rho}{N} \left(1 - \frac{J\rho}{N} \log \frac{kT}{D} \right)^{-1} \right] . \tag{16.41}$$

The magnitude of the localized spin, as given by (16.39), takes the value S near that of the free spin for $T \gg T_{\mathrm{K}}$ and decreases with decreasing temperature. In case $J > 0$, this change is slow and the magnitude becomes $S(1 - J\rho/2N)$ at $T = 0$. Therefore this contraction cancels with the polarization of the conduction-electron spins and the magnitude of the spin becomes S on including the polarization of conduction electrons. That is, the total spins at $T = 0$ retain the value at the starting point. However, in our case $J < 0$, the contraction of the localized spin is large and diverges for $T \leq T_{\mathrm{K}}$ after passing through zero. Here also, we cannot use perturbation theory for $T < T_{\mathrm{K}}$. However, from the results obtained by the perturbation calculation, we expect that the localized spin gradually decreases with decreasing temperature and vanishes at $T = 0$. If $\langle \boldsymbol{S} \rangle$ vanishes, $\langle \boldsymbol{\sigma} \rangle$ also vanishes. As the result, it can be considered that the spin-singlet state is realized as a whole at $T = 0$.

The free energy $-\beta^{-1} \log Z_0$ for zero magnetic field can be also calculated in the perturbation theory. The result in the most divergent approximation is divided into a normal part ΔF_{n} without the $\log T$ term and an anomalous term $\Delta F_{\rm an}$ with the $\log T$ term, as $\Delta F = \Delta F_{\rm n} + \Delta F_{\rm an}$. $\Delta F_{\rm an}$ is calculated as [16.7, 8]

$$\Delta F_{\rm an} = -\left(\frac{J\rho}{N}\right)^3 \frac{\pi^2}{3} S(S+1)kT\left(1 + 3\frac{J\rho}{N}\log\frac{kT}{D} + \ldots\right). \tag{16.42}$$

From this result, by assuming the series in the bracket as $[1-(J\rho/N)\log(kT/D)]^{-3}$, we obtain the excess specific heat $\Delta C_{\rm an}$ as

$$\Delta C_{\rm an} = \left(-\frac{J\rho}{N}\right)^4 \pi^2 S(S+1)k\left(1 + 4\frac{J\rho}{N}\log\frac{kT}{D} + \ldots\right). \tag{16.43}$$

We find from this expression that for $J < 0$ the specific heat increases with decreasing temperature from high temperatures. Since the specific heat vanishes at $T = 0$, it decreases again after showing a peak. Since we can consider that the localized spin vanishes at $T = 0$ from the calculated result for the susceptibility, we can understand that this Schottky specific heat is related to the vanishing of the entropy $k\log(2S+1)$ due to the free localized spin.

17. Ground State of s-d System

In the preceding chapter, we described the perturbation expansions with respect to the s-d exchange interaction for the resistivity, the susceptibility and other physical quantities, and showed that these perturbation expansions form a power series in $(J\rho/N)\log(kT/D)$, if we take the most divergent terms in each order of perturbation. Therefore, this perturbation expansions correspond to the high temperature expansion which is valid for $(J\rho/N)\log(kT/D) < 1$. As a first step in clarifying the low temperature behavior of the s-d system, we consider in this chapter mainly the ground state.

17.1 Theory of the Ground State by the Perturbation Method

The result for the susceptibility obtained by the perturbation calculation suggests that the localized spin falls into a singlet ground state. The wave function for the free state without s-d interaction is written as

$$\chi_\alpha \psi_v, \qquad \chi_\beta \psi_v , \tag{17.1}$$

where ψ_v is the wave function of the Fermi sphere and χ_α and χ_β represent the up and down states of the localized spin. Hereafter, unless otherwise specified, the magnitude of the localized spin S is $1/2$.

The first important problem we encounter in discussing the ground state of the spin-singlet state is that the two degenerate states are not connected with each other by the s-d exchange interaction. Therefore, when there are two degenerate states with localized up and down spins and the matrix element between the states is nonvanishing, the spin-singlet state composed of the linear combination of these states

$$\psi_{\mathrm{singlet}} = \frac{1}{\sqrt{2}}(\psi_\alpha \chi_\alpha - \psi_\beta \chi_\beta) \tag{17.2}$$

has an energy lower than that of the perturbed state of (17.1) and becomes the ground state in place of this state. In order to satisfy this condition, the wave functions representing the states of conduction electrons, ψ_α and

ψ_β, should possess $\mp 1/2$ spin component, as a whole, to compensate the $\pm 1/2$ component of the localized spin. Therefore, for example, ψ_α should contain $N/2$ $\uparrow$ spin electrons and $N/2+1$ $\downarrow$ spin electrons. Since the s-d exchange interaction is a local interaction, one excess electron should be localized around the impurity spin. The simplest representations for ψ_α and ψ_β satisfying this condition are

$$\psi_{0\alpha} = \sum_{\boldsymbol{k}} \Gamma_{\boldsymbol{k}} a^\dagger_{\boldsymbol{k}\downarrow} \psi_v \, , \qquad \psi_{0\beta} = \sum_{\boldsymbol{k}} \Gamma_{\boldsymbol{k}} a^\dagger_{\boldsymbol{k}\uparrow} \psi_v \, . \tag{17.3}$$

The state $\psi_{0\alpha}$ represents the state in which one $\downarrow$ spin electron is excited outside the Fermi sphere and couples with the localized $\uparrow$ spin. As will be confirmed below, these states $\psi_{0\alpha}$ and $\psi_{0\beta}$ are coupled by the *s-d* exchange interaction; however, at the same time by this interaction electrons inside the Fermi sphere are excited outside the sphere to create a new state involving electron-hole pairs. In other words, the state (17.3) is not an eigenstate of the *s-d* interaction Hamiltonian. Thus the eigenvalue and eigenstate of the spin-singlet state should be obtained by the perturbation calculation starting with (17.3).

In our starting wave function (17.3) for the perturbation calculation, there is one electron excited outside the Fermi sphere which is actually bound to the impurity spin. In the general case, however, only opposite-spin electrons amounting to one electron as a whole must be bound (or virtually bound, as noted by Friedel) with the impurity. From this point of view, the calculation by a variational method which describes ψ_α and ψ_β by Slater determinants composed of the scattering waves was presented by *Anderson* [17.1], and another variational calculation using a simpler wave function was done by *Kondo* [17.2]. However, by these variational methods we cannot calculate correctly the ground state energy including both normal and anomalous parts. In this section we consider the ground state on the basis of the perturbation method mentioned above [17.3–8]. First we expand ψ_α and ψ_β starting with (17.3) as follows:

$$\begin{aligned}\psi_\alpha = \Bigg[& \sum_{1} \Gamma^\alpha_1 a^\dagger_{1\downarrow} + \sum_{123} \left(\Gamma^{\alpha\downarrow}_{12,3} a^\dagger_{1\downarrow} a^\dagger_{2\downarrow} a_{3\downarrow} + \Gamma^{\alpha\uparrow}_{12,3} a^\dagger_{1\downarrow} a^\dagger_{2\uparrow} a_{3\uparrow} \right) \\ & + \sum_{12345} \left(\Gamma^{\alpha\downarrow\downarrow}_{123,45} a^\dagger_{1\downarrow} a^\dagger_{2\downarrow} a^\dagger_{3\downarrow} a_{4\downarrow} a_{5\downarrow} + \Gamma^{\alpha\downarrow\uparrow}_{123,45} a^\dagger_{1\downarrow} a^\dagger_{2\downarrow} a^\dagger_{3\uparrow} a_{4\downarrow} a_{5\uparrow} \right. \\ & \left. + \Gamma^{\alpha\uparrow\uparrow}_{123,45} a^\dagger_{1\downarrow} a^\dagger_{2\uparrow} a^\dagger_{3\uparrow} a_{4\uparrow} a_{5\uparrow} \right) + \ldots \Bigg] \psi_v \, . \end{aligned} \tag{17.4}$$

Here we have represented $\boldsymbol{k}_1, \boldsymbol{k}_2, \boldsymbol{k}_3, \ldots$ with $1, 2, 3, \ldots$ for simplicity. We obtain ψ_β by inverting the spin direction in the creation and annihilation operators for the conduction electrons. Now we substitute the linear combination (17.2) of ψ_α and ψ_β, expanded as stated above, into the Schrödinger equation

$$(\mathcal{H} - E)\psi_{\text{singlet}} = 0 \, . \tag{17.5}$$

Here the Hamiltonian is given by the sum of the kinetic energy and $\mathcal{H}_{sd}$,

$$\mathcal{H} = \sum_{k} \varepsilon_k a_{k\sigma}^\dagger a_{k\sigma} - \frac{J}{2N} \sum_{kk'} [(a_{k'\uparrow}^\dagger a_{k\uparrow} - a_{k'\downarrow}^\dagger a_{k\downarrow}) S_z + a_{k'\downarrow}^\dagger a_{k\uparrow} S_+ + a_{k'\uparrow}^\dagger a_{k\downarrow} S_-] \tag{17.6}$$

and E is the energy of the ground state.

In the equation thus obtained, we put to zero all the coefficients of the state possessing one excited electron, the states possessing one excited electron and one excited electron-hole pair and the states possessing one excited electron and two excited electron-hole pairs ..., respectively. Then we obtain a series of equations which connect the amplitude $\Gamma_{[2n+1]}$ of the state having n excited electron-hole pairs with the amplitudes $\Gamma_{[2n-1]}$ and $\Gamma_{[2n+3]}$ of the states having one less or one more electron-hole pairs. The first two equations obtained in this way are

$$(\varepsilon_1 - E)\Gamma_1 + \frac{3J}{4N} \sum_1 \Gamma_1 - \frac{3J}{4N} \sum_{23} \Gamma_{21,3}^{\alpha\uparrow} = 0 \,, \tag{17.7}$$

$$\begin{aligned}(\varepsilon_1 + \varepsilon_2 - \varepsilon_3 - E)\Gamma_{12,3}^{\alpha\uparrow} &- \frac{J}{4N}(\Gamma_1 + 2\Gamma_2) + \frac{J}{4N} \sum_4 (\Gamma_{42,3}^{\alpha\uparrow} - \Gamma_{14,3}^{\alpha\uparrow} \\ &+ \Gamma_{12,4}^{\alpha\uparrow} + 2\Gamma_{[42],3}^{\alpha\uparrow} + 2\Gamma_{21,4}^{\alpha\uparrow}) - \frac{J}{4N} \sum_{45} (\Gamma_{[14]2,53}^{\alpha\downarrow\uparrow} + \Gamma_{1[24],[53]}^{\alpha\uparrow\uparrow} \\ &- 2\Gamma_{[42]1,35}^{\alpha\downarrow\uparrow} = 0 \,.\end{aligned} \tag{17.8}$$

The following notation has been used for the subscript of the amplitude Γ: subscripts before the comma refer to the wave vectors of the excited electrons and those behind the comma refer to the wave vectors of the excited holes; the square bracket attached to subscripts mean antisymmetric sums such as

$$\Gamma_{[12],3}^{\alpha\uparrow} = \Gamma_{12,3}^{\alpha\uparrow} - \Gamma_{21,3}^{\alpha\uparrow} \,. \tag{17.9}$$

In the equations such as (17.7, 8), the following relations hold among the amplitudes of the same order for the spin-singlet state:

$$\begin{aligned}\Gamma_{[12],3}^{\alpha\downarrow} &= \Gamma_{[12],3}^{\alpha\uparrow} \,, \\ \Gamma_{[123],[45]}^{\alpha\downarrow\downarrow} &= \frac{1}{2}\Gamma_{[123],[45]}^{\alpha\downarrow\uparrow} \,, \\ \Gamma_{1[23],[45]}^{\alpha\uparrow\uparrow} &= \Gamma_{[32]1,[45]}^{\alpha\downarrow\uparrow} - \Gamma_{[321],[45]}^{\alpha\downarrow\downarrow} \,.\end{aligned} \tag{17.10}$$

Now we rewrite (17.8) to give $\Gamma_{12,3}^{\alpha\uparrow}$, as

$$\Gamma_{12,3}^{\alpha\uparrow} = \frac{1}{\varepsilon_1 + \varepsilon_2 - \varepsilon_3 - E} \left[\frac{J}{4N}(\Gamma_1 + 2\Gamma_2) - \frac{J}{4N} \sum \Gamma[3] + \frac{J}{4N} \sum \Gamma[5] \right] , \tag{17.11}$$

where $\Gamma[3]$ and $\Gamma[5]$ represent symbolically the third and fourth terms of (17.8), respectively. If we regard the first term as the first approximation for

$\Gamma^{\alpha\uparrow}_{12,3}$, and use the first approximation for the second term, we obtain the second approximation proportional to J^2. The third approximation is given by the sum of two contributions. Namely, one comes from the second approximation in $\Gamma[3]$ of the second term. The other is given by the form $J^3\sum\Gamma[1]$ which arises by using as $\Gamma[5]$ in the third term the first approximation of $\Gamma[5]$. Thus, representing $\Gamma^{\alpha\uparrow}_{12,3}$ in a series in J by successive approximations and substituting this into (17.7), we obtain the integral equation to determine the lowest amplitude Γ_1 :

$$(\varepsilon_1 - \widetilde{E})\Gamma(\varepsilon_1) + \frac{3J}{4N}\sum_2 \Gamma(\varepsilon_2) = \sum_2 K(\varepsilon_1, \varepsilon_2; \widetilde{E})\Gamma(\varepsilon_2) , \tag{17.12}$$

where $\widetilde{E}$ is defined by

$$E = \widetilde{E} + \Delta E , \tag{17.13}$$

$$\Delta E = -\frac{3}{8}\left(\frac{J}{N}\right)^2 \sum_{23} \frac{1}{\varepsilon_2 - \varepsilon_3 + \varepsilon_1 - \widetilde{E}} + \frac{3}{16}\left(\frac{J}{N}\right)^3 \sum_{234} \frac{1}{\varepsilon_2 - \varepsilon_3 + \varepsilon_1 - \widetilde{E}}$$
$$\times \left[\frac{1}{\varepsilon_4 - \varepsilon_3 + \varepsilon_1 - \widetilde{E}} + \frac{1}{\varepsilon_2 - \varepsilon_4 + \varepsilon_1 - \widetilde{E}}\right] + \dots . \tag{17.14}$$

ΔE is the energy obtained by the usual perturbation calculation for the state which is composed of the localized spin and the Fermi sphere plus an electron with energy ε_1, and differs from the perturbed energy of the doubly degenerate state composed of the localized spin and the Fermi sphere at $T = 0$; the difference is that the excess positive quantity $\varepsilon_1 - \widetilde{E}$ is added to the energy denominator of each term of the perturbation. Expanding ΔE with respect to $\varepsilon_1 - \widetilde{E} > 0$, collecting the most divergent terms and neglecting the other terms, we obtain

$$\Delta E = \Delta E_n - \frac{3}{8}\left(\frac{J\rho}{N}\right)^2 (\varepsilon_1 - \widetilde{E}) \log\frac{\varepsilon_1 - \widetilde{E}}{D}\left[1 + \frac{J\rho}{N}\log\frac{\varepsilon_1 - \widetilde{E}}{D}\right.$$
$$\left. + \left(\frac{J\rho}{N}\log\frac{\varepsilon_1 - \widetilde{E}}{D}\right)^2 + \dots\right] . \tag{17.15}$$

Here we neglect the second term and approximate ΔE as ΔE_n; ΔE_n is the perturbation energy of the doubly degenerate state possessing a localized spin $\pm 1/2$. If we keep the second term in (17.15), the first term of (17.12) should be multiplied by the following factor:

$$a = 1 - \frac{3}{8}\left(\frac{J\rho}{N}\right)^2 \log\left(\frac{\varepsilon_1 - \widetilde{E}}{D}\right) \Big/ \left[1 - \frac{J\rho}{N}\log\left(\frac{\varepsilon_1 - \widetilde{E}}{D}\right)\right] , \tag{17.16}$$

where the definition of $\widetilde{E}$ is retained as $E - \Delta E_n$. In the following, the calculation is done thoroughly on the basis of the most divergent approximation. In this approximation we consider $(J\rho/N)\log[(\varepsilon_1 - \widetilde{E})/D]$ as order unity. In this case the second term of (17.16) is order of $J\rho/N \ll 1$ and is negligible compared with unity. That is, the approximation which ignores the second term of (17.15) by putting $\Delta E = \Delta E_n$ can be permitted for the weak coupling case with $J\rho/N \ll 1$. In the approximation including the next divergent terms, the second term of (17.16) should be retained.

In the kernel $K(\varepsilon_1, \varepsilon_2; \widetilde{E})$ on the right-hand side of (17.12), we use the same approximation for the energy denominator in it. Then the kernel is given by a series expansion in J starting with J^2 as

$$K(\varepsilon_1, \varepsilon_2; \widetilde{E}) = \frac{3}{16}\left(\frac{J}{N}\right)^2 \left[\sum_3 \frac{1}{D_{12,3}} + \frac{J}{4N}\sum_{34}\left(\frac{2}{D_{24,3}D_{14,3}} + \frac{1}{D_{12,3}D_{42,3}} + \frac{1}{D_{14,3}D_{12,3}} - \frac{5}{D_{12,3}D_{12,4}}\right) + \dots\right], \quad (17.17)$$

$$D_{12,3} = \varepsilon_1 + \varepsilon_2 - \varepsilon_3 - \widetilde{E}\,. \quad (17.18)$$

The first term (proportional to J^2) in (17.17) gives $\rho \log[(\varepsilon_1 + \varepsilon_2 - \widetilde{E})/D]$ after the integration over ε_3. The four terms proportional to J^3 have the same structure as the diagrams involving electron-hole pair excitations in the intermediate state, namely diagrams (d), (e), (f) and (g) of Fig. 16.2, which appeared in the calculation of the third-order term in the scattering matrix. Therefore the first term possessing the same structure as diagram (g) does not contribute to the most divergent term. All the other terms contribute to the most divergent terms and give $\rho^2 \log^2[(\varepsilon_1 + \varepsilon_2 - \widetilde{E}/D]$ as a whole. Moreover, by calculating the J^4 and J^5 terms, we see that the most divergent terms form a geometric series here also. Thus we obtain the kernel as

$$K(\varepsilon_1, \varepsilon_2; \widetilde{E}) \cong -\frac{3}{16}\frac{J}{N}\frac{J\rho}{N}\log\left(\frac{\varepsilon_1 + \varepsilon_2 - \widetilde{E}}{D}\right) \times \left[1 - \frac{J\rho}{N}\log\left(\frac{\varepsilon_1 + \varepsilon_2 - \widetilde{E}}{D}\right)\right]^{-1}. \quad (17.19)$$

The integral equation to determine the lowest order amplitude $\Gamma(\varepsilon)$ and the energy for the spin-singlet ground state has thus been derived within the weak-coupling approximation, i.e., the most divergent approximation. The next problem is to solve this equation, but before that, as preparation, let us consider a simple case where the term having the kernel is ignored. In this case, the wave function is given by the zeroth order approximation (17.3); the amplitude of the one-electron excitation Γ_k is given by

$$\Gamma(\varepsilon) = \frac{1}{\varepsilon - \widetilde{E}} \quad (17.20)$$

and the energy eigenvalue $\widetilde{E}$ is given for $J < 0$ by

$$\widetilde{E} = -D\exp\left(\frac{4N}{3J\rho}\right). \quad (17.21)$$

That is, the excess electron added outside the Fermi sphere forms a spin-singlet bound state by the coupling with the localized spin. The binding energy of the bound state is given by (17.21). This bound state has the same character as the usual bound state which appears at an energy below the bottom of the energy band, due to the strong attractive potential from the impurity atom. Therefore, if we assume $J > 0$, the spin-triplet state obtained by changing the sign $-$ into $+$ in (17.2) forms a bound state with binding energy

$$\widetilde{E}_t = -D \exp\left(-\frac{4N}{J\rho}\right) . \tag{17.22}$$

There is, however, the following difference: In the usual case of a bound state situated at an energy below the bottom of the band, the energy change of the other electrons below the Fermi surface cancel the binding energy with a singular form in J. On the other hand, the bound state obtained by the zero-th order approximation is composed of plane waves outside the Fermi sphere. Since the Fermi sphere doesn't change at the stage of this approximation, our problem has been reduced to a one-body problem accidentally. However, in principle, we are considering the ground state of the total system which is a many-body problem in nature so that it cannot be reduced to the one-body problem. We may consider that the zeroth order approximation forces one particular electron to carry the character of the many-body spin-singlet state; in the higher order, this burden is gradually shared by all the electrons. This situation will be made clear by actually determining the solution of (17.12).

First, from $\Gamma(\varepsilon)$, we introduce $G(\varepsilon)$ defined by

$$G(\varepsilon) = \int_0^{\varepsilon} \Gamma(\varepsilon) d\varepsilon . \tag{17.23}$$

Using (17.19) as the kernel of (17.12) and carrying out an integration by parts, we obtain as the right-hand side of (17.12)

$$\frac{3}{16}\left(\frac{J\rho}{N}\right)^2 \int_0^D d\varepsilon' \frac{G(\varepsilon')}{\varepsilon + \varepsilon' - \widetilde{E}} \left(1 - \frac{J\rho}{N} \log \frac{\varepsilon + \varepsilon' - \widetilde{E}}{D}\right)^{-2} .$$

Within the most divergent approximation this can be written exactly as

$$\frac{3}{16}\left(\frac{J\rho}{N}\right)^2 \int_{\varepsilon}^D d\varepsilon' \frac{G(\varepsilon')}{\varepsilon' - \widetilde{E}} \left(1 - \frac{J\rho}{N} \log \frac{\varepsilon' - \widetilde{E}}{D}\right)^{-2} .$$

After this replacement, by differentiating (17.12) with respect to ε we can rewrite the integral equation as the following differential equation:

$$(\varepsilon - \widetilde{E})^2 \frac{d^2 G}{d\varepsilon^2} + (\varepsilon - \widetilde{E}) \frac{dG}{d\varepsilon} + \frac{3}{16}\left(\frac{J\rho}{N}\right)^2 G(\varepsilon) \left(1 - \frac{J\rho}{N} \log \frac{\varepsilon - \widetilde{E}}{D}\right)^{-2} = 0 . \tag{17.24}$$

The two independent solutions of this differential equation are

$$G(\varepsilon) = \left(1 - \frac{J\rho}{N}\log\frac{\varepsilon - \widetilde{E}}{D}\right)^{1/4}, \qquad \left(1 - \frac{J\rho}{N}\log\frac{\varepsilon - \widetilde{E}}{D}\right)^{3/4}. \tag{17.25}$$

The coefficients of the linear combination are determined by the two boundary conditions:

$$G(0) = 0\,, \tag{17.26}$$

$$\left[(\varepsilon - \widetilde{E})\frac{dG}{d\varepsilon}\right]_{\varepsilon=D} = -\frac{3J\rho}{4N}G(D)\,. \tag{17.27}$$

The second condition is obtained by putting $\varepsilon = D$ in (17.12). Thus we have

$$G(\varepsilon) = \frac{G(D)}{1 - \frac{1}{3}(1-x)^{1/2}}\left[\left(1 - \frac{J\rho}{N}\log\frac{\varepsilon - \widetilde{E}}{D}\right)^{3/4} - (1-x)^{1/2}\left(1 - \frac{J\rho}{N}\log\frac{\varepsilon - \widetilde{E}}{D}\right)^{1/4}\right]; \tag{17.28}$$

here x is defined by

$$x = \frac{J\rho}{N}\log\frac{-\widetilde{E}}{D}\,. \tag{17.29}$$

Putting $\varepsilon = D$ in (17.28) we obtain the equation to determine the energy eigenvalue $-\widetilde{E}$:

$$1 - \frac{1 - (1-x)^{1/2}}{1 - \frac{1}{3}(1-x)^{1/2}} = 0 \tag{17.30}$$

The only solution of this equation is $x = 1$; from this the energy eigenvalue is obtained as

$$\widetilde{E} = -D\exp\left(\frac{N}{J\rho}\right) = -kT_{\mathrm{K}}\,. \tag{17.31}$$

$\Gamma(\varepsilon)$ is obtained by differentiation of (17.28) with respect to ε; apart for the normalization constant, it is

$$\Gamma(\varepsilon) = \frac{1}{\varepsilon - \widetilde{E}}\left[\left(1 - \frac{J\rho}{N}\log\frac{\varepsilon - \widetilde{E}}{D}\right)^{-1/4} - \frac{1}{3}(1-x)^{1/2}\left(1 - \frac{J\rho}{N}\log\frac{\varepsilon - \widetilde{E}}{D}\right)^{-3/4}\right]. \tag{17.32}$$

In order to determine the normalization constant in the limit of weak coupling we have only to obtain the contributions of the lowest-order amplitude. The contributions from higher-order amplitudes give no terms which are the most divergent. Therefore we obtain

$$
\begin{aligned}
&\langle\psi_{\mathbf{singlet}}|\psi_{\mathbf{singlet}}\rangle \\
&= \frac{1}{2}[\langle\psi_\alpha|\psi_\alpha\rangle + \langle\psi_\beta|\psi_\beta\rangle] = \rho\int_0^D \frac{d\varepsilon}{(\varepsilon-\widetilde{E})^2} \\
&\times\left[\left(1-\frac{J\rho}{N}\log\frac{\varepsilon-\widetilde{E}}{D}\right)^{-1/4} - \frac{1}{3}(1-x)^{1/2}\left(1-\frac{J\rho}{N}\log\frac{\varepsilon-\widetilde{E}}{D}\right)^{-3/4}\right]^2 \\
&= \rho\frac{1}{-\widetilde{E}}\frac{4}{9}(1-x)^{-1/2}\,. \qquad (17.33)
\end{aligned}
$$

If we normalize our wave function with this normalization integral, the amplitude $\Gamma(\varepsilon)$ becomes a singular function such that $1/\sqrt{\rho|\widetilde{E}|}$ for $\varepsilon = 0$ and zero for $\varepsilon \neq 0$.

Thus, when all the most divergent terms are taken into account, the exponent $4/3(N/J\rho)$ in the exponential expression for $\widetilde{E}$ obtained in the zeroth approximation is replaced by $(N/J\rho)$ and $\Gamma(\varepsilon)$ is multiplied by an extra function. As a matter of course, the energy is lower than the value obtained in the zeroth approximation. If we take the spin-triplet state as the starting state, the eigenvalue equation is obtained in place of (17.30) as

$$
1 + \frac{1-(1-x)^{3/2}}{5+(1-x)^{3/2}} = 0\,;
$$

this has no solution other than $x \to \infty$ whether J is positive or negative. Therefore the triplet bound state which appeared in the zeroth approximation disappears in higher order. This result means that in the case $J < 0$ the spin-singlet bound state which exists still at the last stage is a completely different state from the one-body state due to the impurity potential, although the expressions for $\Gamma(\varepsilon)$ and $\widetilde{E}$ look similar.

The preceding discussion shows that the ground state in the system where the localized spin ($S = 1/2$) and the conduction electrons are coupled to each other is the spin-singlet state, and the energy is given by $E = \Delta E_n + \widetilde{E}$. $\widetilde{E}$ is equal to $-kT_\mathrm{K}$ and can be understood as the binding energy of the spin-singlet state. From this result we can see that the Kondo temperature, at which divergences arise in the high-temperature expansion, has the physical meaning of the binding energy of the singlet bound states.

Since the wave function and the energy eigenvalue have been obtained (within the most divergent approximation), it is possible to evaluate various quantities in the ground state. The structure of the conduction band (i.e., the relation between wave vector and energy) is necessary to calculate the spatial distribution of conduction electrons in each wave function of conduction electrons ψ_α, ψ_β, when we decompose the wave function of the singlet state into components with up and down localized spin. In contrast to this, the electron density for each spin at the origin and the total electron number localized

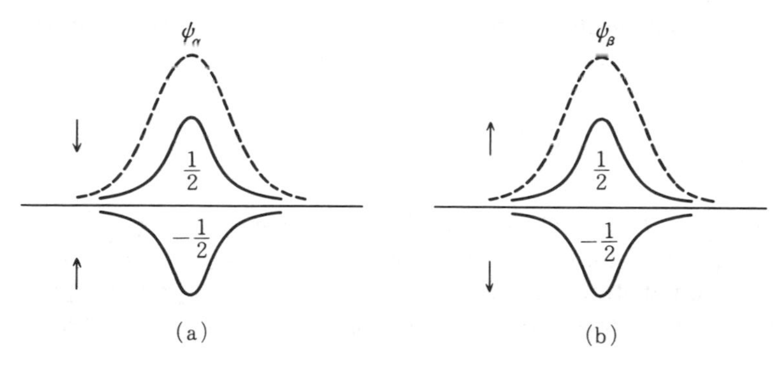

Fig. 17.1. Electron distributions for each spin in ψ_α and ψ_β components. Broken lines show those in the zeroth-order approximation

around the impurity atom can be calculated without detailed information of the band structure. Here we omit the details of the calculation [17.7, 8] and describe only the result. According to this theory, the numbers of the local conduction electrons bound around the localized spin are, as shown in Fig. 17.1, for component ψ_α, $-1/2$ electron with up spin and $+1/2$ electron with down spin. Though the numbers are 0 and 1 at the start of the perturbation calculation, the numbers of localized electrons decrease gradually with the order of the perturbation and take $-1/2$ and $1/2$ at the last stage, by which the total number of localized electrons changes from 1 to 0. Correspondingly, the numbers of the localized electrons for ψ_β are $1/2$ for up-spin electrons and $-1/2$ for down-spin electrons. It is known that the relation between the localized electron number and the phase shift at the Fermi surface holds not only for the one-body problem but also for the general many-body system with the electron correlation [17.9]. By use of this relation, the phase shifts of electrons with $\pm$ spin at the Fermi surface are obtained as

$$\delta_{\alpha\pm} = \mp\frac{\pi}{2}\,, \qquad \delta_{\beta\pm} = \pm\frac{\pi}{2}\,. \tag{17.34}$$

On the other hand, the scattering T-matrix is given by

$$\pi\rho(\varepsilon)T(\varepsilon) = -\mathrm{e}^{\mathrm{i}\delta(\varepsilon)}\sin\delta\,. \tag{17.35}$$

Therefore the T-matrix is determined as

$$T(\varepsilon) = \frac{1}{\mathrm{i}\pi\rho}\,, \qquad \mathrm{Im}\,T(\varepsilon) = -\frac{1}{\pi\rho}\,. \tag{17.36}$$

This value is the highest value that the imaginary part of the T-matrix can take, and is called the *unitarity limit.* Since the localized spin vanishes in the ground state, $T(\varepsilon) = t(\varepsilon), \tau(\varepsilon) = 0$. Thus the electrical resistivity has the maximum value corresponding to the unitarity limit.

Another important result which can be obtained from the theory of the ground state is the magnetic susceptibility due to the impurity. This is given as [17.10]

$$\chi_{\rm imp} = \left(\frac{1}{2}g\mu_{\rm B}\right)^2 \frac{1}{-\widetilde{E}} = \left(\frac{1}{2}g\mu_{\rm B}\right)^2 \frac{1}{kT_{\rm K}} \ . \tag{17.37}$$

The susceptibility at $T = 0$ is finite and does not diverge. It is the binding energy of the singlet state, $|\widetilde{E}|$, which prevents the divergence; due to the existence of the bound state, the temperature dependence of physical quantities near $T = 0$ is expected to include no log-dependence and to behave normally.

Concerning the ground state, the physical picture of the localized spin has been established in this way; however, low-energy excitations from the ground state have not yet been clarified in the above theory. If we cannot calculate these correctly, we cannot obtain the temperature dependences of the specific heat and the electrical resistivity. This problem will be discussed in detail in a later chapter.

17.2 Ground State for Anisotropic Exchange Interaction

The perturbation theory of the singlet ground state is closely related to the perturbation theory of the scattering T-matrix at high temperatures, which was developed by *Abrikosov* [17.11]. This fact can be seen from the structure of the kernel [given by (17.17, 19)] in the integral equation (17.12). The relation between the scattering matrix and the ground state holds also in the case where the exchange interaction is anisotropic.

The s-d exchange interaction can be extended to the anisotropic case as

$$\mathcal{H}_{\rm sd} = -\frac{1}{2N}\sum_i \sum_{\boldsymbol{k}\boldsymbol{k}'\alpha\alpha'} J_i a^\dagger_{\boldsymbol{k}\alpha}\sigma^i_{\alpha\alpha'} a_{\boldsymbol{k}'\alpha'} S_i \ , \tag{17.38}$$

where i denotes the principal axis (x, y and z) of anisotropy. $\sigma^i_{\alpha\alpha'}$ is the $\alpha\alpha'$ component of the Pauli spin matrix σ_i, and α and α' are the spin directions of the electrons with wave vector $\boldsymbol{k}$ and $\boldsymbol{k}'$. For this extended Hamiltonian, the wave function in the zeroth order approximation can be written in term of a linear combination of the following four wave functions:

$$\psi_0^{\alpha\beta} = \sum_{\boldsymbol{k}} \Gamma^{\alpha\beta}_{\boldsymbol{k}} a^\dagger_{\boldsymbol{k}\alpha}\chi_\beta\psi_v \ . \tag{17.39}$$

In this section, α denotes the spin direction of the conduction electron and β denotes the spin direction of the localized electron. Using a linear combination of these wave functions, we obtain the energy eigenvalue and eigenfunction

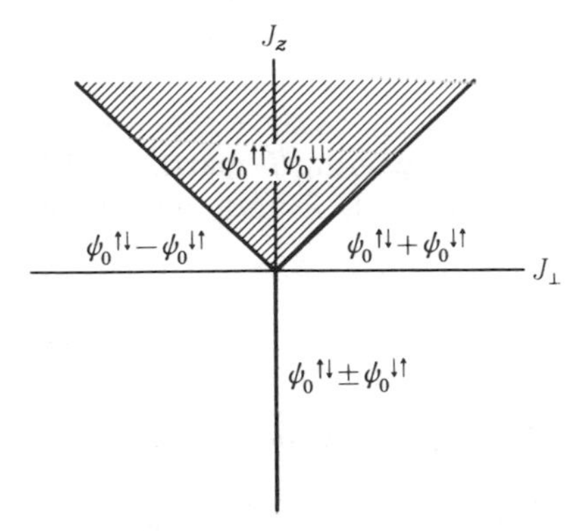

Fig. 17.2. Bound states with the lowest energy and the degeneracy in zeroth order approximation for the anisotropic exchange interaction

in the zeroth order approximation. In this case, there is no degeneracy in the bound state with the lowest energy, if all of J_x, J_y and J_z are different from one another. However, when two of J_x, J_y and J_z are equal, there is a region where a degeneracy remains in the lowest bound state. The region in the case $J_x = J_y = J_\perp$ is the shaded part in Fig. 17.2, which corresponds to $J_z > |J_\perp|$, and the $J_z < 0$ part of the z-axis. The case of isotropic exchange interaction with $J > 0$ is the boundary line in the first zone of this figure. On this line the bound state is triply degenerate. The existence of the boundary line separating two regions in $J_\perp$-J_z space was also pointed out and discussed on the basis of the scaling law by *Anderson* et al. [17.12, 13].

Starting with (17.39) or the zeroth order wave function given by their linear combination, we can develop the perturbation calculation in the same way as for the isotropic exchange interaction. Such theory was developed by *Shiba* [17.14]. Hereafter, we discuss this problem by following his theory.

The integral equation for the lowest order amplitude $\Gamma_{\alpha\beta}(\varepsilon)$ can be written down as

$$(\varepsilon - \widetilde{E})\Gamma_{\alpha\beta}(\varepsilon) - \frac{1}{2N} J_i \sigma^i_{\alpha\alpha'} S^i_{\beta\beta'} \rho \int_0^D d\varepsilon' \Gamma_{\alpha'\beta'}(\varepsilon')$$
$$- \rho \int_0^D d\varepsilon' K_{\alpha\beta\alpha'\beta'}(\varepsilon + \varepsilon' - \widetilde{E}) \Gamma_{\alpha'\beta'}(\varepsilon') = 0 \,. \tag{17.40}$$

Here and henceforth it is assumed that we take sum over repeated subscript i, α and β. In the present case, it is difficult to obtain a closed form of the kernel $K(\omega)$ in the most divergent approximation just by calculating first several terms in the perturbation theory. Therefore we use the theory developed for the scattering matrix by Abrikosov. First, noting that the kernel has the same structure as the vertex function Λ_2, as shown by *Nakajima* [17.15] in the case of isotropic exchange interaction, we put Λ_2 as the kernel. The kernel, (17.19), for the amplitude of the singlet state is given by the singlet component of Λ_2. That is,

$$K_{\text{singlet}} = \Lambda_{2\downarrow\uparrow\downarrow\uparrow} - \Lambda_{2\downarrow\uparrow\uparrow\downarrow} \ . \tag{17.41}$$

The Λ_2 introduced by Abrikosov is related to the scattering T-matrix [which is equal to Abrikosov's vertex, i.e., the spin-flip amplitude $\Gamma^A(\omega)$] in the most divergent approximation by the relation

$$\Lambda_{2\alpha\beta\alpha'\beta'}(\omega) = -\rho \int_{|\omega|}^{D} \frac{d\omega_1}{\omega_1} T_{\alpha\beta''\alpha''\beta'}(\omega_1) T_{\alpha''\beta\alpha'\beta''}(\omega_1) \ . \tag{17.42}$$

The T-matrix satisfies, within the same approximation, the integral equation

$$\begin{aligned} T_{\alpha\beta\alpha'\beta'}(\omega) = &- \frac{J}{2N}(\boldsymbol{\sigma} \cdot \boldsymbol{S})_{\alpha\beta\alpha'\beta'} - \rho \int_{|\omega|}^{D} \frac{d\omega_1}{\omega_1} [T_{\alpha\beta\alpha''\beta''}(\omega_1) T_{\alpha''\beta''\alpha'\beta'}(\omega_1) \\ &- T_{\alpha\beta''\alpha''\beta'}(\omega_1) T_{\alpha''\beta\alpha'\beta''}(\omega_1)] \ . \end{aligned} \tag{17.43}$$

If we put

$$T_{\alpha\beta\alpha'\beta'}(\omega) = \tau(\omega)(\boldsymbol{\sigma} \cdot \boldsymbol{S})_{\alpha\beta\alpha'\beta'} \ , \tag{17.44}$$

we obtain an integral equation for $\tau(\omega)$:

$$\tau(\omega) = -\frac{J}{2N} + 2\rho \int_{|\omega|}^{D} \frac{d\omega_1}{\omega_1} \tau(\omega_1)^2 \ . \tag{17.45}$$

Using the solution of this equation

$$\tau(\omega) = -\frac{J}{2N}\left[1 - \frac{J\rho}{N} \log \frac{|\omega|}{D}\right]^{-1} \ , \tag{17.46}$$

in (17.42), we can derive from (17.41)

$$K_{\text{singlet}} = -\frac{3}{16}\frac{J}{N}\left(\frac{J\rho}{N}\right) \log \frac{|\omega|}{D}\left[1 - \frac{J\rho}{N} \log \frac{|\omega|}{D}\right]^{-1} \ . \tag{17.47}$$

This is nothing but (17.19) which was calculated directly. We note that the spin-flip scattering matrix $\tau(\omega)$ given by (16.26) was actually obtained as the solution of (17.43) by Abrikosov.

For the anisotropic exchange interaction, the inhomogeneous term of (17.43) is replaced by

$$-\frac{1}{2N} J_i \sigma^i_{\alpha\alpha'} S^i_{\beta\beta'} \ . \tag{17.48}$$

Corresponding to this replacement, (17.44) is extended as

$$T_{\alpha\beta\alpha'\beta'}(\omega) = \tau_i(\omega) \sigma^i_{\alpha\alpha'} S^i_{\beta\beta'} \ . \tag{17.49}$$

Similarly the integral equation for $\tau(\omega)$ is changed to

$$\tau_i(\omega) = -\frac{J_i}{2N} + 2N \int_0^t ds \tau_j(s) \tau_k(s) \ , \tag{17.50}$$

where t is defined by

$$t = \frac{\rho}{N} \log \frac{D}{|\omega|} \tag{17.51}$$

and i, j, k represent x, y, z in this order, respectively. Differentiating (17.50) with respect to t, we obtain

$$\frac{d\tau_x^2}{dt} = 4N\tau_x(t)\tau_y(t)\tau_z(t) = \frac{d\tau_y^2}{dt} = \frac{d\tau_z^2}{dt} . \tag{17.52}$$

Integrating both sides of (17.52) and using the initial condition $\tau_i(0) = -J_i/2N$, we obtain the relation

$$[\tau_x(t)]^2 - \left(\frac{J_x}{2N}\right)^2 = [\tau_y(t)]^2 - \left(\frac{J_y}{2N}\right)^2 = [\tau_z(t)]^2 - \left(\frac{J_z}{2N}\right)^2 . \tag{17.53}$$

Now we assume, without loss of generality,

$$J_x^2 \geq J_y^2 \geq J_z^2 ; \tag{17.54}$$

introducing the function $u(t)$, we write (17.53) as

$$\begin{aligned} \frac{1}{(2N)^2}\left[\frac{J_x^2 - J_z^2}{u(t)^2} - J_x^2\right] = \frac{1}{3}\Bigg\{ & [\tau_x(t)^2 + \tau_y(t)^2 + \tau_z(t)^2] \\ & - \frac{1}{(2N)^2}(J_x^2 + J_y^2 + J_z^2)\Bigg\} . \end{aligned} \tag{17.55}$$

Differentiating this with respect to t and eliminating τ_x, τ_y and τ_z by means of (17.52, 53), we obtain

$$\frac{1}{(J_x^2 - J_z^2)}\left(\frac{du}{dt}\right)^2 = (1 - u^2)(1 - k^2u^2) , \tag{17.56}$$

where the parameter k is defined by

$$k = \sqrt{\frac{J_x^2 - J_y^2}{J_x^2 - J_z^2}} \qquad (0 \leq k \leq 1) . \tag{17.57}$$

Integrating (17.56), we obtain

$$t - t_c = \pm\frac{1}{\sqrt{J_x^2 - J_z^2}} \int_0^u \frac{du}{(1 - u^2)^{1/2}(1 - k^2u^2)^{1/2}} , \tag{17.58}$$

where $\pm$ is chosen according to the sign of du/dt, namely, that of $-d\tau_x^2/dt$ or the sign of $(J_xJ_yJ_z)$. Since $u = \sqrt{J_x^2 - J_z^2}/|J_x|$ holds at $t = 0, t_c$ is given by

$$t_c = \pm\frac{1}{\sqrt{J_x^2 - J_z^2}} \int_0^{\sqrt{J_x^2 - J_z^2}/|J_x|} \frac{du}{(1 - u^2)^{1/2}(1 - k^2u^2)^{1/2}} , \tag{17.59}$$

where $+$ and $-$ are the same as the sign of $-J_xJ_yJ_z$. The inverse function of (17.58) is the Jacobi elliptic function $\mathrm{sn}(z, k)$. Therefore we have

$$u = \pm\mathrm{sn}[\sqrt{J_x^2 - J_z^2}(t - t_c), k] \ . \tag{17.60}$$

Substituting this result for u into the left side of (17.55), we obtain the following results:

$$\begin{aligned}
\tau_x(t)^2 &= \frac{J_x^2 - J_z^2}{(2N)^2} \frac{1}{\mathrm{sn}^2[\sqrt{J_x^2 - J_z^2}(t - t_c), k]} \ , \\
\tau_y(t)^2 &= \frac{J_x^2 - J_z^2}{(2N)^2} \frac{\mathrm{dn}^2[\sqrt{J_x^2 - J_z^2}(t - t_c), k]}{\mathrm{sn}^2[\sqrt{J_x^2 - J_z^2}(t - t_c), k]} \ , \\
\tau_z(t)^2 &= \frac{J_x^2 - J_z^2}{(2N)^2} \frac{\mathrm{cn}^2[\sqrt{J_x^2 - J_z^2}(t - t_c), k]}{\mathrm{sn}^2[\sqrt{J_x^2 - J_z^2}(t - t_c), k]} \ .
\end{aligned} \tag{17.61}$$

Here the Jacobi elliptic functions $\mathrm{cn}(z)$ and $\mathrm{dn}(z)$ are related to $\mathrm{sn}(z)$ by

$$\mathrm{cn}^2(z) = 1 - \mathrm{sn}^2(z) \ , \qquad \mathrm{dn}^2(z) = 1 - k^2\mathrm{sn}^2(z) \ . \tag{17.62}$$

According to (17.61) the scattering matrix $\tau(\omega)$ possesses poles at the zeros of $\mathrm{sn}(z)$. Therefore $\tau(\omega)$ generally has the diverging points on the positive real axis of t. The exceptional case where divergence does not occur on the positive real axis of t is the case where the period $4K$ of the sn function becomes infinity and in addition to this t_c is negative; this condition is

$$\begin{aligned}
&J_x J_y J_z \geq 0 \ , \\
&J_x^2 > J_y^2 = J_z^2 \ , \qquad \text{i.e.,} \quad k = 1 \ .
\end{aligned} \tag{17.63}$$

This condition is satisfied just within the shaded region and on the negative side of J_z-axis, and coincides with the condition that the lowest bound states of the zeroth approximation are degenerate.

In the general case, when t is increased from zero, the first zero of the sn function corresponds to $t = t_c$ for $t_c > 0$ and $t = 2K/\sqrt{J_x^2 - J_z^2} + t_c$ for $t_c < 0$. Here K is defined by

$$K = \int_0^{\pi/2} (1 - k^2 \sin^2 \phi)^{-\frac{1}{2}} d\phi \ . \tag{17.64}$$

The value of ω corresponding to this value is given by

$$\begin{aligned}
|\omega_c| &= D\mathrm{e}^{-t_c N/\rho} && (t_c > 0) \ , \\
|\omega_c| &= D\mathrm{e}^{-(2K/\sqrt{J_x^2 - J_z^2} + t_c)N/\rho} && (t_c < 0) \ .
\end{aligned} \tag{17.65}$$

$\tau(\omega)$ diverges at this value of ω_c.

In the case of uniaxial exchange interaction, $k = 0$ holds in the region $J_\perp^2 \geq J_z^2$ in Fig. 17.2, and we obtain

$$\mathrm{sn}(z, 0) = \sin z \ , \quad \mathrm{cn}(z, 0) = \cos z \ , \quad \mathrm{dn}(z, 0) = 1 \ , \tag{17.66}$$

$$t_c = -\frac{\mathrm{sgn} J_z}{(J_\perp^2 - J_z^2)^{1/2}} \,\mathrm{Sin}^{-1} \frac{(J_\perp^2 - J_z^2)^{1/2}}{|\, J_\perp \,|} \ . \tag{17.67}$$

Then the scattering amplitude is given by

$$\tau_\perp(t) = \frac{1}{2N}(J_\perp^2 - J_z^2)^{1/2} \frac{\mathrm{sgn}(J_\perp J_z)}{\sin\{\mathrm{Tan}^{-1}[(J_\perp^2 - J_z^2)^{1/2}/(-J_z)] - (J_\perp^2 - J_z^2)^{1/2} t\}} ,$$

$$\tau_z(t) = \frac{1}{2N}(J_\perp^2 - J_z^2)^{1/2} \cot\left[\mathrm{Tan}^{-1}\left(\frac{(J_\perp^2 - J_z^2)^{1/2}}{-J_z}\right) - (J_\perp^2 - J_z^2)^{1/2} t\right] . \tag{17.68}$$

On the other hand, if $J_x^2 > J_y^2 = J_z^2 = J_\perp^2$ holds, we obtain $k = 1$ so that

$$t_c = \frac{1}{(J_x^2 - J_\perp^2)^{1/2}} \tanh^{-1}\left[\frac{(J_x^2 - J_\perp^2)^{1/2}}{-J_x}\right] , \tag{17.69}$$

$$\tau_\perp(t) = \frac{1}{2N}(J_x^2 - J_\perp^2)^{1/2} \frac{\mathrm{sgn}(J_\perp J_x)}{\sinh\{\tanh^{-1}[(J_x^2 - J_\perp^2)^{1/2}/(-J_x)] - (J_x^2 - J_\perp^2)^{1/2} t\}} ,$$

$$\tau_x(t) = \frac{1}{2N}(J_x^2 - J_\perp^2)^{1/2} \coth\left[\tanh^{-1}\frac{(J_x^2 - J_\perp^2)^{1/2}}{-J_x} - (J_x^2 - J_\perp^2)^{1/2} t\right] . \tag{17.70}$$

Since $\tau_i(t)$ has been obtained in this way, we can solve for the uniaxial *s-d* interaction the integral equation for $\Gamma_{\alpha\beta}(\varepsilon)$, which has the kernel $K(\omega)$ obtained by using the relation (17.42). The results obtained by Shiba are the following: There is no bound state solution in the region where the scattering amplitude has no pole on the real t-axis, in other words, in the shaded region of Fig. 17.2 and on the negative side of J_z- axis. However, bound states exist in the other regions. The binding energy $\widetilde{E}$ is summarized in Table 17.1.

Table 17.1. The binding energy $\widetilde{E}$ for the *s-d* interaction having the uniaxial symmetry

	$J_z < 0$	$J_z > 0$
$\lvert J_\perp\rvert > \lvert J_z\rvert$	$-D\exp\left[\frac{-N}{(J_\perp^2 - J_z^2)^{1/2}\rho} \times \mathrm{Tan}^{-1}\frac{(J_\perp^2 - J_z^2)^{1/2}}{-J_z}\right]$	$-D\exp\left\{\frac{-N}{(J_\perp^2 - J_z^2)^{1/2}\rho} \times \left[\pi + \mathrm{Tan}^{-1}\frac{(J_\perp^2 - J_z^2)^{1/2}}{-J_z}\right]\right\}$
$\lvert J_\perp\rvert < \lvert J_z\rvert$	$-D\exp\left[\frac{-N}{(J_z^2 - J_\perp^2)^{1/2}\rho} \times \tanh^{-1}\frac{(J_z^2 - J_\perp^2)^{1/2}}{-J_z}\right]$	0

17.3 Properties of Scattering T-Matrix

In the case of uniaxial exchange interaction, putting $J_x = J_y = J_\perp$, we can derive from (17.53) the relation

$$\tau_\perp^2(t) - \tau_z^2(t) = \left(\frac{J_\perp}{2N}\right)^2 - \left(\frac{J_z}{2N}\right)^2 . \tag{17.71}$$

If we consider t as a parameter, $\tau_\perp$ and τ_z form the hyperbola A shown in Fig. 17.3 for $|J_\perp| > |J_z|$. Similarly the hyperbola B is obtained for $|J_\perp| < |J_z|$. Here we have considered only the case $\tau_\perp > 0$, since the hyperbola A and B are symmetric with respect to the sign change of $\tau_\perp$.

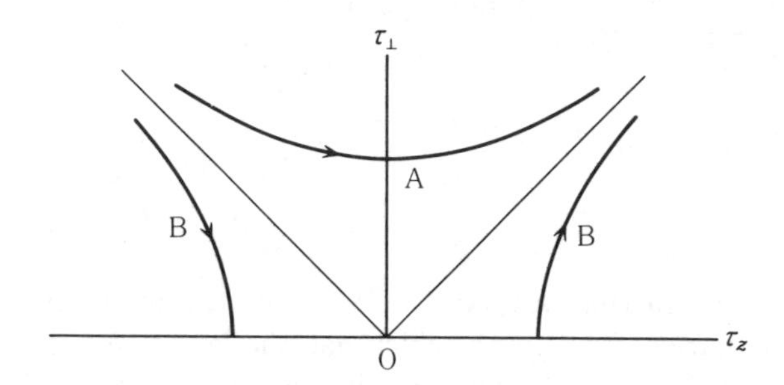

Fig. 17.3. Scaling flow of scattering amplitude τ

From (17.52), τ_z increases with t, since

$$\frac{d\tau_z}{dt} = 2N[\tau_\perp(t)]^2 , \qquad \frac{d\tau_\perp}{dt} = 2N\tau_\perp(t)\tau_z(t) \tag{17.72}$$

hold. Therefore, if we increase t from zero, (that is, decrease ω from D), τ_z and $\tau_\perp$ move from left to right on the hyperbola and approach one of the asymptotes for sufficiently large t in case $|J_\perp| > J_z$; on the other hand we have $\tau_\perp \to 0$ in case $|J_\perp| < |J_z|$ and $J_z > 0$. In the former case, even if the absolute values of τ_z and $\tau_\perp$ at $t = 0$ are very small, the absolute values become large for sufficiently large t .

Since $\tau = -J/2N$ holds at $t = 0$ ($\omega = D$), we write $\tau(t)$ as

$$\tau(t) = -\frac{J_{\text{inv}}(t)}{2N} . \tag{17.73}$$

Then, in the region $|J_\perp| > J_z$, J_{inv} increases in absolute value with increasing t from the value J at $t = 0$ and reaches the strong coupling value at sufficiently large t. The Kondo temperature can be defined as the value of ω at which $\rho|J_{\text{inv}}|/N$ reaches a strong coupling constant around order of unity. J_{inv} corresponds to the invariant coupling constant in the terminology of the renormalization group. When J_{inv} reaches a strong-coupling value, Abrikosov's approach (which is based on the perturbation expansion) is not appropriate.

In this case the spins of conduction electrons are coupled strongly with the localized spin and a singlet state is formed.

Anderson [17.12] and *Anderson* et al. [17.13] considered the problem of the Kondo effect from the point of view of the scaling hypothesis and explained the Kondo effects in terms of the scaling flow of $\tau(\omega)$ at $\omega \to 0$, as mentioned above.

Finally, we return to the consideration of the case of isotropic exchange interaction. By differentiating (17.45) with respect to ω, we obtain

$$\frac{d\tau}{d\omega} = -2\rho\tau^2(\omega)\frac{1}{\omega} . \tag{17.74}$$

Integrating this with respect to ω, and using the boundary condition

$$\tau(\omega)\Big|_{\omega=D} = -\frac{J}{2N} , \tag{17.75}$$

we can derive for $\tau(\omega)$ the result of (16.26). The latter can be obtained by collecting the most divergent terms, as discussed before. Therefore this differential equation combined with the boundary condition is equivalent to (16.26). However, we can derive this differential equation by differentiating $\tau^{(2)}(\omega)$ [given to the second order by (16.26)] with respect to ω and putting $-J/2N = \tau(\omega)$. Therefore, by adding the lowest order contribution in the next divergent terms we can extend it to the form including the next divergent terms. For this purpose, we rewrite (17.74) as

$$\frac{d\omega}{\omega} = -\frac{1}{2\rho}d\tau\left(\frac{1}{\tau^2} + \dots\right) . \tag{17.76}$$

The next divergent term first appears in third order. In this order, the contributions from the perturbation of the wave function should be added to the scattering terms in third order. By combining these contributions as the contribution from the next divergent terms, we obtain $\rho/\tau(\omega)$ which should be added in place of "..." in (17.76). In this way, the differential equation becomes

$$\frac{d\omega}{\omega} = -\frac{1}{2\rho}d\tau\left[\frac{1}{\tau(\omega)^2} + \frac{\rho}{\tau(\omega)}\right] . \tag{17.77}$$

By integrating this differential equation with use of the boundary condition, and defining the Kondo temperature as the value of ω such that $2\rho\tau(\omega) = |J_{\text{inv}}|\rho/N = 1$, we obtain

$$kT_{\text{K}} = \omega_{\text{K}} = D\sqrt{\frac{|J|\rho}{N}}\exp\left[-\frac{N}{|J|\rho}\right] . \tag{17.78}$$

This means that when we take into account the next divergent terms, the value obtained by using only the most divergent terms is multiplied by the factor $\sqrt{|J|\rho/N}$. Owing to this effect, T_{K} becomes smaller by this factor.

Fowler and *Zawadowski* [17.16] and *Abrikosov* and *Migdal* [17.17] derived the differential equation (17.77) (which includes the next divergent terms) as the Lie differential equation for the invariant coupling constant $J_{\rm inv}(\omega)$ in the renormalization group theory. It is possible also to extend in the same way the theory of the singlet ground state so as to include the contributions of the next divergent terms. This extension was actually carried out by *Sakurai* and *Yoshimori* [17.18]; according to their work (17.78) is obtained for the binding energy of the singlet ground state.

18. Anderson's Orthogonality Theorem

This chapter deals with the problems concerning the orthogonality catastrophe, which was first noticed and discussed by Anderson [18.1]. It is related to the Kondo effect, the edge singularities in the X-ray absorption and emission spectra in metals and others. This theorem states that the wave function (normalized) of the ground state of the conduction electrons under the impurity potential ψ is orthogonal to the wave function of the free state in the absence of the impurity potential ψ_0 as $|\langle\psi|\psi_0\rangle| = \exp[-\frac{1}{2}\sum_l(2l+1)(\delta_l/\pi)^2 \log N_l]$: δ_l and N_l are the phase shift at the Fermi surface and the total number of the l-wave electrons. This orthogonality theorem is basic and important as the Friedel sum rule and can be applied widely to various problems.

18.1 Overlap Integral Between Locally Perturbed Wave Functions

It has been already noted in Sect. 16.3 that in the perturbation calculation which starts from the free state composed of the localized spin and conduction electrons and treats the s-d exchange interaction as the perturbation, the normalization integral of the wave function diverges as $J^2 \log N$ in the first-order perturbation, and that this divergence is the origin of Kondo's $\log T$ term appearing in the perturbation calculation at finite temperatures. Though the divergence of the normalization integral occurs also for the impurity potential, in contrast to the s-d exchange interaction, for the case of the potential without any internal degrees of freedom it is cancelled out between the denominator and numerator, giving no direct effects on the calculated results of the expectation values of physical quantities. For the case of the perturbation expansion in the form of (16.34), the overlap integral between the perturbed wave function ψ and the unperturbed wave function, $\langle\psi|\psi_0\rangle$, is normalized to unity. Therefore, the divergence of $\langle\psi|\psi\rangle$ means, in the case where both ψ and ψ_0 are normalized to unity, that the overlap integral between them vanishes. In other words, the perturbed and unperturbed wave functions are orthogonal to each other:

$$\frac{\langle\psi|\psi_0\rangle}{\langle\psi|\psi\rangle^{1/2}\langle\psi_0|\psi_0\rangle^{1/2}} = \langle\psi|\psi\rangle^{-1/2} \to 0 \ .$$

Now let us consider s-wave scattering in the case of the impurity potential and study in what form the wave function diverges as a function of electron number N_e in s-wave. For this purpose, we have only to calculate the higher order of the most divergent terms in (16.38). For the impurity potential, this expansion becomes that of the exponential function of $(\rho V/N)^2 \log N_e$. Here we assume the impurity potential has δ-function form

$$V_i = \frac{V}{N} \sum_{\boldsymbol{k}\boldsymbol{k}'} a_{\boldsymbol{k}'}^{\dagger} a_{\boldsymbol{k}} \tag{18.1}$$

and for simplicity, we ignore the spin of conduction electrons.

Thus the normalization integral is given by the perturbation calculation as

$$\langle\psi|\psi\rangle \to \exp\left[\left(\frac{V\rho}{N}\right)^2 \log N_e\right] . \tag{18.2}$$

If we use a Slater determinant of one-electron wave functions in place of (16.34) as the perturbed wave function of the total electron system, we can easily understand that the normalization integral is given by the above exponential function of $\log N_e$. In this case, ψ is expressed as

$$\psi = \prod_{\boldsymbol{k},\varepsilon_{\boldsymbol{k}}<\varepsilon_{\mathrm{F}}} \left(a_{\boldsymbol{k}}^{\dagger} + \frac{V}{N} \mathrm{P} \sum_{\boldsymbol{k}'} \frac{1}{\varepsilon_{\boldsymbol{k}} - \varepsilon_{\boldsymbol{k}'}} a_{\boldsymbol{k}'}^{\dagger}\right) \Phi_0 , \tag{18.3}$$

where Φ_0 is the vacuum state without any conduction electrons and the operator P means to take the principal part. This normalization integral for ψ is calculated by using the most singular terms as

$$\begin{aligned} \langle\psi|\psi\rangle &= \sum_{n=0}^{\infty} \frac{1}{n!} \left[\left(\frac{V}{N}\right)^2 \sum_{\boldsymbol{k}\boldsymbol{k}'} \frac{f_{\boldsymbol{k}}(1-f_{\boldsymbol{k}'})}{(\varepsilon_{\boldsymbol{k}} - \varepsilon_{\boldsymbol{k}'})^2}\right]^n \\ &= \exp\left[\left(\frac{V}{N}\right)^2 \sum_{\boldsymbol{k}\boldsymbol{k}'} \frac{f_{\boldsymbol{k}}(1-f_{\boldsymbol{k}'})}{(\varepsilon_{\boldsymbol{k}} - \varepsilon_{\boldsymbol{k}'})^2}\right] . \end{aligned} \tag{18.4}$$

This result is correct to order $(\rho V/N)^2$ in the argument of the exponential function. *Anderson* [18.1] pointed out that if we take into account more higher-order terms, then $(\rho V/N)^2$ is replaced by $[\delta(\varepsilon_{\mathrm{F}})/\pi]^2, \delta(\varepsilon_{\mathrm{F}})$ being the phase shift at the Fermi surface. As a result, the overlap integral S between the ground-state wave function for the electron system with the impurity potential and that for the free-electron system is given by

$$S = \exp\left[-\frac{1}{2}\left(\frac{\delta(\varepsilon_{\mathrm{F}})}{\pi}\right)^2 \log N_e\right] = (N_e)^{-(1/2)[\delta(\varepsilon_{\mathrm{F}})/\pi]^2} . \tag{18.5}$$

$S \to 0$ is obtained according to (18.5) in the limit $N_e \to \infty$. This is the orthogonality theorem of *Anderson* [18.1].

From the Friedel sum rule, $\delta(\varepsilon_F)/\pi$ is related to the number of electrons localized around the impurity potential. Anderson's orthogonality theorem holds also between wave functions of total systems representing two states which possess different numbers of localized electrons. In other words, if we assume that the localized electron numbers are n_1 and n_2 in the two states, the condition that the overlap integral between the two states does not vanish is given by

$$n_1 = n_2 . \tag{18.6}$$

This theorem can be extended to the matrix element between ψ_1 and ψ_2, $\langle\psi_2^{(N+1)}|a_0^\dagger|\psi_1^{(N)}\rangle$, of the operators creating one electron at the origin or annihilating it, which are given by

$$a_0^\dagger = \frac{1}{\sqrt{N}}\sum_k a_k^\dagger , \qquad a_0 = \frac{1}{\sqrt{N}}\sum_k a_k . \tag{18.7}$$

One electron added at the origin increases the number of localized electrons to $n_1 + 1$. Therefore, in order for the integral not to vanish, the numbers of localized electrons should be equal, that is, the condition

$$n_2 = n_1 + 1 \tag{18.8}$$

should hold in addition to the condition that the total numbers of electrons of two states $\psi_2^{(N+1)}$ and $a_0^\dagger\psi_1^{(N)}$ are equal. Thus, the Anderson's orthogonality theorem means the conservation law of the localized electron number.

When we wrote down the wave function of the singlet ground state in the *s-d* system in the form of (17.2), it was necessary to couple ψ_α and ψ_β by the *s-d* exchange interaction in order to construct the singlet state . For this reason, we selected $\psi_{0\alpha}$ and $\psi_{0\beta}$ given by (17.3) at the starting point of perturbation. In these states one electron put above the Fermi surface forms a spin bound state antiparallel to the localized spin. Therefore, if we consider the down spin electron, its phase shifts at the Fermi surface are π in ψ_α and 0 in ψ_β, and satisfy the condition (18.8). In the final state, the phase shift of ψ_α is $\pi/2$ and that of ψ_β is $-\pi/2$, and the condition (18.8) is always satisfied. The conservation law of localized electron number, (18.8) was used also in the variational calculation of the ground state energy by *Anderson* [18.1]. In the latter case the phase shift $\delta(\varepsilon)$ of the scattering wave was treated as the variational function.

Since (18.8) is the general relation, it can be considered to hold also for the ground state in the presence of the magnetic field. In this case, if we assume that the localized electron numbers in the two components ψ_α and ψ_β are $n_{\uparrow\downarrow}^\alpha$ and $n_{\uparrow\downarrow}^\beta$, respectively, the following relations hold:

$$n_\uparrow^\alpha = n_\uparrow^\beta - 1 , \qquad n_\downarrow^\alpha = n_\downarrow^\beta + 1 . \tag{18.9}$$

From the condition that the total localized charge vanishes,

$$\frac{[(n_\uparrow^\alpha + n_\downarrow^\alpha)\langle\psi_\alpha|\psi_\alpha\rangle + (n_\uparrow^\beta + n_\downarrow^\beta)\langle\psi_\beta|\psi_\beta\rangle]}{(\langle\psi_\alpha|\psi_\alpha\rangle + \langle\psi_\beta|\psi_\beta\rangle)} = 0 , \tag{18.10}$$

and the relation (16.40)

$$\frac{[(n_\uparrow^\alpha - n_\downarrow^\alpha)\langle\psi_\alpha|\psi_\alpha\rangle + (n_\uparrow^\beta - n_\downarrow^\beta)\langle\psi_\beta|\psi_\beta\rangle]}{(\langle\psi_\alpha|\psi_\alpha\rangle + \langle\psi_\beta|\psi_\beta\rangle)} = \frac{J\rho}{N}\langle S_z\rangle \tag{18.11}$$

which holds for the total localized spin polarization, we obtain the relations

$$\begin{aligned} n_\downarrow^\alpha = -n_\uparrow^\alpha &= \frac{1}{2} - \langle S_z\rangle\left(1 + \frac{J\rho}{2N}\right) , \\ n_\uparrow^\beta = -n_\downarrow^\beta &= \frac{1}{2} + \langle S_z\rangle\left(1 + \frac{J\rho}{2N}\right) . \end{aligned} \tag{18.12}$$

Here we have used the relation

$$\langle S_z\rangle = \frac{1}{2}\left(\frac{\langle\psi_\alpha|\psi_\alpha\rangle - \langle\psi_\beta|\psi_\beta\rangle}{\langle\psi_\alpha|\psi_\alpha\rangle + \langle\psi_\beta|\psi_\beta\rangle}\right) , \tag{18.13}$$

which holds generally. Having the localized electron number of each component from (18.12) and using the Friedel sum rule, we can immediately derive the general relation which holds between the value of resistance $R(H)$ and the localized moment induced by the magnetic field, as [18.2]

$$R(H) = R_0 \cos^2\left[\pi\langle S_z\rangle\left(1 + \frac{J\rho}{2N}\right)\right] . \tag{18.14}$$

Here, R_0 is the value of resistivity in the unitarity limit in the absence of the magnetic field. This result was also derived by H. Ishii and *Hamann* [18.3].

18.2 Nozières-de Dominicis Theory in the Time Representation

Let us consider a spinless conduction-electron system and write the wave function of the free state as $|0\rangle$. If the impurity potential (18.1) is suddenly applied at $t = 0$, the wave function at time t is given by ($\hbar$ is put to 1)

$$\psi(t) = \mathrm{e}^{-\mathrm{i}\mathcal{H}t}|0\rangle , \tag{18.15}$$

where $\mathcal{H}$ is the Hamiltonian in the presence of the impurity potential and is given by the sum of the kinetic energy $\mathcal{H}_0$ and (18.1). Then the overlap integral between this wave function and that at the same time in the absence of impurity potential $\mathrm{e}^{-\mathrm{i}\mathcal{H}_0 t}|0\rangle$ is given by

$$g(t) = \langle 0 \mid e^{i\mathcal{H}_0 t} e^{-i\mathcal{H}t} | 0 \rangle \ . \tag{18.16}$$

Similarly, let us consider a free state $a_0^\dagger|0\rangle$, in which one electron is added at the origin. The overlap integral between the wave function at time t, after the impurity potential is applied to this state suddenly at $t = 0$, and the wave function of the free state at t in the absence of the impurity potential is given by

$$F(t) = \langle 0|e^{i\mathcal{H}_0 t} a_0 e^{-i\mathcal{H}t} a_0^\dagger |0\rangle \ . \tag{18.17}$$

The limiting values of (18.16, 17) at $t \to \infty$ coincide with the two overlap integrals discussed in the previous section, respectively. Here, we discuss the method of *Nozières* and *de Dominicis* [18.4] which derives the integrals of (18.16, 17) assuming finite t and $N \to \infty$.

Inserting the expansion form equivalent to (16.30) ($t > t_0$),

$$e^{i\mathcal{H}_0 t} e^{-i\mathcal{H}(t-t_0)} e^{-i\mathcal{H}_0 t_0} = \mathrm{T}\exp\left[-i\int_{t_0}^{t} V_i(\tau)d\tau\right] = S(t,t_0) \tag{18.18}$$

into (18.16, 17), and taking the ratio of them, we obtain

$$\frac{F(t-t_0)}{g(t-t_0)} = \frac{\langle 0|\mathrm{T}a_0(t)a_0^\dagger(t_0)S(t,t_0)|0\rangle}{\langle 0|S(t,t_0)|0\rangle} \ , \tag{18.19}$$

where $V_i(\tau), a_0(t)$, *etc.* are operators developing by $\mathcal{H}_0$ in the interaction representation and $a_0(t)$ is defined by

$$a_0(t) = e^{i\mathcal{H}_0 t} a_0 e^{-i\mathcal{H}_0 t} \ . \tag{18.20}$$

The expansion form of (18.18) is easily derived by applying an iteration method to the equation

$$e^{i\mathcal{H}_0 t} e^{-i\mathcal{H}(t-t_0)} e^{-i\mathcal{H}_0 t_0} = 1 - i\int_{t_0}^{t} V_i(\tau) e^{i\mathcal{H}_0 \tau} e^{-i\mathcal{H}(\tau - t_0)} e^{-i\mathcal{H}_0 t_0} d\tau \ ; \tag{18.21}$$

we can obtain this relation by differentiating the left-hand side with respect to t and by integrating over t from t_0 to t. The operator T in (18.18, 19) means that the product of functions of t should be arranged always according to the order of time. In (18.19), if we replace $S(t, t_0)$ by $S(\infty, -\infty)$, the equation becomes equal to the electron Green function at the origin in the presence of the impurity potential

$$G(t-t_0) = \langle \mathrm{T}\tilde{a}_0(t)\tilde{a}_0^\dagger(t_0)\rangle \ , \tag{18.22}$$

where $\tilde{a}_0(t)$ is the operator in the Heisenberg representation developing in time by $\mathcal{H}$. Therefore, if we define the following quantity for τ and τ', which satisfy $t > \tau, \tau' > t_0$,

$$\psi(\tau,\tau') = \frac{\langle 0|\mathrm{T}a_0(\tau)a_0^\dagger(\tau')S(t,t_0)|0\rangle}{\langle 0|S(t,t_0)|0\rangle} , \tag{18.23}$$

this is the Green function extended to the case where the impurity potential exists within finite time interval $t - t_0$. After obtaining $\psi(\tau,\tau')$ and taking the limit $\tau' \to t_0, \tau \to t$, we can arrive at (18.19). The Green function of (18.23) satisfies the integral equation (called the Dyson equation)

$$\psi(\tau,\tau') = G_0(\tau - \tau') - \mathrm{i}V \int_{t_0}^{t} d\tau'' G_0(\tau - \tau'')\psi(\tau'',\tau') , \tag{18.24}$$

where $G_0(\tau - \tau')$ is the Green function in the case without the impurity potential,

$$G_0(\tau - \tau') = \langle \mathrm{T}a_0(\tau)a_0^\dagger(\tau')\rangle . \tag{18.25}$$

Using (18.20) for $a_0(\tau)$ and (16.22) for the density of states of the conduction electrons, we obtain $G_0(\tau - \tau')$ as

$$G_0(\tau - \tau') = \frac{\rho}{N}\left[\frac{1-\exp(-\mathrm{i}D|\tau - \tau'|)}{\mathrm{i}(\tau - \tau')}\right] . \tag{18.26}$$

The Dyson equation (18.24) can be derived directly by representing (18.23) in the expansion form with respect to $V_i(\tau)$ and by writing each term in the form of integration of the product of $G_0(\tau)$ with use of Wick's theorem. In order to solve the Dyson equation (18.24), Nozières-de Dominicis approximated $G_0(\tau - \tau')$ by its asymptotic form,

$$G_0(\tau - \tau') \cong -\mathrm{i}\frac{\rho}{N}\frac{1}{\tau - \tau'} , \qquad |\tau - \tau'| > D^{-1} . \tag{18.27}$$

Since the integration over τ'' on the right-hand side of (18.24) passes through $\tau = 0$ in $G_0(\tau)$, we take the principal part to avoid the singularity. In this approximation, (18.24) can be written as

$$\psi(\tau,\tau') = -\mathrm{i}\frac{\rho}{N}\mathrm{P}\left(\frac{1}{\tau - \tau'}\right) - \frac{\rho V}{N}\int_{t_0}^{t} d\tau''\mathrm{P}\left(\frac{1}{\tau - \tau''}\right)\psi(\tau'',\tau') . \tag{18.28}$$

In this approximation, the Fourier transform $G_0(\varepsilon)$ of $G_0(\tau)$ is given by

$$G_0(\varepsilon) = \int_{-\infty}^{\infty} G_0(\tau)\mathrm{e}^{\mathrm{i}\varepsilon\tau}d\tau = \frac{\pi\rho}{N}\mathrm{sgn}(\varepsilon) . \tag{18.29}$$

The Green function $G(\tau)$ in the presence of the impurity potential is given by the solution of the integral equation obtained by putting $t \to \infty, t_0 \to -\infty$ in (18.28). This integral equation can be easily solved by Fourier transformation; the solution is

$$G(\varepsilon) = \frac{G_0(\varepsilon)}{1 + \mathrm{i}VG_0(\varepsilon)} = \frac{\pi\rho}{N}\mathrm{sgn}(\varepsilon)\frac{1}{1 + \mathrm{i}\pi\dfrac{\rho V}{N}\mathrm{sgn}(\varepsilon)} . \tag{18.30}$$

Therefore, by the inverse transformation, $G(\tau)$ is given by

$$G(\tau) = -\mathrm{i}\frac{\rho}{N}\frac{1}{1+\pi^2(\rho V/N)^2}\left(\mathrm{P}\frac{1}{\tau} + \pi^2\frac{\rho V}{N}\delta(\tau)\right) . \tag{18.31}$$

On the other hand, the phase shift of the conduction electron $\delta(\varepsilon)$ in the presence of the impurity potential is

$$\tan\delta(\varepsilon) = -\pi\frac{\rho V}{N}\Bigg/\left(1 - \frac{\rho V}{N}\mathrm{P}\int_{-D}^{D}\frac{d\varepsilon'}{\varepsilon-\varepsilon'}\right) . \tag{18.32}$$

Therefore the phase shift at the Fermi surface ($\varepsilon = \varepsilon_\mathrm{F} = 0$) is given by

$$\tan\delta(\varepsilon_\mathrm{F}) = -\pi\frac{\rho V}{N} . \tag{18.33}$$

Here we have assumed the band is symmetric with respect to ε_F. Hereafter, the phase shift at Fermi surface is simply written as δ.

Now we return to the integral equation (18.28). This equation is solved by following the procedure in "Singular Integral Equation" by *Muskhelishvili* [18.5], as follows: First we define the following Cauchy integral:

$$\Phi(z) = \frac{1}{2\pi\mathrm{i}}\int_{t_0}^{t} d\tau''\frac{1}{\tau''-z}\left[\psi(\tau'',\tau') - \mathrm{i}\pi\frac{\rho}{N}\cot\delta\cdot\delta(\tau''-\tau')\right] . \tag{18.34}$$

Denoting the value continued to the real axis from the upper part of the complex plane as $\Phi^+(\tau)$ and the value continued from the lower part as $\Phi^-(\tau)$, (where $t > \tau > t_0$), we obtain

$$\Phi^+(\tau) - \Phi^-(\tau) = \psi(\tau,\tau') - \mathrm{i}\pi\frac{\rho}{N}\cot\delta\cdot\delta(\tau-\tau') \tag{18.35}$$

and

$$\Phi^+(\tau) + \Phi^-(\tau) = \frac{1}{\pi\mathrm{i}}\int_{t_0}^{t} d\tau''\mathrm{P}\left(\frac{1}{\tau''-\tau}\right)\psi(\tau'',\tau') - \frac{\rho}{N}\cot\delta\cdot\mathrm{P}\left(\frac{1}{\tau'-\tau}\right) . \tag{18.36}$$

Using (18.35, 36) in the integral equation (18.28), eliminating $\psi(\tau,\tau')$ and noting (18.33), we obtain the relation

$$\Phi^+(\tau) = \mathrm{e}^{-2\mathrm{i}\delta}\Phi^-(\tau) - \mathrm{i}\pi\frac{\rho}{N}\frac{\cos^2\delta}{\sin\delta}\mathrm{e}^{-\mathrm{i}\delta}\delta(\tau-\tau') . \tag{18.37}$$

Now we look for a function $X(z)$, which is analytic in the complex plane except for the real axis $t - t_0$ on which it satisfies the boundary condition

$$X^+(\tau) = \mathrm{e}^{-2\mathrm{i}\delta}X^-(\tau) . \tag{18.38}$$

On taking the logarithm of both sides of (18.38), we obtain

$$\log X^+(\tau) - \log X^-(\tau) = -2\mathrm{i}\delta .$$

In this way $X(z)$ is given by

$$X(z) = \exp\left(-\frac{\delta}{\pi}\int_{t_0}^{t}\frac{d\tau'}{\tau' - z}\right) . \tag{18.39}$$

Thus we obtain

$$X^{\pm}(\tau) = \mathrm{e}^{\mp \mathrm{i}\delta} \exp\left[-\frac{\delta}{\pi}\int_{t_0}^{t} \mathrm{P}\left(\frac{1}{\tau' - \tau}\right) d\tau'\right] . \tag{18.40}$$

By eliminating δ in the first term of the right-hand side of (18.37) by use of (18.38), (18.37) can be written as

$$\frac{\Phi^+(\tau)}{X^+(\tau)} - \frac{\Phi^-(\tau)}{X^-(\tau)} = -\mathrm{i}\pi\frac{\rho}{N}\frac{\cos^2\delta}{\sin\delta}\mathrm{e}^{-\mathrm{i}\delta}\frac{1}{X^+(\tau)}\delta(\tau - \tau') . \tag{18.41}$$

Here we must again solve the Hilbert problem to obtain an analytic function in the complex plane other than the real axis $t - t_0$; the solution possesses the discontinuity on the real axis $t - t_0$. Since this function is given by

$$-\frac{\rho}{2N}\frac{\cos^2\delta}{\sin\delta}\mathrm{e}^{-\mathrm{i}\delta}\int_{t_0}^{t}\frac{d\tau''}{\tau'' - z}\frac{\delta(\tau'' - \tau')}{X^+(\tau'')} ,$$

$\Phi(z)$ is given by

$$\Phi(z) = -\frac{\rho}{2N}\frac{\cos^2\delta}{\sin\delta}\mathrm{e}^{-\mathrm{i}\delta}X(z)\int_{t_0}^{t}\frac{d\tau''}{\tau'' - z}\frac{\delta(\tau'' - \tau')}{X^+(\tau'')} . \tag{18.42}$$

From this and using (18.38), we obtain

$$\Phi^+(\tau) - \Phi^-(\tau) = -\mathrm{i}\frac{\rho}{N}\cos^2\delta\left[\mathrm{P}\left(\frac{1}{\tau - \tau'}\right) + \pi\cot\delta\cdot\delta(\tau - \tau')\right]\frac{X^+(\tau)}{X^+(\tau')} . \tag{18.43}$$

From this result and (18.35), we obtain the solution of Dyson's equation, which reproduces $\psi(\tau) = -\mathrm{i}(\rho/N)\mathrm{P}(1/\tau)$ at $\delta = 0$ as

$$\psi(\tau, \tau') = G(\tau - \tau')\frac{X^+(\tau)}{X^+(\tau')} . \tag{18.44}$$

Here $G(\tau - \tau')$ is the Green function in the presence of the impurity potential and is given by (18.31) in our approximation. By using (18.40) for $X^{\pm}(\tau)$, we obtain the factor multiplying $G(\tau - \tau')$ in (18.44) as

$$\frac{X^+(\tau)}{X^+(\tau')} = \left[\frac{(t - \tau')(\tau - t_0)}{(\tau' - t_0)(t - \tau)}\right]^{\delta/\pi} . \tag{18.45}$$

Equation (18.19), which is our goal, is given from (18.44) by taking the limit $\tau' \to t_0$ and $\tau \to t$. In this limit the denominator in (18.45) vanishes. This is due to our approximation of replacing $G_0(\tau - \tau')$ with the asymptotic form

(18.27). The original equation (18.26) for $G_0(\tau)$ converges at $\tau \to 0$ and $1/\tau$ in the asymptotic form is replaced by $\mathrm{i}D\mathrm{sgn}(\tau)$ in the correct form. Therefore, by replacing $1/(\tau - t_0)$ and $1/(t-\tau)$ with $\mathrm{i}D$ in (18.45), we obtain (18.19) as

$$\frac{F(t-t_0)}{g(t-t_0)} = \frac{\rho D}{N}\cos^2\delta\left[\mathrm{i}D(t-t_0)\right]^{-1+2\delta/\pi} ; \tag{18.46}$$

here we have used (18.31) for $G(t-t_0)$.

The overlap integral $g(t-t_0)$ can be calculated with use of $\psi(\tau,\tau')$ as follows: By changing the impurity potential V to λV and taking the logarithmic derivative of (18.16 or 18) with respect to λ, we obtain the relation

$$\frac{1}{g_\lambda(t-t_0)}\frac{\partial g_\lambda(t-t_0)}{\partial\lambda} = \mathrm{i}V\int_{t_0}^{t}\psi_\lambda(\tau,\tau_+)d\tau . \tag{18.47}$$

Here τ_+ is the value which tends to τ from the side larger than τ. If we put $\tau' \to \tau$ in (18.44) and expand (18.44) with respect to $\tau' - \tau$, the first term diverges due to $G(\tau-\tau')$ and the second term becomes constant. Since the divergence of the first term originates again from the asymptotic approximation, we replace $G(\tau-\tau')$ by the correct value of the Green function $G(\tau-\tau_+)$. By using this technique, we obtain

$$\psi_\lambda(\tau,\tau_+) = G_\lambda(\tau-\tau_+) - \mathrm{i}\frac{\rho}{N}\cos^2\delta_\lambda\frac{1}{X_\lambda^+(\tau)}\frac{dX_\lambda^+(\tau)}{d\tau} . \tag{18.48}$$

Therefore the right-hand side of (18.47) becomes

$$\mathrm{i}VG_\lambda(\tau-\tau_+)(t-t_0) + \frac{\rho V}{N}\cos^2\delta_\lambda[\log X_\lambda^+(t) - \log X_\lambda^+(t_0)] .$$

$-VG_\lambda(\tau-\tau_+)$ is just the average value of the impurity potential divided by λ. On the other hand, it is well known that the relation

$$\lambda\frac{\partial E_n(\lambda)}{\partial\lambda} = \langle\psi_n|\lambda\mathcal{H}'|\psi_n\rangle \tag{18.49}$$

holds when the Hamiltonian is given by the sum of $\mathcal{H}_0$ and a perturbation term $\lambda\mathcal{H}'$ and the eigenfunctions and eigenvalues of the Hamiltonian are given by ψ_n and E_n, respectively. This relation can be derived from the fact that E_n is stationary with respect to changes in ψ_n. Therefore, the integral of $-VG_\lambda(\tau-\tau_+)$ over λ from 0 to 1 gives the energy shift ΔE due to the impurity potential. With use of (18.40) and the relation

$$-\pi\frac{\rho V}{N}\cos^2\delta_\lambda d\lambda = d\delta_\lambda ,$$

and by integration over λ from 0 to 1, (or by integration over δ_λ from 0 to δ), the second term becomes

$$\frac{1}{2}\left(\frac{\delta}{\pi}\right)^2 \int_{t_0}^{t} \left[\mathrm{P}\left(\frac{1}{\tau' - t}\right) - \mathrm{P}\left(\frac{1}{\tau' - t_0}\right) \right] d\tau' = -\left(\frac{\delta}{\pi}\right)^2 \log[\mathrm{i}D(t - t_0)] \; .$$

Here we have used again the technique of replacing $t_+ - t$ and $t_{0+} - t_0$ with $1/\mathrm{i}D$. Thus, by integrating (18.47) over λ from 0 to 1, we arrive at the result

$$g(t - t_0) = \mathrm{e}^{-\mathrm{i}\Delta E(t-t_0)}[\mathrm{i}D(t - t_0)]^{-(\delta/\pi)^2} \; . \tag{18.50}$$

Using this result in (18.46), we obtain

$$F(t - t_0) = \mathrm{e}^{-\mathrm{i}\Delta E(t-t_0)} \frac{\rho D}{N} \cos^2 \delta [\mathrm{i}D(t - t_0)]^{-(1-\delta/\pi)^2} \; . \tag{18.51}$$

These two results are the correct asymptotic forms of the overlap integrals in the limit of $t - t_0 \to \infty$, as derived by Nozières-de Dominicis. This calculation can be extended to the case in which the impurity potential V_1 exists in the initial state and it changes suddenly at $t = t_0$ to V_2. In this case δ in the exponents of (18.50, 51) is replaced by the difference of phase shifts due to the two potentials, $\delta_2 - \delta_1$. Therefore the condition for the overlap integral not to vanish is given by (18.6, 8). That is, two wave functions which do not have the same number of local electrons are orthogonal to each other.

The calculation of the overlap integral by Nozières-de Dominicis, which we have described in this section, was carried out in the theory of the edge singularities in the soft X-ray absorption and emission spectra in metals; for example, in the process of soft X-ray absorption, core electrons absorb the X-rays and are excited to the levels above the Fermi level in the conduction band. In this case, while the conduction electrons are in completely free states before the transition, in the final state the conduction electrons experience the local potential due to the hole created in the core state. The transition probability is directly related to the overlap integral (18.51) derived here. In particular, the absorption intensity for the case where p-electrons in the core states are excited to the conduction s band is given by the Fourier transform of (18.51) in the region $t - t_0 = \tau$ ($\tau > 0$). Therefore a singularity like $(\omega)^{-2\delta/\pi+(\delta/\pi)^2}$ occurs at the absorption edge ($\omega \to 0$).

We just mentioned above that the orthogonality theorem about the overlap integral between two wave functions with different local potentials is closely related to the edge singularities of the soft X-ray absorption and emission spectra in metals. The relation of the orthogonality theorem to Kondo effect was remarked in Sect. 18.1 and can also be seen in the next section in connection with the Anderson-Yuval theory.

However, in order to apply the orthogonality theorem to other problems, it is required to extend the theorem to the cases in which the local potential has no spherical symmetry. The overlap integral for general local potentials without spherical symmetry can be expressed in terms of the scattering S-matrix as

$$S = \exp\left[\frac{1}{8\pi^2} \mathrm{Tr}\left\{ \log[\widehat{S}_\mathrm{f}(\mu) \cdot \widehat{S}_\mathrm{in}(\mu)^{-1}] \right\}^2 \cdot \log N_e \right] \tag{18.52}$$

where Tr means to take trace; $S_{\rm in}$ and $S_{\rm f}$ are, respectively, the scattering S-matrices in the initial and final states at the Fermi energy μ [18.6]. The corresponding expression for the Friedel sum rule is given by

$$\Delta Z = \frac{1}{2\pi \mathrm{i}} \operatorname{Tr} \log[\widehat{S}_{\rm f}(\mu)\widehat{S}_{\rm in}(\mu)^{-1}] . \tag{18.53}$$

This expression was obtained by *Langer* and *Ambegaokar* [18.7].

18.3 Anderson-Yuval Theory

Anderson and *Yuval* [18.8] applied the Nozières-de Dominicis theory described in the previous section to the problem of the s-d interaction and discussed the structure of the partition function of the s-d system. They included the longitudinal component of the s-d exchange interaction in the unperturbed term of the Hamiltonian and treated the transverse component as the perturbation:

$$\mathcal{H} = \mathcal{H}_0 + \mathcal{H}' ,$$

$$\mathcal{H}_0 = \sum_{k\alpha} \varepsilon_k a_{k\alpha}^\dagger a_{k\alpha} - \frac{J_z}{2N} S_z \sum_{kk'\alpha} a_{k\alpha}^\dagger \sigma_{\alpha\alpha}^z a_{k'\alpha} , \tag{18.54}$$

$$\mathcal{H}' = -\frac{J_\perp}{4N} \sum_{kk'\alpha\alpha'} a_{k\alpha}^\dagger a_{k'\alpha'} (\sigma_{\alpha\alpha'}^- S_+ + \sigma_{\alpha\alpha'}^+ S_-) . \tag{18.55}$$

Here, for the moment, we distinguish between J_z and $J_x = J_y = J_\perp$; σ is the Pauli spin matrix and α, α' are the spin states of the conduction electrons. The ground states of the unperturbed Hamiltonian $\mathcal{H}_0$ are degenerate with respect to the direction (up and down) of the localized spin. Taking the ground-state wave function of the up spin as $\psi_{0\uparrow}$, we consider the integral

$$F(t) = \langle \psi_{0\uparrow} | \mathrm{e}^{-\mathrm{i}\mathcal{H}t} | \psi_{0\uparrow} \rangle = \left\langle \psi_{0\uparrow} \left| \mathrm{e}^{-\mathrm{i}\mathcal{H}_0 t} \mathrm{T} \exp\left[-\mathrm{i} \int_0^t \mathcal{H}'(\tau) d\tau \right] \right| \psi_{0\uparrow} \right\rangle . \tag{18.56}$$

This quantity reduces to $\exp(-\beta E_g)$ by putting $t \to -\mathrm{i}\beta$ and taking the limit $\beta \to \infty$; E_g is the ground state energy. Though (18.56) itself gives the overlap integral at $t \to \infty$ between the eigenstate of $\mathcal{H}_0$ and the eigenstate of $\mathcal{H}$ and vanishes because of the orthogonality theorem, E_g can be extracted from the oscillating parts of (18.50, 51), as far as t remains finite.

Now in the power expansion of (18.56) with respect to transverse component $J_\perp$ the $2n$th order term (the odd order terms vanish) can be written as

$$\int_0^t dt_{2n} \ldots \int_0^{t_3} dt_2 \int_0^{t_2} dt_1 \langle \psi_{0\uparrow} | \mathrm{i}\mathcal{H}' \mathrm{e}^{-\mathrm{i}\mathcal{H}_0(t_{2n}-t_{2n-1})} \mathrm{i}\mathcal{H}'$$

$$\ldots \mathrm{i}\mathcal{H}' \mathrm{e}^{-\mathrm{i}\mathcal{H}_0(t_2-t_1)} \mathrm{i}\mathcal{H}' | \psi_{0\uparrow} \rangle , \tag{18.57}$$

here we have put energy of $\psi_{0\uparrow}, E_{0\uparrow}$, to zero. Using (18.55) for $\mathcal{H}'$, we can rewrite the integrand into the product of the up and down spin parts as

$$\begin{aligned}&(-1)^n\left(\frac{J_\perp}{2}\right)^{2n}\big[\langle\psi_{0\uparrow}|a_{0\uparrow}(t_{2n})S_\uparrow(t_{2n},t_{2n-1})a_{0\uparrow}^\dagger(t_{2n-1})a_{0\uparrow}(t_{2n-2})\\&\times S_\uparrow(t_{2n-2},t_{2n-3})\dots S_\uparrow(t_2,t_1)a_{0\uparrow}^\dagger(t_1)a_{0\downarrow}^\dagger(t_{2n})S_\downarrow(t_{2n},t_{2n-1})\\&\times a_{0\downarrow}(t_{2n-1})\dots S_\downarrow(t_2,t_1)a_{0\downarrow}(t_1)|\psi_{0\uparrow}\rangle\big]\,.\end{aligned} \tag{18.58}$$

Here $a_{0\uparrow}^\dagger(t)$ and $a_{0\downarrow}^\dagger(t)$ are the operators developing in time by the Hamiltonian

$$\mathcal{H}_{0\alpha}=\sum_k\varepsilon_k a_{k\alpha}^\dagger a_{k\alpha}\mp\frac{J_z}{4N}\sum_{kk'}a_{k\alpha}^\dagger a_{k'\alpha}\,. \tag{18.59}$$

$S_\alpha(t_{2n},t_{2n-1})$ is given by

$$S_\alpha(t_{2n},t_{2n-1})=\mathrm{T}\exp\left[\mp\mathrm{i}\frac{J_z}{2N}\int_{t_{2n-1}}^{t_{2n}}\sum_{kk'}a_{k'\alpha}^\dagger(t)a_{k\alpha}(t)dt\right]\,, \tag{18.60}$$

where we choose $-$ for $\alpha=\uparrow$ and $+$ for $\alpha=\downarrow$. Since the term with $\alpha=\downarrow$ in (18.58) is obtained by changing the sign of J_z and exchanging $a_0^\dagger$ and a_0 at the same time, it becomes equal to the term with $\alpha=\uparrow$ and (18.58) can be rewritten as

$$\begin{aligned}&(-1)^n\left(\frac{J_\perp}{2}\right)^{2n}\Bigg[\sum_{\mathrm{P}}(-1)^{\mathrm{P}}G(t_{2\mathrm{P}[n]},t_{2n-1})G(t_{2\mathrm{P}[n-1]},t_{2n-3})\\&\qquad\dots G(t_{2\mathrm{P}[1]},t_1)\langle\psi_{0\uparrow}|S_\uparrow(t_{2n},t_1)|\psi_{0\uparrow}\rangle\Bigg]^2\,.\end{aligned} \tag{18.61}$$

P is the permutation operator for $1,2,\dots,n$. $G(t_2,t_1)$ is the extended Green function of Nozières-de Dominicis. In our case, the product of $S_\uparrow$ from 1 to n, $S_\uparrow(t_{2n},t_1)$ appears in place of $S(t,t_0)$. Therefore $G(t_2,t_1)$ is the Green function in the case where the extra potential works between t_2 and t_1, t_4 and t_3 .... By extending the Nozières-de Dominicis theory to this case, (18.61) is, as easily guessed from (18.51), approximated as

$$\begin{aligned}&(-1)^n\left(\frac{J_\perp}{2}\right)^{2n}\Bigg[\sum_{\mathrm{P}}(-1)^{\mathrm{P}}G_0(t_{2\mathrm{P}[n]},t_{2n-1})G_0(t_{2\mathrm{P}[n-1]},t_{2n-3})\dots\\&G_0(t_{2\mathrm{P}[1]},t_1)\Bigg]^{2(1-\varepsilon/2)}=(-1)^n\left(\frac{J_\perp}{2}\right)^{2n}\left[\det G_0(t_{2l},t_{2k-1})\right]^{(2-\varepsilon)}\,.\end{aligned} \tag{18.62}$$

Here $G_0(t_2,t_1)$ is the Green function at $T=0$ in the case $S(t_{2n},t_1)=1$ and the asymptotic form (18.27) is assumed. If $\delta_\uparrow$ is the phase shift for $\mathcal{H}_{0\uparrow}$ in (18.59), then ε is given by

$$\frac{\varepsilon}{2} = -4\frac{\delta_\uparrow}{\pi} - 4\left(\frac{\delta_\uparrow}{\pi}\right)^2 \cong -\frac{J_z\rho}{N} - \frac{1}{4}\left(\frac{J_z\rho}{N}\right)^2 . \tag{18.63}$$

The determinant appearing in (18.62) is called Cauchy's determinant, for which the following formula [18.9] holds:

$$\det\left|\frac{1}{\tau_{2i} - \tau_{2j-1}}\right| = \frac{\prod\limits_{i<j}(\tau_{2i} - \tau_{2j})(\tau_{2i-1} - \tau_{2j-1})}{\prod\limits_{i,j}(\tau_{2i} - \tau_{2j-1})} . \tag{18.64}$$

Using this formula and putting $t = -\mathrm{i}\beta$ in $F(t)$, we obtain

$$\begin{aligned} F(-\mathrm{i}\beta) = &\sum_{n=0}^{\infty}\left(\frac{J_\perp}{2}\right)^{2n}\int d\beta_{2n}\int d\beta_{2n-1}\cdots\int d\beta_1 \\ &\times \exp\left[(2-\varepsilon)\sum_{n>n'}(-1)^{n-n'}\log[D(\beta_n - \beta_{n'})]\right] . \end{aligned} \tag{18.65}$$

Here, the region near the logarithmic singularity $\beta_n - \beta_{n'} \le D^{-1}$ is excluded from the region of integration because of the asymptotic approximation. Equation (18.65) is equivalent to the partition function of hard rods with + and − charges which are arranged in one dimension and interacting with each other via a two-dimensional Coulomb force. In this expression, the temperature is $(2-\varepsilon)^{-1}$, $J_\perp/2$ is $\mathrm{e}^{-\mu}$ (μ is the chemical potential), and the length is β.

On the other hand, rewriting the exponent of the exponential function in (18.65) as

$$\int_0^\beta d\beta' \int_0^\beta d\beta'' \delta(\beta' - \beta_n)\delta(\beta'' - \beta_{n'})$$

and carrying out partial integrations over β' and β'', we can rewrite (18.65) as

$$\begin{aligned} F(-\mathrm{i}\beta) &= \langle\psi_{0\uparrow}|\mathrm{e}^{-\beta\mathcal{H}}|\psi_{0\uparrow}\rangle \\ &= \int d(\text{paths})\exp\left[\frac{2-\varepsilon}{2}\int_0^\beta d\beta' \int_0^\beta d\beta'' \frac{S_z(\beta')S_z(\beta'')}{(\beta'-\beta'')^2}\right. \\ &\quad \left. + \left(\log\frac{|J_\perp|}{2}\right) \times (\text{number of jumps})\right] , \end{aligned} \tag{18.66}$$

where $S_z(\beta)$ is the step function taking the values $\pm 1/2$ with $2n$ jumps. In this form F is equivalent to the partition function of a one-dimensional Ising spin system with interaction proportional to the inverse square of distance. Since the exact solution of the one-dimensional Ising system has not yet been obtained, we cannot obtain any new knowledge about the Kondo effect from the result of (18.66). On the contrary, from the conclusion obtained about the ground state of the *s*-*d* system, (that is, from the conclusion that though the localized spin exists for $J > 0$, the localized spin vanishes to make a singlet state for $J < 0$), we can expect that Ising system with interaction proportional to the inverse square of distance has a critical point at $\varepsilon = 0$; it is nonmagnetic at temperatures higher than this point and has ferromagnetic long-range order at lower temperatures. Equation (18.65), as stated in Chap. 10, is equivalent to the partition function of two-dimensional XY model; Kosterlitz studied the behavior near the critical point of the two-dimensional XY model using the scaling method due to *Anderson* et al. [18.10].

Anderson-Yuval theory expresses the partition function (at $T = 0$) of the *s*-*d* system as a power expansion with respect to the transverse component of the *s*-*d* exchange interaction on the basis of the asymptotic approximation. Even if we calculate each term of the perturbation expansion one by one, it is difficult to calculate the singular part of the ground state energy from this result. This fact was stated in the discussion of the perturbation expansion. However, from the structure of this expansion formula, it is possible to make some conclusions regarding the ground state of the *s*-*d* system. Using the expansion formula of (18.65), *Anderson* et al. [18.10] derived the scaling law with respect to the coupling constants J_z and $J_\perp$ of the exchange interaction, and on the basis of this law they discussed the separation into two regions by the boundary line $J_z = |J_\perp|$ in $J_z - J_\perp$ plane, as stated in Sect. 17.3.

19. Behavior of the *s-d* System at Low Temperatures

Although the essential nature of the localized spin has been made clear by the singlet ground state theory, the extension of this theory to finite temperatures was difficult. On the other hand, Anderson discussed the *s-d* problem on the basis of the scaling law and Anderson-Yuval applied the Nozières-de Dominicis theory to the *s-d* problem and derived the partition function in an expansion form with respect to the transverse component of the *s-d* exchange. These theoretical developments led us to a deeper understanding of the Kondo effect. However, the low temperature behavior of the *s-d* system remained unsolved yet. In this chapter we turn our consideration to the low temperature behavior of the *s-d* system. We start with a survey of the Wilson theory.

19.1 Wilson Theory

Wilson [19.1] made full use of the renormalization group theory and the scaling law, and showed very clearly the transition from weak coupling to strong coupling with respect to the invariant coupling constant J_{inv} and actually carried out the calculation of the specific heat and the magnetic susceptibility at low temperatures ($T \ll T_{\text{K}}$). In the following, we outline the Wilson theory.

Simplifying the problem, Wilson started with the Hamiltonian

$$\mathcal{H} = \int_{-1}^{1} k u_k^\dagger u_k dk - J A^\dagger \sigma A \cdot \tau \quad ; \tag{19.1}$$

here and henceforth we follow Wilson's notation. If we use the partial-wave representation for the conduction electrons rather than the plane-wave representation, it is only the *s*-wave part which is scattered by the *s-d* exchange interaction given by (17.6). Therefore we have only to consider this component and so we can treat the problem as one-dimensional; the states of the conduction electrons are specified only by the energy values. We write the energy (measured from the Fermi energy) as k, assume half the band width as the energy unit; the energy eigenvalues are distributed uniformly within the region from −1 to 1. The first term of (19.1) represents the kinetic energy of the s-electrons. Throughout this section, we omit the suffices for spin components and the corresponding summation. The second term is the s-d exchange interaction between the localized spin situated at the origin and the s-electrons; A is defined, corresponding to the Fermi operator a_0 at the origin, as

$$A = \int_{-1}^{1} a_k dk \quad ; \tag{19.2}$$

$\boldsymbol{\sigma}$ and $\boldsymbol{\tau}$ denote the Pauli spin operators of the s-electrons and the localized spin, respectively.

At this stage we replace the states of the electron system (which are distributed uniformly with respect to energy k) with a discrete spectrum equally separated in the logarithmic scale. That is, as shown in Fig. 19.1, we change to the discrete energy distribution $k = 1, 1/\Lambda, 1/\Lambda^2, \ldots 1/\Lambda^n, \ldots$; Λ is adopted to be a value around 2 or 3. If we let Λ tend to 1, we recover the original spectrum.

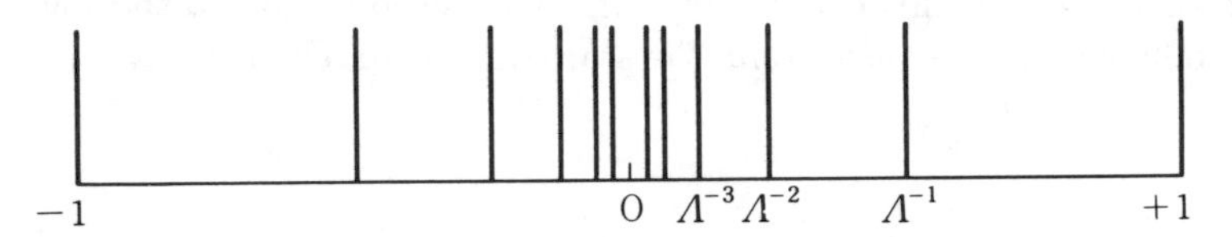

Fig. 19.1. Discrete energy levels of conduction electrons

Let us consider one of the discrete levels, for example Λ^{-n}. Since this represents the continuous distribution of k within $\Lambda^{-(n+1)} < k < \Lambda^{-n}$, the states numbering $\Lambda^{-n}(1-\Lambda^{-1})$ in this range are replaced by one discrete level. Therefore, if we represent Fermi operator of discrete level as a_n, since a_k and a_n are the amplitudes, we can put all a_k in the region $\Lambda^{-(n+1)} < k < \Lambda^{-n}$ uniformly as

$$a_k = \frac{\Lambda^{n/2}}{\sqrt{1-\Lambda^{-1}}} a_n \ . \tag{19.3}$$

Here the coefficient of a_n is a normalized function which is independent of k for k between Λ^{-n} and $\Lambda^{-(n+1)}$ and zero outside the region. The operator for the discrete level is written as b_n for negative energies. With use of the relation (19.3), the Hamiltonian (19.1) becomes

$$\mathcal{H} = \frac{1}{2}(1+\Lambda^{-1}) \sum_{n=0}^{\infty} \Lambda^{-n}(a_n^\dagger a_n - b_n^\dagger b_n) - JA^+ \boldsymbol{\sigma} A \cdot \boldsymbol{\tau} \ , \tag{19.4}$$

$$A = \sqrt{1-\Lambda^{-1}} \sum_{n=0}^{\infty} \Lambda^{-n/2}(a_n + b_n) \ . \tag{19.5}$$

The above is the first stage of the transformation and is the most ingenious contrivance. Let us now make the second step of the transformation. This transformation is that from $\{a_n^\dagger, a_n; b_n^\dagger, b_n\}$ to new complete orthonormal set $\{f_n^\dagger, f_n\}$. The relations between a_n, b_n and f_n are given by

$$a_n = \sum_l u_{nl} f_l \ , \qquad b_n = \sum_l v_{nl} f_l \ , \tag{19.6}$$

$$f_l = \sum_n (u_{nl} a_n + v_{nl} b_n) \ . \tag{19.7}$$

f_0 for $l = 0$ is assigned to the normalized A (19.5), namely

$$u_{n0} = v_{n0} = c_0 \Lambda^{-n/2} \ ,$$
$$c_0 = \sqrt{1 - \Lambda^{-1}}/\sqrt{2} \ .$$

Using (19.6) for the kinetic energy term H_{K} in the Hamiltonian, we represent it as

$$\frac{H_{\mathrm{K}}}{\frac{1}{2}(1+\Lambda^{-1})} = \sum_{n=0}^{\infty} \Lambda^{-n} \left(a_n^\dagger \sum_l u_{nl} f_l - b_n^\dagger \sum_l v_{nl} f_l \right) \ . \tag{19.8}$$

If we pick up the $l = 0$ term, we obtain

$$\sum_{n=0}^{\infty} \Lambda^{-n} (u_{n0} a_n^\dagger - v_{n0} b_n^\dagger) f_0 \ .$$

The quantity by which f_0 is multiplied is orthogonal to f_0, and is defined as $f_1^\dagger$ after normalization;

$$f_1^\dagger = c_1 \sum_{n=0}^{\infty} \Lambda^{-n} (u_{n0} a_n^\dagger - v_{n0} b_n^\dagger) \ , \tag{19.9}$$

where c_1 is a normalization constant. For $l = 1$, the coefficient by which f_1 is multiplied is orthogonal to f_1. The coefficient is composed of $f_0^\dagger$ and the part orthogonal to f_1 and f_0. Normalizing the latter, we define it as $f_2^\dagger$. Hereafter, by proceeding in the same way, Hamiltonian (19.4) divided by the factor $(1 + \Lambda^{-1})/2$ is transformed to

$$\mathcal{H} = \sum_{n=0}^{\infty} \Lambda^{-n/2} \varepsilon_n (f_n^\dagger f_{n+1} + f_{n+1}^\dagger f_n) - \widetilde{J} f_0^\dagger \boldsymbol{\sigma} f_0 \cdot \boldsymbol{\tau} \ , \tag{19.10}$$

where ε_n is of order unity and approaches 1 in the limit of $n \to \infty$. Therefore, all ε_n are put to 1. $\widetilde{J}$ is the product of original J and $4(1 + \Lambda^{-1})^{-1}$, in which factor 2 originates from $A = \sqrt{2} f_0$ and another factor originates from the fact that the total Hamiltonian was divided by $(1+\Lambda^{-1})/2$. The Hamiltonian (19.10) represents the hopping motion of electrons among the lattice points $0, 1, 2, \ldots, n, \ldots$.

At the final stage, we stop summing the first term of (19.10) over n at $N - 1$ and multiply the total Hamiltonian by factor $\Lambda^{(N-1)/2}$, to define

$$\mathcal{H}_N = \Lambda^{(N-1)/2}\left[\sum_{n=0}^{N-1} \Lambda^{-n/2}(f_n^\dagger f_{n+1} + f_{n+1}^\dagger f_n) - \tilde{J} f_0^\dagger \sigma f_0 \cdot \tau\right] . \quad (19.11)$$

This is the Hamiltonian discussed by Wilson. The relation between the real Hamiltonian $\mathcal{H}$ and $\mathcal{H}_N$ is given by

$$\mathcal{H} = \lim_{N\to\infty} \Lambda^{-(N-1)/2}\mathcal{H}_N . \quad (19.12)$$

The kinetic energy part of $\mathcal{H}_N$ except for the factor $\Lambda^{(N-1)/2}$ is one electron Hamiltonian; the one electron energy eigenvalues distribute discretely from around zero to 1. With increasing N, new levels are created near zero. The factor $\Lambda^{(N-1)/2}$ expands the low energy levels near the Fermi surface; by multiplying this factor, low energy levels become constant independently of N for large enough values of N. According to the numerical calculation, one-body energy levels, which is the kinetic energy part of $\mathcal{H}_N$, are given for $\Lambda = 2$ as follows:

N_{odd} ; $\varepsilon = \pm 0.6555,\ \pm 1.976,\ \pm 4,\ \pm 8, \ldots,\ \pm 2^l, \ldots$

N_{even} ; $\varepsilon = 0,\ \pm 1.297,\ \pm 2.827,\ \pm 4\sqrt{2},\ \ldots,\ \pm 2^{l-1}\sqrt{2}, \ldots$

Compared with this result, the s-d exchange interaction increases more and more with increasing N due to the factor $\Lambda^{(N-1)/2}$. Thus, for the low energy excitation the exchange interaction increases up to $-\infty$ and arrives at the strong coupling limit. However, for the high energy excitation, the factor $\Lambda^{(N-1)/2}$ works similarly for both of the kinetic energy part and the exchange interaction so that the exchange interaction works still as weak coupling. Thus we can see that even if the exchange interaction is weak coupling for the high energy excitations, it becomes strong coupling in the limit of low energy excitation tending to zero and the fixed point of J_{inv} is $-\infty$.

Now, if we put $J = 0$ in $\mathcal{H}_N$, we obtain

$$\mathcal{H}_N = \Lambda^{(N-1)/2} \sum_{n=0}^{N-1} \Lambda^{-n/2}(f_n^\dagger f_{n+1} + f_{n+1}^\dagger f_n) . \quad (19.13)$$

On the other hand, if we put $J = -\infty$, then since f_0 couples with τ to form the spin-singlet state and this state is never destroyed, $\mathcal{H}_N$ is given by

$$\mathcal{H}_N = \Lambda^{(N-1)/2} \sum_{n=1}^{N-1} \Lambda^{-n/2}(f_n^\dagger f_{n+1} + f_{n+1}^\dagger f_n) . \quad (19.14)$$

This differs from the Hamiltonian in case $J = 0$ only in that the term $n = 0$ is excluded in the sum over n. Therefore, if we shift site indices by one as $f_1 \to f_0, f_2 \to f_1$, we obtain (19.13) with N replaced by $N - 1$. Thus, if we assume N_{odd} for $J = 0$, we obtain for $J = -\infty$ the excitation spectrum in case $J = 0$ and N_{even}. If we assume N_{even}, we obtain the spectrum in case $J = 0$ and N_{odd}. From this result, when $|J| \ll 1$ is satisfied, the energy spectrum for small N (but large enough for the one-electron energy levels in $J=0$ to be independent of N) is close to that for $J=0$ and N_{even}; however, they shift to the energy levels for $J=0$ and N_{odd} with increasing N, as far as low-energy levels are concerned. Actually, the numerical calculation ($\Lambda=2, \widetilde{J}=-0.053$) shows that if we denote the first two many-body energy levels of the total system by E_1 and E_2, these levels are found to change as presented in Table 19.1.

Table 19.1. First two energy levels of total system

N	20	22 ...	44	46 ...	66 ...	78
E_1	0.0945	0.1007	0.296	0.3418	0.6537	0.6555
E_2	0.1269	0.1353	0.4177	0.4929	1.3037	1.3109

Here E_1 and E_2 at $N=20$ correspond to the energies of the states in which two electrons occupy the one-body energy level $\varepsilon=0$ for $J=0, N_{\text{even}}(S_{\text{tot}}=1/2)$ and one electron occupies it ($S_{\text{tot}}=1$), respectively. For $J=0$, both E_1 and E_2 are zero. At $N=78, E_1$ corresponds to the state where one electron occupies the level 0.6555 for $J=0, N_{\text{odd}}$; another electron couples with the localized spin to form the singlet state. E_2 corresponds to the state where two electrons (electron and hole) occupy the same level ($S_{\text{tot}}=1$). This even-odd crossover occurs near $N\sim 40$.

In Wilson's Hamiltonian $\mathcal{H}_N$ the larger value of N is needed for the lower energy excitations; therefore, the lower the temperature, the larger the value of N which is needed. From this result we find that the excitation energy and temperature are scaled by $\Lambda^{-N/2}$.

In order to calculate the specific heat and the susceptibility at low temperatures, we have only to calculate the low-energy excitation spectrum as $N\to\infty$. Though the excitation energy in the limit $N=\infty$ is given by the one-electron energy, it is necessary to know how to approach this limit in order to calculate the specific heat and the susceptibility . For this purpose, Wilson considers the effective Hamiltonian

$$\mathcal{H}_N=\mathcal{H}^*+[\lambda(f_1^\dagger f_2+f_2^\dagger f_1)+\eta(f_1^\dagger f_1-1)^2]\Lambda^{-N/2}\ , \tag{19.15}$$

where $\mathcal{H}^*$ is the Hamiltonian at the fixed point $N\to\infty, \widetilde{J}\to-\infty$. The additional term in (19.15) tends to zero as $\Lambda^{-N/2}$ with increasing N. Such terms are called irrelevant in the renormalization group theory. Wilson calculates the excitation energy analytically up to first order in λ and η, and makes the linear term in λ, η equal to the difference between energy level E_{iN} and $E_{i\infty}$. He calculates E_{iN} numerically for the actual Hamiltonian $\mathcal{H}_N$ and determines the values of λ and η so that this relation holds. By numerical calculations he finds that the values of λ and η determined for E_1 and E_2 hold also for the other low-energy excitations. That is,

$$
\begin{aligned}
E_{1N} - E_{1\infty} &= \widetilde{\lambda}\Lambda^{-N/2}\,, & \widetilde{\lambda} &= 2^{32} \times (-3.668 \times 10^{-3}) \\
E_{2N} - E_{2\infty} &= 2\widetilde{\lambda}\Lambda^{-N/2} - \widetilde{\eta}\Lambda^{-N/2}\,, & \widetilde{\eta} &= 2^{32} \times (7.2496 \times 10^{-3}) \\
E_{3N} - E_{3\infty} &= 2\widetilde{\lambda}\Lambda^{-N/2} + \widetilde{\eta}\Lambda^{-N/2}\,, & &(N = 64, \Lambda = 2, \widetilde{J} = -0.053)\,,
\end{aligned}
\tag{19.16}
$$

where $\widetilde{\lambda}$ and $\widetilde{\eta}$ differ from λ and η by factors of order unity. As shown here, he introduces the pseudo-potential so that it reproduces the numerical results, and calculates the linear term in $\widetilde{\lambda}$ and $\widetilde{\eta}$ for the specific heat and the susceptibility on the basis of the Hamiltonian (19.15). The results are

$$
\begin{aligned}
\chi_{\text{imp}} &= \frac{g^2\mu_{\text{B}}^2}{kT}(-0.834\widetilde{\lambda} + 0.422\widetilde{\eta})2^{-N/2}\,, \\
C_{\text{imp}} &= k(-10.97\widetilde{\lambda})2^{-N/2}\,,
\end{aligned}
\tag{19.17}
$$

where $\Lambda = 2, \widetilde{J} = -0.053$, and $T = 2 \times 2^{-N/2}$.

The striking point is that while the specific heat involves only $\widetilde{\lambda}$, the susceptibility contains both $\widetilde{\lambda}$ and $\widetilde{\eta}$. Moreover, with use of the values given for $\widetilde{\lambda}$ and $\widetilde{\eta}$ in (19.16), the two terms of this susceptibility have the same numerical value $10^{-3} \times 2^{32} \times 3.059$. Therefore, if we take the ratio of the susceptibility χ_{imp} to the specific heat C_{imp}, we obtain

$$
\frac{T\chi_{\text{imp}}}{C_{\text{imp}}} = \frac{g^2\mu_{\text{B}}^2}{k^2} 2 \times \frac{0.834}{10.97} = \frac{g^2\mu_{\text{B}}^2}{k^2} \cdot 0.1521 = \frac{g^2\mu_{\text{B}}^2}{k^2} \cdot \frac{3}{2\pi^2}\,. \tag{19.18}
$$

This ratio is twice the ratio of the susceptibility to the specific heat in the noninteracting conduction-electron system. A half of the contribution comes from the $\widetilde{\eta}$ term. The fact that the ratio of the susceptibility to the specific heat is twice that in the usual case is one of the characteristic properties of the *s-d* system; this ratio is called the Wilson ratio.

As noted above, Wilson succeeded in calculating the specific heat proportional to T at low temperatures and the susceptibility at $T = 0$, by making full use of numerical calculations. In this process, he introduced a pseudo-potential with two parameters and derived the final result with use of this pseudo-potential. Since there is a relation between parameter $\widetilde{\lambda}$ and $\widetilde{\eta}$, the number of independent parameter is actually only one. This reflects the fact that all physical quantities in the *s-d* system are scaled by the Kondo temperature T_{K}.

19.2 Local Fermi-Liquid Theory of Nozières

One of the important conclusions from Wilson theory is that the ratio of the susceptibility to the specific heat proportional to T, namely the Wilson ratio, is twice that in the noninteracting conduction-electron system. Another important conclusion of this theory is that the low-temperature properties of the localized spin can be described by the effective Hamiltonian possessing two parameters λ and η. This effective Hamiltonian includes no longer the localized spin. The localized spin couples strongly with the spin of the conduction electrons to form the singlet state and does not appear explicitly. The localized spin coupled in the singlet state leads to two terms. One is the one electron term which represents the transfer of conduction electrons from lattice site 1 to lattice site 2; the coefficient is λ. Another one is the term with coefficient η, which is written as

$$(f_1^\dagger f_1 - 1)^2 = (f_{1\uparrow}^\dagger f_{1\uparrow} + f_{1\downarrow}^\dagger f_{1\downarrow} - 1)^2 = 2n_{1\uparrow}n_{1\downarrow} + [1-(n_{1\uparrow}+n_{1\downarrow})] \,. \quad (19.19)$$

As seen from this expression, this term represents the repulsive force between two electrons with antiparallel spins which occupy the same lattice point. Therefore it can be seen that when lattice point 1 is considered as the position of impurity atom, Wilson's effective Hamiltonian possesses the same structure as the Anderson Hamiltonian. In this case the term with λ corresponds to the *s-d* mixing term of V, and η to the repulsive force U. Wilson obtained the excitation energy up to the first order in λ and η and showed that this reproduces the low energy excitations from the ground state of the real *s-d* system.

Nozières [19.2] noticed this fact and developed a local Fermi-liquid theory. In the scattering problem of the conduction electrons due to the impurity potential, the most important physical quantity is the phase shift δ of the conduction electrons near the Fermi surface. All the information of the scattering problem is included in δ. If we include the localized spin in the number of localized conduction electrons as in the Anderson model, the phase shift at the Fermi surface is $\delta(\varepsilon_{\mathrm{F}}) = \pi/2$; the reason is that the localized electron numbers are 1/2 for both of up and down spin. Following the spirit of Landau's Fermi-liquid theory [19.3], Nozières considers that the state of the *s-d* system can be described by the distribution function of quasi-particles with spin σ, n_σ. In this case, the phase shift of scattered quasi-particles depends on the energy ε_σ and the distribution $n_{\sigma'}$. Namely, the phase shift can be written as

$$\delta_\sigma(\varepsilon_\sigma, n_{\sigma'}) \,. \quad (19.20)$$

Expanding δ_σ with respect to the deviation of distribution $n_{\sigma'}$ from the ground state, $\delta n_{\sigma'} = n_{\sigma'} - n_{\sigma' 0}$ and taking up to the first order, he expresses δ_σ as

$$\delta_\sigma(\varepsilon) = \delta_0(\varepsilon) + \sum_{\varepsilon'\sigma'} \phi_{\sigma\sigma'}(\varepsilon,\varepsilon')\delta n_{\sigma'}(\varepsilon') \,. \quad (19.21)$$

Next, he expands δ_0 and $\phi_{\sigma\sigma'}(\varepsilon,\varepsilon')$ with respect to ε (measured from chemical potential μ) as follows:

$$\begin{aligned} &\delta_0(\varepsilon) = \delta_0 + \alpha\varepsilon + \beta\varepsilon^2 + \dots \\ &\phi_{\sigma\sigma'}(\varepsilon,\varepsilon') = \phi_{\sigma\sigma'} + \psi_{\sigma\sigma'}(\varepsilon+\varepsilon') + \dots \,. \end{aligned} \quad (19.22)$$

If we confine ourselves to low temperatures and low magnetic fields, we have only to include linear terms in ε, T and H. In this case we are left with four parameters, $\delta_0, \alpha, \phi_{\sigma,\pm\sigma} = \phi^s \pm \phi^a$. If we assume that the total electron number is constant, then ϕ^s does not appear explicitly. Denoting the electron numbers with up and down spin as $n_\uparrow$ and $n_\downarrow$, respectively, and putting $n_\uparrow - n_\downarrow = m$, we can express the phase shift δ_σ as

$$\delta_\sigma(\varepsilon) = \delta_0 + \alpha\varepsilon + \sigma\phi^a m \ , \tag{19.23}$$

where σ is 1 for $\uparrow$ and -1 for $\downarrow$. Since δ_0 is $\pi/2$, this expression has two parameters, α and ϕ^a. In the case where the repulsive force acts only between antiparallel spins, the relation

$$\phi^s + \phi^a = 0 \tag{19.24}$$

holds. The above is the basis of the local Fermi-liquid theory of Nozières; α corresponds to Wilson's λ, and ϕ^a to η.

Since δ_σ/π is the extra number of electrons localized around the impurity, the new energy $\tilde{\varepsilon}_\sigma$ which is shifted from the electron energy ε by the quantity of δ_σ/π divided by ρ (density of states of the conduction electrons without impurity) is defined as

$$\tilde{\varepsilon}_\sigma = \varepsilon - \delta_\sigma(\varepsilon)/\pi\rho \ . \tag{19.25}$$

In this case, we assume that electrons occupy the energy levels up to $\tilde{\varepsilon}_\sigma = \mu = 0$. In terms of the original ε, electrons occupy the energy levels up to $\varepsilon = \delta_\sigma/\pi\rho$. In this case the electron number increases by δ_σ/π compared with $\delta_\sigma = 0$. For $H = 0$ and $m = 0$, the change $\delta\rho$ in the density of states is given by

$$\delta\rho = \rho\left(\frac{d\varepsilon}{d\tilde{\varepsilon}} - 1\right) = \frac{\alpha}{\pi} \tag{19.26}$$

and the change in the specific heat is

$$\delta C_v/C_v = \alpha/\pi\rho \ . \tag{19.27}$$

In the presence of the magnetic field H the new energies $\tilde{\varepsilon}_\uparrow$ and $\tilde{\varepsilon}_\downarrow$ are given by

$$\tilde{\varepsilon}_\uparrow = \varepsilon_\uparrow - \frac{1}{2}g\mu_B H - \frac{1}{\pi\rho}(\delta_0 + \alpha\varepsilon_\uparrow + \phi^a m) \ ,$$
$$\tilde{\varepsilon}_\downarrow = \varepsilon_\downarrow + \frac{1}{2}g\mu_B H - \frac{1}{\pi\rho}(\delta_0 + \alpha\varepsilon_\downarrow - \phi^a m) \ ,$$

at $T = 0$. In equilibrium we have

$$\tilde{\varepsilon}_\uparrow = \tilde{\varepsilon}_\downarrow = \mu = 0 \ ;$$

dividing the magnetization $(1/2)g\mu_B m$ by H, we obtain the susceptibility as

$$\chi = \frac{1}{2}\frac{g\mu_{\rm B}m}{H} = \frac{1}{2}\rho(g\mu_{\rm B})^2\left(1 + \frac{\alpha}{\pi\rho} + \frac{2\phi^a}{\pi}\right) ; \tag{19.28}$$

note that the terms α and ϕ^a are order of $1/N$ compared with 1. From (19.27, 28), the Wilson ratio is obtained as

$$(\delta\chi/\chi)/(\delta C_v/C_v) = 1 + (2\rho\phi^a/\alpha) . \tag{19.29}$$

Nozières reasons that the second term in (19.29) (which arises from the electron interaction) actually gives unity as shown by Wilson, as follows: In case of weak *s*-*d* exchange interaction, the Kondo temperature is small compared with the band width D ($T_{\rm K} \ll D$). In this case, the change in the conduction-electron states is confined to the neighborhood of the Fermi surface. Now, if ε and the Fermi energy μ are shifted by the same amount, the phase shift $\delta[\varepsilon, n_0(\mu)]$ remains invariant. This condition becomes, from (19.21),

$$\alpha + 2\rho\phi^s = 0 . \tag{19.30}$$

From this relation and (19.24), one finds immediately that $2\rho\phi^a/\alpha = 1$.

Nozières showed that the low temperature behavior of the localized spin (which Wilson derived on the basis of renormalization group theory, using numerical calculations) can be described phenomenologically on the basis of the local Fermi-liquid theory. Moreover, Nozières derived also the expression for the temperature dependence of the electrical resistivity at low temperatures which cannot be obtained by Wilson theory. However, we do not discuss this problem here, leaving it to later discussions.

19.3 Kondo Effect Viewed from the Anderson Hamiltonian

So far we have considered the Kondo effect on the basis of the *s*-*d* model, which is derived from the Anderson model in the limit of large U. In this section, we consider this problem directly on the Anderson model without passing through the *s*-*d* model.

19.3.1 Viewpoint of the Anderson Model

On the basis of the *s*-*d* model, the localized spin couples strongly with the conduction electron spins at zero temperature to form the singlet bound state; the freedom of localized spin vanishes. According to Wilson, the low-energy excitations from this ground state can be described correctly by the effective Hamiltonian without localized spin, which contains two parameters. On the other hand, Nozières showed that the result which Wilson derived from this effective Hamiltonian can be reproduced phenomenologically by the local Fermi-liquid theory.

The *s-d* model can be derived from the Anderson Hamiltonian in the limiting case where the repulsive potential U between two electrons with antiparallel spins occupying the lattice point of the impurity ion is large compared with the matrix element of mixing term V. On the contrary, in the case where U is smaller than V, the *d*-orbital localized at the impurity ion is occupied by the same number of up-and down-spin electrons; the spins of the *d*-electrons are cancelled and the impurity ion is in the nonmagnetic state without a localized moment. In the Hartree-Fock approximation for the Anderson Hamiltonian, when U becomes larger than π times the broadening of the *d*-electron energy level $\Delta(= \pi\rho|V|^2)$, up- and down-spin electrons strongly repel each other (owing to U) and the impurity ion possesses a localized moment. This case can be described by the *s-d* model. However, the localized spin thus created also couples with the conduction electrons at low enough temperatures via the antiferromagnetic *s-d* exchange interaction and forms the singlet state; the impurity ion is again in the nonmagnetic state. Is the latter nonmagnetic state different from the former nonmagnetic state, which is realized in the case of small U in the Anderson model? Actually, when U increases and becomes equal to $\pi\Delta$, the Hartree-Fock approximation predicts that the nonmagnetic state becomes unstable and the localized moment appears. However, in the case of the Anderson model (which describes the problem of a single impurity ion), the phase transition should not take place, when U increases; the phase transition arises only in the limit of the particle number $N \to \infty$. Therefore, the phase transition resulting from the Hartree approximation for a small system, is an artifact due to the approximation and should not occur in real systems. It means that even if U is increased in the Anderson Hamiltonian, the nonmagnetic ground state realized for $U = 0$ remains to be the ground state. From this point of view, the singlet state of the localized spin arising in the *s-d* model at low temperatures should be the same as the nonmagnetic state in the large U limit of the Anderson model. Although the *s-d* model is assumed on the basis of existence of the localized spin, the localized spin does not exist actually at low temperatures ($T < T_K$). It is only in the region $T > T_K$ that the localized spin can rotate its direction freely as assumed at the starting point.

Yamada and *Yosida* [19.4] developed the perturbation theory of the Anderson Hamiltonian on the basis of the consideration mentioned above. Following this theory, we describe how the Kondo effect can be understood on the basis of the Anderson model. In this case we do not use the *s-d* model at the intermediate stage so that our description becomes explicit and can approach the essence of the Kondo effect without any sophisticated considerations.

19.3.2 Perturbation of the Anderson Model

The Anderson Hamiltonian is given by (15.54). To make clear the essence of the problem, we consider the case where electrons and holes are symmetric with respect to the Fermi surface (*i.e.*, $E_{\mathrm{d}} = -U/2$); we rewrite the second and the fourth terms in (15.54) as

$$\mathcal{H}' = -\frac{U}{4} + U\left(n_{\mathrm{d}\uparrow} - \frac{1}{2}\right)\left(n_{\mathrm{d}\downarrow} - \frac{1}{2}\right) . \tag{19.31}$$

In the case where the relation $E_{\mathrm{d}} = -U/2$ does not hold, the term $(E_{\mathrm{d}} + U/2)(n_{\mathrm{d}\uparrow} + n_{\mathrm{d}\downarrow})$ should be added; this is the term corresponding to the impurity potential to the localized d-electron and is not considered to be so essential to the problem of the electron correlation.

If we neglect this term, the Anderson Hamiltonian is equivalent to the s-d model in the limit $U \to \infty$; in that case, the relation between J and U is given by

$$\frac{J}{N} = -\frac{8|V|^2}{U} \tag{19.32}$$

from (15.103). Here the value of J is twice that in (15.103). The second term in (19.31) represents the deviation of the Coulomb interaction between two d-electrons from the average value in the Hartree-Fock unperturbed state, and is treated as a perturbation to the unperturbed Hamiltonian

$$\mathcal{H}_0 = \sum_{ks} \varepsilon_k a_{ks}^\dagger a_{ks} + V \sum_{ks} (a_{ks}^\dagger a_{\mathrm{d}s} + a_{\mathrm{d}s}^\dagger a_{ks}) - \frac{U}{4} , \tag{19.33}$$

where the last (constant) term is the first term of (19.31). In other words, the one-body Hamiltonian given by the Hartree-Fock approximation is considered as the unperturbed state and the fluctuation term ignored in this approximation is considered as the perturbation. The one-body Hamiltonian $\mathcal{H}_0$ is easily diagonalized; if the one-electron eigenstate is denoted as $|n\rangle$, Fermi operators of the electron in this state are given by (15.59). The ground state of the total system is the state in which up- and down-spin electrons occupy the one-electron states up to the Fermi energy. In order to describe the behavior of the unperturbed state at finite temperatures, it is convenient to introduce the d-electron temperature Green function. This Green function is defined by

$$G_{ij} = G(\tau_i - \tau_j) = -\langle\langle \mathrm{T}_\tau a_{\mathrm{d}\sigma}(\tau_i) a_{\mathrm{d}\sigma}^\dagger(\tau_j) \rangle\rangle , \tag{19.34}$$

where $a_{\mathrm{d}\sigma}(\tau)$ is given by

$$a_{\mathrm{d}\sigma}(\tau) = \mathrm{e}^{\mathcal{H}_0 \tau} a_{\mathrm{d}\sigma} \mathrm{e}^{-\mathcal{H}_0 \tau} \tag{19.35}$$

and T_τ orders the operators $a_{\mathrm{d}\sigma}(\tau_i)$ and $a_{\mathrm{d}\sigma}^\dagger(\tau_j)$ with larger value of τ at the left; a minus sign is introduced in case of changing the order. The symbol $\langle\langle A \rangle\rangle$

means the thermal average $\mathrm{Tr}\{\mathrm{e}^{-\beta\mathcal{H}_0}A\}/\mathrm{Tr}\{\mathrm{e}^{-\beta\mathcal{H}_0}\}$. This Green function is decomposed into the Fourier components as

$$G_{ij} = \frac{1}{\beta}\sum_n G(\omega_n)\mathrm{e}^{-\mathrm{i}\omega_n(\tau_i-\tau_j)} \tag{19.36}$$

$$\omega_n = \frac{\pi(2n+1)}{\beta}, \qquad \beta = \frac{1}{kT} \tag{19.37}$$

where n is an integer. For $\mathcal{H}_0$ in (19.33) the Green function $G(\omega)$ is calculated as

$$G(\omega_n) = \frac{1}{\mathrm{i}(\omega_n + \Delta\mathrm{sgn}\omega_n)}, \tag{19.38}$$

where Δ is the width of the d-level given by (15.70) and sgn is the sign function. Thus $G(\omega_n)$ has a discontinuity at $\omega_n = 0$. By the analytic continuation of $\mathrm{i}\omega_n$ from the upper half of the complex plane to the real axis, *i.e.*, by the change of $\mathrm{i}\omega \to \omega + \mathrm{i}\eta$, the thermal Green function is transformed into the retarded Green function ($z = \omega + \mathrm{i}\eta$). G_{ij} changes its sign under exchange of i and j.

Expanding $\exp(-\beta\mathcal{H})$ as (16.30) or (18.18) and using the d-electron temperature Green function, we can write the free energy F with the perturbation expansion in which the second term of $\mathcal{H}'$ is treated as the perturbation. The result is

$$F = F_0 - \frac{U}{4} - \sum_{n=1}^{\infty}\frac{U^{2n}}{(2n)!}\frac{1}{\beta}\int_0^\beta \cdots \int d\tau_1 d\tau_2 \ldots d\tau_{2n}\left[D^{2n}(1,2,\ldots,2n)\right]^2_{\mathrm{conn}}, \tag{19.39}$$

where $D^n(1,2,\ldots,n)$ is an n-row, n-column antisymmetric determinant constructed by the elements G_{ij} in which the diagonal elements are zero. Since such antisymmetric determinants vanish identically in odd order, odd-order terms with respect to U do not appear in the free energy F.

The local susceptibility of d-electrons is given by the integral of the d-electron spin correlation function:

$$\chi = \left(\frac{g\mu_\mathrm{B}}{2}\right)^2 \frac{1}{\beta}\int_0^\beta\int_0^\beta d\tau d\tau' \langle\!\langle \mathrm{T}_\tau[n_{\mathrm{d}\uparrow}(\tau) - n_{\mathrm{d}\downarrow}(\tau)][n_{\mathrm{d}\uparrow}(\tau') - n_{\mathrm{d}\downarrow}(\tau')]\rangle\!\rangle . \tag{19.40}$$

If we divide the susceptibility into two parts as

$$\chi_{\uparrow\uparrow} = \frac{(g\mu_\mathrm{B})^2}{2}\frac{1}{\beta}\int_0^\beta\int_0^\beta d\tau d\tau' \langle\!\langle \mathrm{T}_\tau \tilde{n}_{\mathrm{d}\uparrow}(\tau)\tilde{n}_{\mathrm{d}\uparrow}(\tau')\rangle\!\rangle , \tag{19.41a}$$

$$\chi_{\uparrow\downarrow} = -\frac{(g\mu_\mathrm{B})^2}{2}\frac{1}{\beta}\int_0^\beta\int_0^\beta d\tau d\tau' \langle\!\langle \mathrm{T}_\tau \tilde{n}_{\mathrm{d}\uparrow}(\tau)\tilde{n}_{\mathrm{d}\downarrow}(\tau')\rangle\!\rangle , \tag{19.41b}$$

we obtain the following expansion formulae:

$$\chi_{\uparrow\uparrow} = \frac{(g\mu_{\mathrm{B}})^2}{2} \sum_{n=0}^{\infty} \frac{U^{2n}}{(2n)!} \frac{1}{\beta} \int_0^\beta \cdots \int d\tau_1 d\tau_2 \ldots d\tau_{2n+2} \times \left[D^{2n+2}(1,2,\ldots,2n+2) D^{2n}(1,2,\ldots,2n)\right]_{\mathrm{conn}} , \tag{19.42}$$

$$\chi_{\uparrow\downarrow} = \frac{(g\mu_{\mathrm{B}})^2}{2} \sum_{n=0}^{\infty} \frac{U^{2n+1}}{(2n+1)!} \frac{1}{\beta} \int_0^\beta \cdots \int d\tau_1 d\tau_2 \ldots d\tau_{2n+3} \times \left[D^{2n+2}(1,2,\ldots,2n+2) D^{2n+2}(2,3,\ldots,2n+3)\right]_{\mathrm{conn}} . \tag{19.43}$$

Here $\tilde{n}$ is the difference between n and its average value 1/2. For the same reason as in F given by (19.39), $D^n(1,2,\ldots,n)$ with n odd do not appear here also. $\chi_{\uparrow\uparrow}$ is given by a power series with only even-order terms in U, and $\chi_{\uparrow\downarrow}$ is given by one with only odd-order terms.

The symbol conn in (19.39, 42, 43) means that only the terms in which all τ are connected should be retained, when the two determinants are expanded in the form of sums of products of the G_{ji}. In this case, for the integral over last τ (left after the integrals over other τ have been done one by one), the integrand does not depend on τ, and it gives the factor β. This factor cancels β in front of the integrand and gives a finite value in the limit $\beta \to \infty$. In contrast to this, in the case of disconnected terms such as the product of two connected terms, the factor β appears for each term in the last integration so that the product of two connected terms gives the term proportional to β. In the same way we can see that "connected" means the operation which omits all the terms possessing the factor β^n $(n \geq 1)$.

Next we consider $\chi_{\uparrow\uparrow}$ in (19.42). The integrand over τ is given by the product of two determinants, but τ_{2n+1} and τ_{2n+2} are not included in the right determinant. Therefore we expand the left determinant in the sum of cofactors and carry out the integration over τ_{2n+1} and τ_{2n+2}. We take the limit $T \to 0$ $(\beta \to \infty)$ in the result obtained, replace the sum over ω_n with an integration and carry out a partial integration. Then, by using of the properties of determinant, we can write $\chi_{\uparrow\uparrow}$ at $T = 0$ in the following form:

$$\chi_{\uparrow\uparrow} = \frac{(g\mu_{\mathrm{B}})^2}{2} \frac{1}{\pi\Delta} \widetilde{\chi}_{\uparrow\uparrow} \tag{19.44}$$

$$\widetilde{\chi}_{\uparrow\uparrow} = 1 - \lim_{\beta\to\infty} \sum_{n=1}^{\infty} \frac{U^{2n}}{(2n)!} \frac{1}{\beta} \int_0^\beta \cdots \int_0^\beta d\tau_1 d\tau_2 \ldots d\tau_{2n} \times \sum_{ji} [(\tau_j - \tau_i) D_{ji}^{2n} D^{2n}]_{\mathrm{conn}} . \tag{19.45}$$

Here D_{ji}^{2n} is the minor determinant (cofactor) derived by removing the jth row and ith column from D^{2n}. By a similar operation, $\chi_{\uparrow\downarrow}$ at $T = 0$ can be rewritten as

$$\chi_{\uparrow\downarrow} = \frac{(g\mu_{\mathrm{B}})^2}{2} \frac{1}{\pi\Delta} \widetilde{\chi}_{\uparrow\downarrow} \tag{19.46}$$

$$\widetilde{\chi}_{\uparrow\downarrow} = \frac{1}{\pi\Delta} \lim_{\beta\to\infty} \sum_{n=0}^{\infty} \frac{U^{2n+1}}{(2n+1)!} \frac{1}{\beta} \int_0^\beta \cdots \int_0^\beta d\tau_1 d\tau_2 \ldots d\tau_{2n+1} \,,$$
$$\times \Bigg(\sum_{\substack{ji \\ \text{incl.} j=i}} D_{ji}^{2n+1} \Bigg)^2_{\text{conn}} . \tag{19.47}$$

One of the striking properties of the perturbation expansion of the Anderson Hamiltonian with respect to U is that every order of the perturbation can be written in a closed form, as shown above. Therefore, by comparing perturbation terms of arbitrary order for various physical quantities with each other, we can find relations between physical quantities. For this purpose, here we carry out the perturbation expansion of the temperature Green function for d-electrons

$$G_\sigma^{\mathrm{P}}(\tau,\tau') = -\langle\langle \mathrm{T}_\tau \tilde{a}_{\mathrm{d}\sigma}(\tau) \tilde{a}_{\mathrm{d}\sigma}^\dagger(\tau') \rangle\rangle \,, \tag{19.48}$$

which is an important physical quantity. Here $\tilde{a}_{\mathrm{d}\sigma}(\tau)$ is the operator in the Heisenberg representation and evolves by following the total Hamiltonian including the interaction. This can be written in the interaction representation as

$$G_\sigma^{\mathrm{P}}(\tau,\tau') = -\frac{\left\langle \mathrm{T}_\tau a_{\mathrm{d}\sigma}(\tau) a_{\mathrm{d}\sigma}^\dagger(\tau') \exp\left[-\int_0^\beta \mathcal{H}'(\tau')d\tau' \right] \right\rangle}{\left\langle \mathrm{T}_\tau \exp\left[-\int_0^\beta \mathcal{H}'(\tau')d\tau' \right] \right\rangle} . \tag{19.49}$$

Here $\langle \quad \rangle$ means the thermal average in the unperturbed state. The perturbation $\mathcal{H}'$ is the second term in (19.31). Equation (19.49) is expanded in powers of U, similarly to the free energy and susceptibility, as

$$G_\sigma^{\mathrm{P}}(\tau,\tau') = G_{\tau\tau'} + \sum_{n=1}^{\infty} \frac{U^n(-1)^n}{n!} \int_0^\beta \cdots \int d\tau_1 \ldots d\tau_n$$
$$\times \, [\widetilde{D}^{n+1}(\tau\tau';1,2,\ldots,n) D^n(1,2,\ldots,n)]_{\text{conn}} \,, \tag{19.50}$$

where

$$\widetilde{D}^{n+1}(\tau\tau';1,2,\ldots,n) = \begin{vmatrix} 0 & G_{\tau 1} & \cdots\cdots & G_{\tau n} \\ G_{1\tau'} & 0 & G_{12}\cdots & G_{1n} \\ \vdots & G_{21} & & \vdots \\ \vdots & \vdots & & \vdots \\ G_{n\tau'} & G_{n1} & \cdots\cdots & 0 \end{vmatrix}$$
$$= - \sum_{\substack{ji \\ \text{incl.} j=i}} G_{\tau i} G_{j\tau'} D_{ji}^n . \tag{19.51}$$

Since $D^n(1,2,\ldots,n)$ vanishes identically for odd n, the sum over n in (19.50) is done for even n. For $\tau' \to \tau + 0$ the determinant (19.51) vanishes, since it becomes an odd order antisymmetric determinant for even n. Therefore we obtain

$$\langle\langle n_{\mathrm{d}\sigma}\rangle\rangle = \lim_{\tau' \to \tau+0} G_\sigma^{\mathrm{P}}(\tau, \tau') = \lim_{\tau' \to \tau+0} G_{\tau\tau'} = \frac{1}{2} \,. \tag{19.52}$$

The localized d-electron number is always fixed to the unperturbed value 1/2 even if the perturbation is switched on. This means that the phase shift of conduction electrons at the Fermi surface is fixed to $\pi/2$. This result arises from the condition of electron-hole symmetry, *i.e.*, $E_{\mathrm{d}} = -U/2$. By using the minor-determinant expansion of $\widetilde{D}^{n+1}$ in the relation of (19.50), we obtain the Fourier component of $G_\sigma^{\mathrm{P}}(\tau, \tau')$ as

$$\begin{aligned} G^{\mathrm{P}}(\omega_n) = G(\omega_n) - \sum_{n=1}^{\infty} \frac{U^{2n}}{(2n)!} \frac{1}{\beta} \int_0^\beta \cdots \int_0^\beta \mathrm{e}^{\mathrm{i}\omega_n(\tau-\tau')} d\tau d\tau' d\tau_1 \dots d\tau_{2n} \\ \times \sum_{ji} [G_{\tau i} G_{j\tau'} D_{ji}^{2n}(1, 2, \dots, 2n) D^{2n}]_{\mathrm{conn}} \,. \end{aligned} \tag{19.53}$$

From the relation

$$G^{\mathrm{P}}(\omega_n) = G(\omega_n) + G(\omega_n)\Sigma'(\omega)G(\omega_n) \,, \tag{19.54}$$

the improper self-energy $\Sigma'(\omega)$ is given by

$$\begin{aligned} \Sigma'(\omega_n) = -\sum_{n=1}^{\infty} \frac{U^{2n}}{(2n)!} \frac{1}{\beta} \int_0^\beta \cdots \int_0^\beta d\tau_1 \dots d\tau_{2n} \\ \times \sum_{ji} [\mathrm{e}^{-\mathrm{i}\omega_n(\tau_j - \tau_i)} D_{ji}^{2n} D^{2n}]_{\mathrm{conn}} \,. \end{aligned} \tag{19.55}$$

Here, if we put $\omega_n \to 0$, we find that $\Sigma'(\omega \to 0) = 0$ by the relation $D_{ji}^{2n} = -D_{ij}^{2n}$. Therefore, when $\Sigma'(\omega_n)$ is expanded with respect to ω_n, the first term is linear in ω_n. This linear term is obtained from the differentiation of (19.55) by ω_n as

$$\begin{aligned} \Delta^{(1)}\Sigma'(\omega_n) = \mathrm{i}\omega_n \sum_{n=1}^{\infty} \frac{U^{2n}}{(2n)!} \frac{1}{\beta} \int_0^\beta \cdots \int_0^\beta d\tau_1 \dots d\tau_{2n} \\ \times \sum_{ji} [(\tau_j - \tau_i) D_{ji}^{2n} D^{2n}]_{\mathrm{conn}} \,. \end{aligned} \tag{19.56}$$

From comparison of this relation with the expression (19.45) for $\widetilde{\chi}_{\uparrow\uparrow}$, the relation

$$\Delta^{(1)}\Sigma'(\omega_n) = -\mathrm{i}\omega_n(\widetilde{\chi}_{\uparrow\uparrow} - 1) \tag{19.57}$$

or

$$\frac{\partial \Sigma'(\omega_n)}{\partial(\mathrm{i}\omega_n)} = -(\widetilde{\chi}_{\uparrow\uparrow} - 1) \tag{19.58}$$

holds at $T = 0$. This is one of the relations between physical quantities. This is nothing but a special form of Ward identities.

On the other hand, the proper self-energy $\Sigma(\omega_n)$ is defined by

$$G^{\mathrm{P}}(\omega_n) = G(\omega_n) + G(\omega_n)\Sigma(\omega_n)G^{\mathrm{P}}(\omega_n) \ . \tag{19.59}$$

Therefore $\Sigma'(\omega_n)$ and $\Sigma(\omega_n)$ are connected by the relation

$$\Sigma(\omega_n) = \frac{\Sigma'(\omega_n)}{1 + G(\omega_n)\Sigma'(\omega_n)} \ . \tag{19.60}$$

Since $\Sigma'(\omega_n)$ vanishes as $\omega_n \to 0$, $\Sigma(\omega_n)$ also vanishes as $\omega_n \to 0$ and the derivatives of both self-energies with respect to ω_n are equal to each other. Therefore Σ' in (19.57, 58) can be written as Σ. Since $\Sigma(\omega_n)$ with ω_n replaced by $-\mathrm{i}\omega$ gives the self-energy for the retarded Green function $\Sigma^{\mathrm{R}}(\omega)$, the relation (19.57) can be written as

$$\Delta^{(1)}\Sigma^{\mathrm{R}}(\omega) = -\omega(\widetilde{\chi}_{\uparrow\uparrow} - 1) \ , \tag{19.61}$$

$$\left.\frac{\partial \Sigma^{\mathrm{R}}(\omega)}{\partial \omega}\right|_{\omega=0} = -(\widetilde{\chi}_{\uparrow\uparrow} - 1) \ . \tag{19.62}$$

$\widetilde{\chi}_{\uparrow\uparrow}$ is real so that the quantity on the left-hand side is naturally real.

Now we return to the expression for the free energy, (19.39). Rewriting D^{2n} in this equation by the minor-determinant expansion of D^{2n} as

$$D^{2n} = \frac{1}{2n}\sum_{ji} G_{ji}D^{2n}_{ji} = \frac{1}{2n}\frac{1}{\beta}\sum_{ji}\sum_{n} G(\omega_n)\mathrm{e}^{-\mathrm{i}\omega_n(\tau_j - \tau_i)}D^{2n}_{ji} \ ,$$

we obtain the $2n$th order term as

$$F^{(2n)} = \frac{1}{2n}\frac{1}{\beta}\sum_{n} G(\omega_n)\Sigma'^{(2n)}(\omega_n) \ . \tag{19.63}$$

Let us consider the sum of an arbitrary function $F(\omega_n)$ over ω_n. At $T = 0$ this sum is replaced by an integral over ω. At low temperatures, the difference between the sum and the integral is expanded in a power series in T as

$$\frac{1}{\beta}\sum_{n} F(\omega_n) = \frac{1}{2\pi}\int_{-\infty}^{\infty} F(\omega)d(\omega) - \frac{1}{6}\left(\frac{\pi}{\beta}\right)^2 \frac{1}{2\pi}[F'(0_-) - F'(0_+)] + \dots \ . \tag{19.64}$$

Using this relation in (19.63), we pick up the T^2-term from $F^{(2n)}$. It is noted in this case that only $G(\omega_n)$ has a discontinuity at $\omega_n = 0$, while the derivative of $G(\omega_n)$, $\Sigma'(\omega_n)$, and its derivative are continuous there. The discontinuity of $G(\omega_n)$ at $\omega_n \to 0$ is given from (19.38) by

$$G(0_+) - G(0_-) = -\frac{2}{\Delta}\mathrm{i} \ . \tag{19.65}$$

Moreover, when the $2n$th order term of the free energy (19.63) is represented with the diagram, there are $4n$ electron lines; we can choose any of them as $C(\omega_n)$ in (19.63). Therefore we have to multiply by an extra factor $4n$. Taking this point into account, we obtain, as the T^2-term of $F^{(2n)}$,

$$F^{(2n)} \rightarrow \frac{1}{3}\left(\frac{\pi}{\beta}\right)^2 \frac{1}{\pi\Delta} \lim_{\beta\to\infty} \left.\frac{\partial \Sigma'^{(2n)}(\omega_n)}{\partial(\mathrm{i}\omega_n)}\right|_{\omega_n=0} . \tag{19.66}$$

Since the specific heat is given by $-T(\partial^2 F/\partial T^2)$, the coefficient γ of the T-linear specific heat is given by

$$\gamma = \frac{2\pi^2 k^2}{3}\frac{1}{\pi\Delta}\tilde{\gamma} , \tag{19.67}$$

$$\tilde{\gamma} = 1 - \left.\frac{\partial \Sigma^{\mathrm{R}}(\omega)}{\partial\omega}\right|_{\omega=0} , \tag{19.68}$$

where $\Sigma^{\mathrm{R}}(\omega)$ is the self-energy of the retarded Green function and is obtained by replacing $\mathrm{i}\omega_n$ by ω, as noted above. The result (19.68) is the relation between the specific heat and the self-energy. With use of the relation (19.62), we can show $\tilde{\gamma}$ is nothing but $\widetilde{\chi}_{\uparrow\uparrow}$. Thus, the Wilson ratio is obtained as

$$\frac{T\chi}{C_v} = \frac{(g\mu_{\mathrm{B}})^2/2}{2(\pi k)^2/3}\left(1 + \frac{\widetilde{\chi}_{\uparrow\downarrow}}{\widetilde{\chi}_{\uparrow\uparrow}}\right) . \tag{19.69}$$

Now we consider the case where the signs of E_{d} and U are changed. In this case, the energy levels of the d-electron are situated above the Fermi surface; when two electrons occupy the d-orbital, the energy level is situated below the Fermi surface. Therefore the states with zero or two d-electrons are stable; these states have no localized spins. Thus the susceptibility in this case is given by $\chi_{\uparrow\uparrow} - \chi_{\uparrow\downarrow}$ and decreases with increasing U, vanishing at $U = \infty$. From this result, $\chi_{\uparrow\uparrow} = \chi_{\uparrow\downarrow}$ is obtained for $U = \infty$. $\chi_{\uparrow\uparrow} - \chi_{\uparrow\downarrow}$ is called the charge susceptibility, and we can say that the charge susceptibility vanishes at $U = \infty$. In this case, the quantity in the bracket in the expression for the Wilson ratio becomes 2. In the case $U = 0$, the ratio is unity; it increases monotonically with increasing U and approaches 2 as $U \rightarrow \infty$. This is the value of the s-d limit obtained by Wilson and Nozières. For general values of U, there are two parameters, $\widetilde{\chi}_{\uparrow\uparrow}$ and $\widetilde{\chi}_{\uparrow\downarrow}$, corresponding to α and $2\rho\phi^a$ in the theory of Nozières.

19.3.3 Temperature Dependence of Electrical Resistivity

Now we consider the electrical resistivity. The inverse of the average lifetime $\tau(\omega)$ due to the scattering of the conduction electrons by the impurity, namely, the transition probability for the s-d model, is given by

$$\frac{1}{\tau(\omega)} = \frac{2\pi}{\hbar}\rho(\omega)\left[|t(\omega)|^2 + S(S+1)|\tau(\omega)|^2\right] = -\frac{2}{\hbar}\,\mathrm{Im}\ t_{\boldsymbol{kk}}(\omega) . \tag{19.70}$$

The second relation holds not only for the s-d model but also in general. Here ω is the electron energy measured from the Fermi energy and $\rho(\omega)$ is the density of states. The T-matrix is related to the retarded Green function of the d-electron by

$$t_{\boldsymbol{kk}}(\omega) = V^2 G^{\mathrm{R}}(\omega) = V^2 \frac{1}{\omega + \mathrm{i}\Delta - \Sigma^{\mathrm{R}}(\omega)} , \tag{19.71}$$

which can be understood easily from the $\boldsymbol{kk}'$ component of the Green function, (15.74); $\Sigma^{\mathrm{R}}(\omega)$ is the self-energy of the d-electron. Denoting its real part as $R(\omega)$ and its imaginary part as $I(\omega)$, we obtain

$$\begin{aligned} -\operatorname{Im} t_{\boldsymbol{kk}}(\omega) &= \frac{1}{\pi\rho} \frac{\Delta^2 - \Delta I}{(\omega - R)^2 + (\Delta - I)^2} \\ &= \pi\rho|t|^2 - \frac{1}{\pi\rho} \frac{\Delta I}{(\omega - R)^2 + (\Delta - I)^2} . \end{aligned} \tag{19.72}$$

For small ω and T, $R(\omega)$ is proportional to ω, as seen from (19.61); $I(\omega)$ is proportional to ω^2 and T^2, as shown below. Thus, for small ω and T, the second term of (19.72) is approximated by $-(1/\pi\rho)(I/\Delta)$. The existence of the extra term means that there exists inelastic scattering channel in addition to the elastic scattering given by the first term. The lowest-order term of the inelastic scattering arises from two-body scattering. If we denote the scattering amplitude from $k_1 \uparrow k_2 \uparrow$ to $k_3 \uparrow k_4 \uparrow$ as $A_{\uparrow\uparrow}$ and that from $k_1 \uparrow k_2 \downarrow$ to $k_3 \downarrow k_4 \uparrow$ as $A_{\uparrow\downarrow}$, the lifetime τ_{in} of the $k \uparrow$ electron due to the two-body scattering can be obtained from the relation

$$\begin{aligned} \left(\frac{\partial f(\omega)}{\partial t}\right)_{\mathrm{coll}} &= -\frac{f(\omega) - f_0(\omega)}{\tau_{\mathrm{in}}(\omega)} \\ &= \frac{2\pi}{\hbar}\rho^3 \int d\omega_1 d\omega_2 d\omega_3 \left(|A_{\uparrow\downarrow}|^2 + \frac{1}{2}|A_{\uparrow\uparrow}|^2\right) \delta(\omega + \omega_3 - \omega_2 - \omega_1) \\ &\quad \times [(1-f)(1-f_{30}) f_{10} f_{20} - f f_{30} (1-f_{10})(1-f_{20})] . \end{aligned} \tag{19.73}$$

This relation represents the time development of the distribution function $f(\omega)$ due to the two-body scattering. $f_0(\omega)$ is that in the thermal equilibrium, namely, the Fermi distribution function. Carrying out the integration, we obtain $1/\tau_{\mathrm{in}}(\omega)$ as

$$\frac{1}{\tau_{\mathrm{in}}(\omega)} = \frac{\pi\rho^3}{\hbar}(\pi^2 T^2 + \omega^2)\left(|A_{\uparrow\downarrow}|^2 + \frac{1}{2}|A_{\uparrow\uparrow}|^2\right) = -\frac{2}{\pi\rho\hbar}\frac{I}{\Delta} , \tag{19.74}$$

where all four electron energies and also T in $A_{\uparrow\downarrow}$ and $A_{\uparrow\uparrow}$ can be put to zero. The fact that the scattering probability of the conduction electrons in metals due to the repulsive force between electrons is proportional to T^2 and ω^2 is the basis of *Landau*'s Fermi-liquid theory [19.3]. Though the electron interaction in our case is a local interaction, this property is unchanged even for a nonlocal interaction.

The scattering amplitudes for two-body scattering, $A_{\uparrow\downarrow}$ and $A_{\uparrow\uparrow}$, can be obtained from the vertex part of the two-body temperature Green function. This vertex can be expanded in a similar way as for the free energy and the susceptibility. At this point, if we put the four energies to zero and set $T = 0$, the perturbation expansion for $A_{\uparrow\downarrow}$ possesses the same expansion as that for $\widetilde{\chi}_{\uparrow\downarrow}$ in (19.47), and the relation

$$A_{\uparrow\downarrow} = -\frac{1}{\pi\rho^2\Delta}\widetilde{\chi}_{\uparrow\downarrow} \qquad (19.75)$$

is found. Moreover, it is also found that $A_{\uparrow\uparrow}$ vanishes at zero energy. Thus the imaginary part of the d-electron self-energy is obtained at low energy and low temperatures, with use of (19.74, 75), as

$$I(\omega) = -\frac{1}{2}\Delta\widetilde{\chi}^2_{\uparrow\downarrow}\left(\frac{\pi^2T^2}{\Delta^2} + \frac{\omega^2}{\Delta^2}\right) . \qquad (19.76)$$

This result can be derived directly from $\Sigma(\omega)$. In the local Fermi-liquid theory of Nozières, $A_{\uparrow\uparrow} = 0$ and $A_{\uparrow\downarrow} = -2\phi^a/\pi\rho$ are assumed. For the real part of Σ^{R}, (19.61) holds. Using these two relations in the first equation (19.72) and substituting this result into the expression for the current density (16.3), we obtain the electrical resistivity at low temperatures as

$$R = R_0\left[1 - \frac{\pi^2}{3}\left(\frac{T}{\Delta}\right)^2(2\widetilde{\chi}^2_{\uparrow\downarrow} + \widetilde{\chi}^2_{\uparrow\uparrow})\right] , \qquad (19.77)$$

where R_0 is the resistivity at $T = 0$ and the value of resistivity corresponding to the unitarity limit given by $t_{\boldsymbol{kk}}(0) = 1/i\pi\rho$.

As explained above, it has been shown that for the Anderson Hamiltonian the specific heat, susceptibility and electrical resistivity at low temperatures can be written in terms of two physical quantities $\widetilde{\chi}_{\uparrow\uparrow}$ and $\widetilde{\chi}_{\uparrow\downarrow}$. This fact holds also for other physical quantities, as far as low temperatures and low energies are concerned. For example, the density of states for the d-electron is obtained from the imaginary part of the retarded Green function as

$$\begin{aligned}\rho_{\mathrm{d}}(\omega) &= -\frac{1}{\pi}\,\mathrm{Im}\,G^{\mathrm{R}}(\omega)\\ &= \frac{1}{\pi\Delta}\left[1 - \left(\frac{\omega}{\Delta}\right)^2\left(\frac{1}{2}\widetilde{\chi}^2_{\uparrow\downarrow} + \widetilde{\chi}^2_{\uparrow\uparrow}\right) - \frac{1}{2}\frac{\pi^2T^2}{\Delta^2}\widetilde{\chi}^2_{\uparrow\downarrow}\right] . \quad \left(\frac{\omega}{\Delta} \ll 1\right)\end{aligned} \qquad (19.78)$$

From this result, we arrive at the following picture on the density of states for the d-orbital: It is given by Lorentz type of distribution with half width Δ and possessing a peak at the Fermi surface at $U = 0$. Then the peak becomes narrow with increasing U; the peak width decreases in proportion to the inverse of the susceptibility and becomes of order T_{K} in the limit of large U.

Now we consider the two parameters. The two essential physical quantities $\widetilde{\chi}_{\uparrow\uparrow}$ and $\widetilde{\chi}_{\uparrow\downarrow}$ become equal to each other in the limit $U \to \infty$; they should be equal to a half of the susceptibility for the *s*-*d* model, namely (17.37) obtained by the theory of the singlet ground state,

$$\widetilde{\chi}_{\uparrow\uparrow} = \widetilde{\chi}_{\uparrow\downarrow} = \frac{\pi\Delta}{4T_{\mathrm{K}}} \; . \tag{19.79}$$

On the other hand, for small U, we have only to carry out the perturbation calculation and take the terms of low order in U. The perturbation expansion of the susceptibility is given by

$$\widetilde{\chi}_{\uparrow\uparrow} = 1 + \left(3 - \frac{\pi^2}{4}\right)\left(\frac{U}{\pi\Delta}\right)^2 + 0.0551\left(\frac{U}{\pi\Delta}\right)^4 + \dots , \tag{19.80}$$

$$\widetilde{\chi}_{\uparrow\downarrow} = \frac{U}{\pi\Delta} + \left(15 - \frac{3\pi^2}{2}\right)\left(\frac{U}{\pi\Delta}\right)^3 + \dots . \tag{19.81}$$

It has been made clear that the physical properties of the *s*-*d* system at low temperatures ($T \ll T_{\mathrm{K}}$) can be easily derived, at least in principle, by the perturbation calculation with respect to U on the basis of the Anderson model. On the other hand, at high temperatures ($T \gg T_{\mathrm{K}}$) we have to use the inverse perturbation expansion. That is, U is included in the unperturbed term and the *s*-*d* mixing term V is treated as the perturbation. By this procedure, the *s*-*d* model is derived and the high-temperature expansion in the *s*-*d* model becomes a good approximation. In the high-temperature region, the *d*-electron behaves as a localized spin. The transition from the high-temperature state to the nonmagnetic state at low temperature occurs rapidly around $T \sim T_{\mathrm{K}}$; this transition is continuous and during it the localized spin loses most of the entropy $k \log(2S+1)$ which it possesses.

Here we have assumed for simplicity that the localized *d*-orbital has no orbital degeneracy. The perturbation theory taking into account the 5-fold orbital degeneracy has been developed by *Shiba* [19.5] and *Yoshimori* [19.6].

19.4 Exact Solution on the Basis of Bethe Ansatz

Through the work of Wilson and Nozières, and by the perturbation expansion of the Anderson model, the low-temperature behavior of the localized spin has been well explained and the essence of the Kondo effect has been made clear. In particular, the understanding of the concept of localized spin has been made deeper than before. The magnetic ion in the nonmagnetic metal has no localized moment at $T = 0$ when the external magnetic field $H = 0$. However, because of the strong repulsive force between the electrons occupying the localized *d*-orbital, the spin fluctuation appears locally with increasing temperature and this fluctuation grows up to become a localized spin. However, even at $T = 0$, localized spins are induced in the presence of magnetic field or in the case of strong RKKY interaction between localized spins.

We have arrived at the above understanding of the essence of the Kondo effect. At this point, further development of the theory on the Kondo effect was brought through the exact solution of the Kondo effect by *Andrei* et al. [19.7] and *Tsvelik* and *Wiegmann* [19.8]. As treated by Wilson, the Kondo problem can be reduced to the one-dimensional problem by taking into consideration only s-wave scattering among partial waves. Andrei and Wiegmann showed that the method of the Bethe hypothesis can be applied to the s-wave scattering problem where the conduction electrons are scattered by the localized spin. Thus the Kondo effect can be included in the group of exactly solvable problems consisting of one-dimensional ferromagnetism, antiferromagnetism, and one-dimensional Hubbard model. After Andrei and Wiegmann obtained the exact solution of the s-d model, Wiegmann and *Okiji* and *Kawakami* [19.9] derived the exact solution for the Anderson model.

These exact solutions neither contradicted the conclusions obtained previously regarding the Kondo effect nor added any corrections. Rather they confirmed the results obtained before for high and low temperatures, and for other special cases. Since the mathematics used there is rather complicated and not so easy, we do not enter the details of the exact solution here; we show only the result for the susceptibility at $T = 0$ for the Anderson model with electron-hole symmetry ($E_\mathrm{d} = -U/2$) and present the consideration of *Zlatić* and *Horvatić* [19.10] on the results.

The dimensionless values of the spin and charge susceptibilities (derived by Okiji and Kawakami) are expressed as

$$\widetilde{\chi}_\mathrm{s} = \widetilde{\chi}_{\uparrow\uparrow} + \widetilde{\chi}_{\uparrow\downarrow} = \sqrt{\frac{\pi}{2u}} \exp\left(\frac{1}{8}\pi^2 u - \frac{1}{2u}\right) + \frac{1}{\sqrt{2\pi u}} \int_{-\infty}^{\infty} \frac{\exp(-x^2/2u)}{1 + \left(\frac{1}{2}\pi u + \mathrm{i}x\right)^2} dx \, , \tag{19.82}$$

$$\widetilde{\chi}_\mathrm{c} = \widetilde{\chi}_{\uparrow\uparrow} - \widetilde{\chi}_{\uparrow\downarrow} = \frac{1}{\sqrt{2\pi u}} \int_{-\infty}^{\infty} \frac{\exp(-x^2/2u)}{1 + \left(\frac{1}{2}\pi u + x\right)^2} dx \, , \tag{19.83}$$

respectively. Here u is defined by $u = U/\pi\Delta$. Since the second term in (19.82) vanishes in the limit $u \to \infty$, the asymptotic form of the spin susceptibility is given by

$$\lim_{u\to\infty} \widetilde{\chi}_\mathrm{s} = \sqrt{\frac{\pi}{2u}} \exp\left(\frac{1}{8}\pi^2 u\right) . \tag{19.84}$$

If we use the relation (19.32) between U and J together with (17.78) as the Kondo temperature, this expression for $\widetilde{\chi}_s$ is just twice (19.79) multiplied by $(D/U)(4/\sqrt{\pi})$. If we put $D = U$, the susceptibility obtained in the exact solution has an extra factor $4/\sqrt{\pi}$. In the limit $u \to \infty$, $\widetilde{\chi}_c$ given by (19.83) naturally goes to zero.

Next we consider the small-u region. The first term of (19.82) possesses an essential singularity at $u = 0$, and therefore $\widetilde{\chi}_s$ is apparently singular at $u = 0$. But, the second term is also singular at $u = 0$; these singularities are cancelled with each other and $\widetilde{\chi}_s$ is completely analytic at $u = 0$. This fact was shown by *Zlatić* and *Horvatić* [19.10]. In order to prove this, we denote first the integral of the second term of (19.82) as $I_s(u)$ and rewrite it as

$$I_s(u) = \frac{1}{\sqrt{2\pi u}} \mathrm{Re} \int_{-\infty}^{\infty} \frac{\mathrm{e}^{-x^2/2u}}{x - z'} dx \ , \qquad z' = -1 + \mathrm{i}\frac{1}{2}\pi u \ .$$

The integrand has a first-order pole at $z = z'$ in the complex plane; this pole is situated on the real axis at $u \to 0$ and makes $I_s(u)$ singular for $u = 0$. Zlatić and Horvatić evaluated the integral by the aid of the contour integration shown in Fig. 19.2 and showed that $I_s(u)$ is divided into a term singular at $u = 0$ and an analytic term, and that the former cancels the first term in (19.82) completely. Thus, $\widetilde{\chi}_s$ is reduced to

$$\widetilde{\chi}_s = \mathrm{e}^{\pi^2 u/8} \sqrt{\frac{2}{\pi u}} \int_0^{\infty} \mathrm{e}^{-x^2/2u} \frac{\cos\left(\frac{1}{2}\pi x\right)}{1 - x^2} dx \ . \tag{19.85}$$

On the other hand, $\widetilde{\chi}_c(u)$ can be rewritten as

$$\widetilde{\chi}_c(u) = \mathrm{e}^{-\pi^2 u/8} \sqrt{\frac{2}{\pi u}} \int_0^{\infty} \mathrm{e}^{-x^2/2u} \frac{\cosh\left(\frac{1}{2}\pi x\right)}{1 + x^2} dx \ . \tag{19.86}$$

In (19.85, 86), expanding $\cos[(1/2)\pi x]/(1 - x^2)$ and $\cosh[(1/2)\pi x]/(1 + x^2)$ in power series in x, multiplying them by $\exp(-x^2/2u)$ and integrating term by term, we obtain

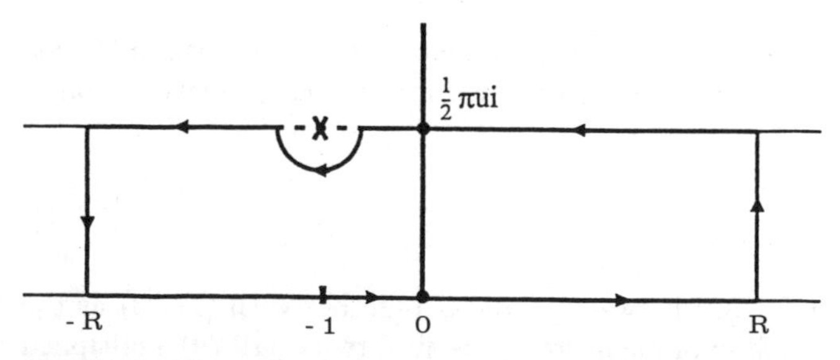

Fig. 19.2. The path of the contour integration for $I_s(u)$

$$\widetilde{\chi}_{\rm s}(u) = \exp\left(\frac{1}{8}\pi^2 u\right)\psi(u) , \tag{19.87}$$

$$\widetilde{\chi}_{\rm c}(u) = \exp\left(-\frac{1}{8}\pi^2 u\right)\psi(-u) , \tag{19.88}$$

$$\psi(u) = \sum_{n=0}^{\infty}(-1)^n \zeta_n u^n , \tag{19.89}$$

where ζ_n satisfies the recurrence formula

$$\zeta_n = \left[\left(\frac{1}{8}\pi^2\right)^n \Big/ n!\right] - (2n-1)\zeta_{n-1} \tag{19.90}$$

with $n \geq 1, \zeta_0 = 1$. The general term of ζ_n is given by

$$\zeta_n = \frac{\left(\frac{1}{8}\pi^2\right)^{n+1}}{(2n+1)(n+1)!}\sum_{k=0}^{\infty}(-1)^k \frac{[2(n+1)]!}{[2(n+k+1)]!}\left(\frac{\pi}{2}\right)^{2k} . \tag{19.91}$$

Since the sum of the series changes from $8/\pi^2$ at $n = 0$ to 1 at $n = \infty$, the asymptotic form of ζ_n for large n is given by the prefactor. Therefore the power series for $\psi(u)$ converges absolutely for $|u| < \infty$ and $\psi(u)$ is analytic for finite u. At $u = 0, \psi(u) = 1$ holds and $\psi(u)$ reaches the asymptotic form $\psi^a(u) = \sqrt{\pi/2u}\exp(-1/2u)$ already around $u = 2$. Thus, at values even near $u \sim 2, \widetilde{\chi}_{\rm s}$ is well approximated by the first term in (19.82).

The power-series expansions of $\widetilde{\chi}_{\rm s}$ and $\widetilde{\chi}_{\rm c}$ with respect to u are obtained by expanding the prefactors, namely, the exponential functions in (19.87, 88). Thus we obtain

$$\widetilde{\chi}_{\rm s}(u) = \sum_{n=0}^{\infty} C_n u^n , \tag{19.92}$$

$$\widetilde{\chi}_{\rm c}(u) = \sum_{n=0}^{\infty}(-1)^n C_n u^n . \tag{19.93}$$

Here the expansion coefficients C_n satisfy the recurrence formula

$$C_n = (2n-1)C_{n-1} - \left(\frac{\pi}{2}\right)^2 C_{n-2} , \tag{19.94}$$

where $n \geq 2, C_0 = C_1 = 1$. The solution of this recurrence formula for general n is given by

$$C_n = \left[\left(\frac{1}{2}\pi\right)^{2n+1} \Big/ (2n+1)!!\right] P_n , \tag{19.95}$$

$$P_n = \sum_{k=0}^{\infty}\frac{(-1)^k}{k!}\frac{(2n+1)!!}{[2(n+k)+1]!!}\left(\frac{\pi^2}{8}\right)^k . \tag{19.96}$$

Since P_n satisfies $(2/\pi) = P_0 < P_n < P_\infty = 1$, the power series for $\widetilde{\chi}_s$ and $\widetilde{\chi}_c$ converge absolutely and both are analytic functions of u. Moreover, since C_n decreases rapidly with increasing n, we obtain sufficiently good approximate values for $\widetilde{\chi}_s$ and $\widetilde{\chi}_c$ by taking the first several terms with respect to u. Taking up to $n = 4$ in (19.92), we recover the results obtained by the direct perturbation calculation, *i.e.*, the sum of (19.80, 81).

As described above, by the exact solution obtained for the Anderson model on the basis of the Bethe hypothesis it has been rigorously proved that various physical quantities such as the susceptibility and the specific heat change smoothly and continuously as a function of u from $u = 0$ to $u \leq \infty$ and there is no singularity at any intermediate value of u. Thus we can consider that for the perturbation U added adiabatically, the philosophy of the Landau Fermi-liquid theory has been completely proved in the Anderson model, though it is local. Even in this one point, the role of the exact solution is important.

Here we close our description of the Kondo effect. In summary, the Kondo problem is that of a localized spin in metals, which is a simple but essential problem; it is of fundamental importance in the history of the electron theory of solids. At present, high-density Kondo states discovered in the rare-earth metallic compounds, the formation of heavy-fermion systems at low temperatures and other problems closely related to the Kondo effect are still actively studied.

References

Chapter 1

1.1 J.C. Slater: Phys. Rev. **34**, 1293 (1929)
1.2 B. Bleaney, K.W.H. Stevens: Rep. Prog. Phys. **16**, 108 (1953)

Chapter 2

2.1 J. van Kranendonk: *7th International Conference on Low Temperature Physics* (Univ. of Toronto Press, Toronto 1961) Proc., p. 9

Chapter 3

3.1 B. Bleaney, K.W.H. Stevens: Rep. Prog. Phys. **16**, 108 (1953)
3.2 K.W.H. Stevens: Proc. Phys. Soc. (London) A **65**, 209 (1952)
3.3 R. Kubo, Y. Obata: J. Phys. Soc. Jpn. **11**, 547 (1956)
3.4 A. Clogston, A.G. Gossard, V. Jaccarino, Y. Yafet: Phys. Rev. Lett. **9**, 262 (1962)
3.5 K. Yosida, M. Tachiki: Prog. Theor. Phys. **17**, 331 (1957)
3.6 U. Öpik, M.H.L. Pryce: Proc. Roy. Soc. (London) A **238**, 425 (1957)
3.7 A.D. Liehr, C.J. Ballhausen: Ann. Phys. (N.Y.) **3**, 304 (1958)
3.8 H. Ohnishi, T. Teranishi, S. Miyahara: J. Phys. Soc. Jpn. **14**, 106 (1959)
3.9 H.F. McMurdie, B.M. Sullivan, F.A. Mauer: J. Res. Natl. Bur. Stand. **45**, 35 (1950)
3.10 J. Kanamori: J. Appl. Phys. Suppl. **31**, 14S (1960)
3.11 M.A. Hepworth, K.H. Jack: Acta Cryst. **10**, 345 (1957)
3.12 A. Okazaki, Y. Suemune: J. Phys. Soc. Jpn. **16**, 176 (1961)
3.13 A. Okazaki: J. Phys. Soc. Jpn. **26**, 870 (1969); *ibid.* **27**, 518 (1969)
3.14 K. Hirakawa, Y. Kurogi: Prog. Theor. Phys. Suppl. **46**, 147 (1970)
3.15 J.D. Dunitz, L.E. Orgel: J. Phys. Chem. Solids **3**, 20 (1957)
3.16 J. Kanamori: Prog. Theor. Phys. **17**, 177, 197 (1957)
3.17 J.C. Slonczewski: Phys. Rev. **110**, 1341 (1958)
3.18 M. Tachiki: Prog. Theor. Phys. **23**, 1055 (1960)
3.19 A. Abragam, M.H.L. Pryce: Proc. Roy. Soc. (London) A **206**, 164 and 173 (1951)

Chapter 4

4.1 W. Heisenberg: Z. Physik **49**, 619 (1928)
4.2 P.W. Anderson: Phys. Rev. **115**, 2 (1959); *Solid State Physics*, **14**, 99 (Academic, New York 1963)

4.3 E.A. Harris, J. Owen: Phys. Rev. Lett. **11**, 9 (1963)
4.4 A. Yoshimori, S. Inagaki: J. Phys. Soc. Jpn. **44**, 101 (1978); *ibid.* **50**, 769 (1981)
4.5 K. Yosida, S. Inagaki: J. Phys. Soc. Jpn. **50**, 3268 (1981)
4.6 H.A. Kramers: Physica **1**, 182 (1934)
4.7 P.W. Anderson: Phys. Rev. **79**, 350 (1950)
4.8 J.B. Goodenough: Phys. Rev. **100**, 564 (1955); J. Phys. Chem. Solids **6**, 287 (1958)
4.9 J. Kanamori: J. Phys. Chem. Solids **10**, 87 (1959)
4.10 E.O. Wollan, H.R. Child, W.C. Koehler, M.K. Wilkinson: Phys. Rev. **112**, 1132 (1958)
4.11 T. Moriya, K. Yosida: Prog. Theor. Phys. **9**, 663 (1953)
4.12 T. Nakamura, H. Taketa: Prog. Theor. Phys. **13**, 129 (1955)
4.13 I. Dzyaloshinsky: J. Phys. Chem. Solids **4**, 241 (1958)
4.14 T. Moriya: Phys. Rev. **120**, 91 (1960); Weak ferromagnetism, in *Magnetism*, ed. by G. Rado and H. Suhl (Academic, New York 1963) Vol. 1, p. 85
4.15 A.S. Borovik-Romanov, M.P. Orlova: JETP **31**, 579 (1956) [Engl. transl.: Sov. Phys. – JETP **4**, 531 (1957)]
4.16 A.S. Borovik-Romanov: JETP **36**, 766 (1959) [Engl. transl.: Sov. Phys. – JETP **9**, 539 (1959)]
4.17 W.C. Koehler, E.O. Wollan: J. Phys. Chem. Solids **2**, 100 (1957)
4.18 W.C. Koehler, E.O. Wollan, M.K. Wilkinson: Phys. Rev. **118**, 58 (1960)
4.19 T. Moriya: Phys. Rev. **117**, 635 (1960)
4.20 A.J. Heeger, O. Beckman, A.M. Portis: Phys. Rev. **123**, 1652 (1961)
4.21 W. Heitler, F. London: Z. Phys. **44**, 455 (1927)
4.22 Y. Sugiura: Z. Phys. **45**, 484 (1927)
4.23 C. Herring: Rev. Mod. Phys. **34**, 631 (1962); Direct exchange between well-separated atoms, in *Magnetism*, ed. by G. Rado and H. Suhl (Academic, New York 1966) Vol. **2B**, p. 1
4.24 D.J. Thouless: Proc. Phys. Soc. (London) **86**, 893, 905 (1965)
M. Roger, J.H. Hetherington, J.M. Delrieu: Rev. Mod. Phys. **55**, 1 (1983)

Chapter 5

5.1 A. Yoshimori: J. Phys. Soc. Jpn. **14**, 807 (1959)
5.2 T.A. Kaplan: Phys. Rev. **116**, 888 (1959)
5.3 J. Villain: Phys. Chem. Solids **11**, 303 (1959)
5.4 T. Nagamiya: *Solid State Physics* **20**, 305 (Academic, New York 1967)
5.5 T.A. Kaplan: Phys. Rev. **119**, 1460 (1960)
5.6 D.H. Lyons, T.A. Kaplan: Phys. Rev. **120**, 1580 (1960)
5.7 T.A. Kaplan, K. Dwight, D.H. Lyons, N. Menyuk: J. Appl. Phys. **32**, 13S (1961); Phys. Rev. **126**, 540 (1962); *ibid.* **127**, 1983 (1962); J. Appl. Phys. **36**, 1129 (1965)

Chapter 6

6.1 L. Néel: Ann. de Physique **5**, 232 (1936)
6.2 C.J. Gorter, J. Haantjes: Physica **18**, 285 (1952)
6.3 T. Nagamiya: Prog. Theor. Phys. **11**, 309 (1954)

Chapter 7

7.1 P.R. Weiss: Phys. Rev. **74**, 1493 (1948)
7.2 T. Oguchi: Prog. Theor. Phys. **13**, 148 (1955)
7.3 P.W. Kasteleijn, J. van Kranendonk: Physica **22**, 317 (1956)
7.4 P.G. de Gennes: In *Magnetism*, ed. by G. Rado and H. Suhl (Academic, New York 1963) Vol. 3, p. 115
7.5 H. Nyquist: Phys. Rev. **32**, 110 (1928)
7.6 H.B. Callen, T.A. Welton: Phys. Rev. **83**, 34 (1951)
7.7 R. Kubo: J. Phys. Soc. Jpn. **12**, 570 (1957)
7.8 L. Van Hove: Phys. Rev. **95**, 1374 (1954)
7.9 H. Mori, K. Kawasaki: Prog. Theor. Phys. **27**, 529 (1962)
7.10 L. Van Hove: Phys. Rev. **95**, 249 (1954)
7.11 C. Domb, M.F. Sykes: Proc. Roy. Soc. (London) A **240**, 214 (1957)
7.12 C. Domb, M.F. Sykes: Phys. Rev. **128**, 168 (1962)
7.13 G. Rushbrooke, P.J. Wood: Mol. Phys. **1**, 257 (1958)
7.14 L. Onsager: Phys. Rev. **65**, 117 (1944)
7.15 Shang-keng Ma: *Modern Theory of Critical Phenomena* (Benjamin, New York 1976) p. 103

Chapter 8

8.1 F. Bloch: Z. Physik **61**, 206 (1930)
8.2 J.C. Slater: Phys. Rev. **35**, 509 (1930)
8.3 C. Kittel: Phys. Rev. **71**, 270 (1947); *ibid.* **73**, 155 (1948)
8.4 R.L. White, I.H. Solt Jr.: Phys. Rev. **104**, 56 (1956)
8.5 J.E. Mercereau, R.P. Feynman: Phys. Rev. **104**, 63 (1956)
8.6 L.R. Walker: Phys. Rev. **105**, 390 (1957)
8.7 H. Bethe: Z. Phys. **71**, 205 (1931)
8.8 F.J. Dyson: Phys. Rev. **102**, 1217, 1230 (1956)
8.9 M. Wortis: Phys. Rev. **132**, 85 (1963)
8.10 T. Holstein, H. Primakoff: Phys. Rev. **58**, 1094 (1940)
8.11 P.W. Anderson: *Solid State Physics* **14**, 99 (Academic, New York, 1963)
8.12 F. Keffer, R. Loudon: J. Appl. Phys. **32**, 2S (1961)
8.13 T. Oguchi: Phys. Rev. **117**, 117 (1960)
8.14 J. Kanamori, M. Tachiki: J. Phys. Soc. Jpn. **17**, 1384 (1962)

Chapter 9

9.1 P.W. Anderson: Phys. Rev. **83**, 1260 (1951)
9.2 P.W. Anderson: Phys. Rev. **86**, 694 (1952)
9.3 T. Nagamiya: Prog. Theor. Phys. **6**, 342, 350 (1951)
9.4 C. Kittel: Phys. Rev. **82**, 565 (1951)
9.5 F. Keffer, C. Kittel: Phys. Rev. **85**, 329 (1952)
9.6 R. Kubo: Phys. Rev. **87**, 568 (1952)
9.7 J. Kanamori, K. Yosida: Prog. Theor. Phys. **14**, 423 (1955)
9.8 J.A. Eisele, F. Keffer: Phys. Rev. **96**, 929 (1954)
9.9 J. Kanamori, M. Tachiki: J. Phys. Soc. Jpn. **17**, 1384 (1962)
9.10 L. Hulthén: Arkiv Mat. Astron. Fysik **26**A, No.11 (1938)
9.11 R. Orbach: Phys. Rev. **112**, 309 (1958)
9.12 L.R. Walker: Phys. Rev. **116**, 1089 (1959)
9.13 C.N. Yang, C.P. Yang: Phys. Rev. **150**, 321, 327 (1966)

9.14 E. Lieb, T. Schultz, D. Mattis: Ann. Phys. (N.Y.) **16**, 407 (1961)
9.15 S. Katsura: Phys. Rev. **127**, 1508 (1962)
9.16 J. des Cloizeaux, J.J. Pearson: Phys. Rev. **128**, 2131 (1962)
9.17 R.B. Griffiths: Phys. Rev. **133**, A768 (1964)
9.18 C.N. Yang, C.P. Yang: Phys. Rev. **151**, 258 (1966)
9.19 J.C. Bonner, M.E. Fisher: Phys. Rev. **135**, A640 (1964)
9.20 C.N. Yang: Phys. Rev. Lett. **19**, 1312 (1967)
9.21 E.H. Lieb, F.Y. Wu: Phys. Rev. Lett. **20**, 1445 (1968)

Chapter 10

10.1 J.M. Kosterlitz, D.J. Thouless: J. Phys. C **6**, 1181 (1973)
10.2 J.M. Kosterlitz: J. Phys. C **7**, 1046 (1974)
10.3 J.B. Kogut: Rev. Mod. Phys. **51**, 659 (1979)
10.4 F. Spitzer: *Principles of Random Walk* (Van Nostrand, Princeton 1964) pp. 148–151
10.5 P.W. Anderson, G. Yuval: J. Phys. C **4**, 607 (1971)
10.6 P.W. Anderson, G. Yuval, D.R. Hamann: Phys. Rev. B **1**, 4464 (1970)
10.7 D.R. Nelson, J.M. Kosterlitz: Phys. Rev. Lett. **39**, 1201 (1977)

Chapter 11

11.1 L. Landau: Z. Physik **64**, 629 (1930)
11.2 J. Bardeen, L.N. Cooper, J.R. Schrieffer: Phys. Rev. **108**, 1175 (1957)

Chapter 12

12.1 F. Bloch: Z. Physik **57**, 545 (1929)
12.2 E.P. Wigner: Phys. Rev. **46**, 1002 (1934)
12.3 W. Macke: Z. Naturforsch. **5a**, 192 (1950)
12.4 D. Bohm, D. Pines: Phys. Rev. **92**, 609 (1953)
12.5 D. Pines: Phys. Rev. **92** 626 (1953)
12.6 M. Gell-Mann, K.A. Brueckner: Phys. Rev. **106**, 364 (1957)
12.7 K. Sawada: Phys. Rev. **106**, 372 (1957)
12.8 K. Sawada, K.A. Brueckner, N. Fukuda, R. Brout: Phys. Rev. **108**, 507 (1957)
12.9 D. Pines: Phys. Rev. **95**, 1090 (1954)
12.10 K.A. Brueckner, K. Sawada: Phys. Rev. **112**, 328 (1958)
12.11 B.S. Shastry: Phys. Rev. Lett. **38**, 449 (1977)

Chapter 13

13.1 W. Kohn, L.J. Sham: Phys. Rev. **140**, A1133 (1965)
13.2 J.C. Slater: Phys. Rev. **49**, 537 (1936)
13.3 J.C. Slater: Phys. Rev. **34**, 1293 (1929)
13.4 J.C. Slater, H. Statz, G. F. Koster: Phys. Rev. **91**, 1323 (1953)
13.5 J. Kanamori: Prog. Theor. Phys. **30**, 275 (1963)
13.6 K.A. Brueckner, C.A. Levinson: Phys. Rev. **97**, 1344 (1955)
13.7 K.A. Brueckner, J.L. Gammel: Phys. Rev. **109**, 1023, 1040 (1958)
13.8 V.M. Galitskii: J. Exp. Theor. Phys. (USSR) **34**, 151 (1958); Sov. Phys. – JETP **7**, 104 (1958)

13.9 J. Hubbard: Proc. Roy. Soc. (London) A **276**, 238 (1963); ibid. A **277**, 238 (1964); A **281**, 401 (1964)
13.10 M.C. Gutzwiller: Phys. Rev. Lett. **10**, 159 (1963); Phys. Rev. **134**, A923 (1964); **137**, A1726 (1965)
13.11 T. Ogawa, K. Kanda: Z. Physik B **30**, 355 (1978);
T. Ogawa, K. Kanda, T. Matsubara: Prog. Theor. Phys. **53**, 614 (1975)
13.12 D. Vollhardt: Rev. Mod. Phys. **56**, 99 (1984)
13.13 W. Metzner, D. Vollhardt: Phys. Rev. Lett. **59**, 121 (1987); Phys. Rev. B **37**, 7382 (1988)
13.14 F. Gebhard, D. Vollhardt: Phys. Rev. Lett. **59**, 1472 (1987); Phys. Rev. B **38**, 6911 (1988)
13.15 W.F. Brinkman, T.M. Rice: Phys. Rev. B **2**, 4302 (1970)
13.16 P.W. Anderson, W.F. Brinkman: Theory of anisotropic superfluidity in He, in *The Helium Liquids*, ed. by J.G.M. Armitage and I.E. Farqhar (Academic, New York 1975);
P.W. Anderson, W.F. Brinkman: In *The Physics of Liquid Helium and Solid Helium, Part II*, ed. by K.H. Bennemann and J.B. Ketterson (Wiley, New York, 1978)

Chapter 14

14.1 C. Herring, C. Kittel: Phys. Rev. **81**, 869 (1951)
14.2 C. Herring: Phys. Rev. **85**, 1003 (1952)
14.3 H. Nyquist: Phys. Rev. **32**, 110 (1928)
14.4 H.B. Callen, T.R. Welton: Phys. Rev. **83**, 34 (1951)
14.5 R. Kubo: J. Phys. Soc. Jpn. **12**, 570 (1957)
14.6 T. Izuyama, D.-J. Kim, R. Kubo: J. Phys. Soc. Jpn. **18**, 1025 (1963)
14.7 T. Moriya: J. Phys. Soc. Jpn. **18**, 516 (1963)
14.8 S. Doniach, S. Engelsberg: Phys. Rev. Lett. **17**, 750 (1966)
14.9 N.F. Berk, J.R. Schrieffer: Phys. Rev. Lett. **17**, 433 (1966)
14.10 K.K. Murata, S. Doniach; Phys. Rev. Lett. **29**, 285 (1972)
14.11 T. Moriya, A. Kawabata: J. Phys. Soc. Jpn. **34**, 639 (1973)
14.12 T. Moriya: *Spin Fluctuations in Itinerant Electron Magnetism,* Springer Ser. Solid-State Sci., Vol. 56 (Springer, Berlin, Heidelberg 1985)

Chapter 15

15.1 W. Kohn, C. Majumdar: Phys. Rev. **138**, A1617 (1965)
15.2 J. Friedel: Nuovo Cimento Supplement **7**, 287 (1958)
15.3 H. Fröhlich, F.R.N. Nabarro: Proc. Roy. Soc. (London) A **175**, 382 (1940)
15.4 C. Zener: Phys. Rev. **81**, 440 (1951)
15.5 M.A. Ruderman, C. Kittel: Phys. Rev. **96**, 99 (1954)
15.6 T. Kasuya: Prog. Theor. Phys. **16**, 45 (1956)
15.7 K. Yosida: Phys. Rev. **106**, 893 (1957); ibid. **107**, 396 (1957)
15.8 A. Blandin: J. Phys. Radium **22**, 507 (1961)
15.9 J. Owen, M.E. Browne, W.D. Knight, C. Kittel: Phys. Rev. **102**, 1501 (1956)
15.10 J. Owen, M.E. Browne, V. Arp, A.F. Kip: J. Phys. Chem. Solids **2**, 85 (1957)
15.11 P.W. Anderson: Phys. Rev. **124**, 41 (1961)
15.12 J.R. Schrieffer, P.A. Wolff: Phys. Rev. **149**, 491 (1966)

Chapter 16

16.1 J. Kondo: Prog. Theor. Phys. **32**, 37 (1964)
16.2 A.A. Abrikosov: Physics **2**, 5 (1965)
16.3 H. Suhl: Phys. Rev. **138**, A515 (1965); Physics **2**, 39 (1965); Phys. Rev. **141**, 483 (1966)
16.4 Y. Nagaoka: Phys. Rev. **138**, A1112 (1965)
16.5 H. Miwa: Prog. Theor. Phys. **34**, 1040 (1965)
16.6 K. Yosida, A. Okiji: Prog. Theor. Phys. **34**, 505 (1965)
16.7 J. Kondo: Prog. Theor. Phys. **40**, 683 (1968)
16.8 K. Yosida, H. Miwa: Prog. Theor. Phys. **41**, 1416 (1969)

Chapter 17

17.1 P.W. Anderson: Phys. Rev. **164**, 352 (1967)
17.2 J. Kondo: Prog. Theor. Phys. **36**, 429 (1966)
17.3 K. Yosida: Phys. Rev. **147**, 223 (1966); Prog. Theor. Phys. **36**, 875 (1966)
17.4 A. Okiji: Prog. Theor. Phys. **36**, 712 (1966)
17.5 A. Yoshimori: Phys. Rev. **168**, 493 (1968)
17.6 A. Yoshimori, K. Yosida: Prog. Theor. Phys. **39**, 1413 (1968)
17.7 K. Yosida, A. Yoshimori: Prog. Theor. Phys. **42**, 753 (1969)
17.8 K. Yosida, A. Yoshimori: In *Magnetism*, ed. by H. Suhl (Academic, New York 1973) Vol. 5, p. 253
17.9 J.S. Langer, V. Ambegaokar: Phys. Rev. **121**, 1090 (1961)
17.10 H. Ishii, K. Yosida: Prog. Theor. Phys. **38**, 61 (1967); H. Ishii: Prog. Theor. Phys. **40**, 201 (1968)
17.11 A.A. Abrikosov: Physics **2**, 5 (1965)
17.12 P.W. Anderson: J. Phys. C **3**, 2436 (1970)
17.13 P.W. Anderson, G. Yuval, D.R. Hamann: Phys. Rev. B **1**, 4464 (1970)
17.14 H. Shiba: Prog. Theor. Phys. **43**, 601 (1970)
17.15 S. Nakajima: Prog. Theor. Phys. **39**, 1402 (1968)
17.16 M. Fowler, A. Zawadowski: Solid State Commun. **9**, 471 (1971)
17.17 A.A. Abrikosov, A.A. Migdal: J. Low Temp. Phys. **3**, 519 (1970)
17.18 A. Sakurai, A. Yoshimori: Prog. Theor. Phys. **49**, 1840 (1973)

Chapter 18

18.1 P.W. Anderson: Phys. Rev. Lett. **18**, 1049 (1967); Phys. Rev. **164**, 352 (1967)
18.2 K. Yosida, A. Yoshimori: In *Magnetism*, ed. by H. Suhl (Academic, New York 1973) Vol. 5, p. 253
18.3 D.R. Hamann: Phys. Rev. B **2**, 1373 (1970)
18.4 P. Nozières, C.T. de Dominicis: Phys. Rev. **178**, 1097 (1969)
18.5 N.I. Muskhelishvili: *Singular Integral Equations* (Noordhoff 1953)
18.6 K. Yamada, K. Yosida: Prog. Theor. Phys. **68**, 1504 (1982)
18.7 J.S. Langer, V. Ambegaokar: Phys. Rev. **121**, 1090 (1961)
18.8 P.W. Anderson, G. Yuval: Phys. Rev. Lett. **23**, 89 (1969); G. Yuval, P.W. Anderson: Phys. Rev. B **1**, 1522 (1970)
18.9 G. Pólya, G. Szegö: *Aufgaben und Lehrsätze aus der Analysis II*, (Springer, Berlin, Heidelberg 1964) p. 98
18.10 P.W. Anderson, G. Yuval, D.R. Hamann: Phys. Rev. B **1** 4464 (1970)

Chapter 19

19.1 K. Wilson: *Nobel Symposia* **24**, 68 (Academic, New York 1974); Rev. Mod. Phys. **47**, 773 (1975)
19.2 P. Nozières: J. Low Temp. Phys. **17**, 31 (1974)
19.3 L.D. Landau: JETP **30**, 1058 (1956) [Sov. Phys. – JETP **3**, 920 (1957)]; JETP **32**, 59 (1957) [Sov. Phys. – JETP **5**, 101 (1957)]; JETP **35**, 97 (1958) [Sov. Phys. – JETP **8**, 70 (1959)];
P. Nozières: *Theory of Interacting Fermi Systems*, (Benjamin, New York 1964) Chap. 1
19.4 K. Yosida, K. Yamada: Prog. Theor. Phys. Suppl. No.46, 244 (1970);
K. Yamada: Prog. Theor. Phys. **53**, 970 (1975);
K. Yosida, K. Yamada: Prog. Theor. Phys. **53**, 1286 (1975);
K. Yamada: Prog. Theor. Phys. **54**, 316 (1975)
19.5 H. Shiba: Prog. Theor. Phys. **54**, 967 (1975)
19.6 A. Yoshimori: Prog. Theor. Phys. **55**, 67 (1976)
19.7 N. Andrei, K. Furuya, J.H. Lowenstein: Rev. Mod. Phys. **55**, 331 (1983)
19.8 A.M. Tsvelick, P.B. Wiegmann: Adv. Phys. **32**, 453 (1983)
19.9 N. Kawakami, A. Okiji: Phys. Lett. A **86**, 483 (1981);
A. Okiji: *Fermi Surface Effects*, ed. by J. Kondo, A. Yoshimori, Springer Ser. Solid-State Sci., Vol. 77 (Springer, Berlin, Heidelberg 1988) p. 63
19.10 V. Zlatić, B. Horvatić: Phys. Rev. B **28**, 6904 (1983)

Subject Index

Springer Series in Solid-State Sciences

Editors: M. Cardona P. Fulde K. von Klitzing H.-J. Queisser

1 **Principles of Magnetic Resonance** 3rd Edition By C. P. Slichter
2 **Introduction to Solid-State Theory** By O. Madelung
3 **Dynamical Scattering of X-Rays in Crystals** By Z. G. Pinsker
4 **Inelastic Electron Tunneling Spectroscopy** Editor: T. Wolfram
5 **Fundamentals of Crystal Growth I** Macroscopic Equilibrium and Transport Concepts By F. E. Rosenberger
6 **Magnetic Flux Structures in Superconductors** By R. P. Huebener
7 **Green's Functions in Quantum Physics** 2nd Edition By E. N. Economou
8 **Solitons and Condensed Matter Physics** Editors: A. R. Bishop and T. Schneider
9 **Photoferroelectrics** By V. M. Fridkin
10 **Phonon Dispersion Relations in Insulators** By H. Bilz and W. Kress
11 **Electron Transport in Compound Semiconductors** By B. R. Nag
12 **The Physics of Elementary Excitations** By S. Nakajima, Y. Toyozawa, and R. Abe
13 **The Physics of Selenium and Tellurium** Editors: E. Gerlach and P. Grosse
14 **Magnetic Bubble Technology** 2nd Edition By A. H. Eschenfelder
15 **Modern Crystallography I** Fundamentals of Crystals Symmetry, and Methods of Structural Crystallography 2nd Edition By B. K. Vainshtein
16 **Organic Molecular Crystals** Their Electronic States By E. A. Silinsh
17 **The Theory of Magnetism I** Statics and Dynamics By D. C. Mattis
18 **Relaxation of Elementary Excitations** Editors: R. Kubo and E. Hanamura
19 **Solitons** Mathematical Methods for Physicists By. G. Eilenberger
20 **Theory of Nonlinear Lattices** 2nd Edition By M. Toda
21 **Modern Crystallography II** Structure of Crystals 2nd Edition By B. K. Vainshtein, V. L. Indenbom, and V. M. Fridkin
22 **Point Defects in Semiconductors I** Theoretical Aspects By M. Lannoo and J. Bourgoin
23 **Physics in One Dimension** Editors: J. Bernasconi and T. Schneider
24 **Physics in High Magnetics Fields** Editors: S. Chikazumi and N. Miura
25 **Fundamental Physics of Amorphous Semiconductors** Editor: F. Yonezawa
26 **Elastic Media with Microstructure I** One-Dimensional Models By I. A. Kunin
27 **Superconductivity of Transition Metals** Their Alloys and Compounds By S. V. Vonsovsky, Yu. A. Izyumov, and E. Z. Kurmaev
28 **The Structure and Properties of Matter** Editor: T. Matsubara
29 **Electron Correlation and Magnetism in Narrow-Band Systems** Editor: T. Moriya
30 **Statistical Physics I** Equilibrium Statistical Mechanics 2nd Edition By M. Toda, R. Kubo, N. Saito
31 **Statistical Physics II** Nonequilibrium Statistical Mechanics 2nd Edition By R. Kubo, M. Toda, N. Hashitsume
32 **Quantum Theory of Magnetism** 2nd Edition By R. M. White
33 **Mixed Crystals** By A. I. Kitaigorodsky
34 **Phonons: Theory and Experiments I** Lattice Dynamics and Models of Interatomic Forces By P. Brüesch
35 **Point Defects in Semiconductors II** Experimental Aspects By J. Bourgoin and M. Lannoo
36 **Modern Crystallography III** Crystal Growth By A. A. Chernov
37 **Modern Chrystallography IV** Physical Properties of Crystals Editor: L. A. Shuvalov
38 **Physics of Intercalation Compounds** Editors: L. Pietronero and E. Tosatti
39 **Anderson Localization** Editors: Y. Nagaoka and H. Fukuyama
40 **Semiconductor Physics** An Introduction 6th Edition By K. Seeger
41 **The LMTO Method** Muffin-Tin Orbitals and Electronic Structure By H. L. Skriver
42 **Crystal Optics with Spatial Dispersion, and Excitons** 2nd Edition By V. M. Agranovich and V. L. Ginzburg
43 **Structure Analysis of Point Defects in Solids** An Introduction to Multiple Magnetic Resonance Spectroscopy By J.-M. Spaeth, J. R. Niklas, and R. H. Bartram
44 **Elastic Media with Microstructure II** Three-Dimensional Models By I. A. Kunin
45 **Electronic Properties of Doped Semiconductors** By B. I. Shklovskii and A. L. Efros
46 **Topological Disorder in Condensed Matter** Editors: F. Yonezawa and T. Ninomiya

Springer Series in Solid-State Sciences

Editors: M. Cardona P. Fulde K. von Klitzing H.-J. Queisser

47 **Statics and Dynamics of Nonlinear Systems**
Editors: G. Benedek, H. Bilz, and R. Zeyher

48 **Magnetic Phase Transitions**
Editors: M. Ausloos and R. J. Elliott

49 **Organic Molecular Aggregates**
Electronic Excitation and Interaction Processes
Editors: P. Reineker, H. Haken, and H. C. Wolf

50 **Multiple Diffraction of X-Rays in Crystals**
By Shih-Lin Chang

51 **Phonon Scattering in Condensed Matter**
Editors: W. Eisenmenger, K. Laßmann, and S. Döttinger

52 **Superconductivity in Magnetic and Exotic Materials** Editors: T. Matsubara and A. Kotani

53 **Two-Dimensional Systems, Heterostructures, and Superlattices**
Editors: G. Bauer, F. Kuchar, and H. Heinrich

54 **Magnetic Excitations and Fluctuations**
Editors: S. W. Lovesey, U. Balucani, F. Borsa, and V. Tognetti

55 **The Theory of Magnetism II** Thermodynamics and Statistical Mechanics By D. C. Mattis

56 **Spin Fluctuations in Itinerant Electron Magnetism** By T. Moriya

57 **Polycrystalline Semiconductors**
Physical Properties and Applications
Editor: G. Harbeke

58 **The Recursion Method and Its Applications**
Editors: D. G. Pettifor and D. L. Weaire

59 **Dynamical Processes and Ordering on Solid Surfaces** Editors: A. Yoshimori and M. Tsukada

60 **Excitonic Processes in Solids**
By M. Ueta, H. Kanzaki, K. Kobayashi, Y. Toyozawa, and E. Hanamura

61 **Localization, Interaction, and Transport Phenomena** Editors: B. Kramer, G. Bergmann, and Y. Bruynseraede

62 **Theory of Heavy Fermions and Valence Fluctuations** Editors: T. Kasuya and T. Saso

63 **Electronic Properties of Polymers and Related Compounds**
Editors: H. Kuzmany, M. Mehring, and S. Roth

64 **Symmetries in Physics** Group Theory Applied to Physical Problems 2nd Edition
By W. Ludwig and C. Falter

65 **Phonons: Theory and Experiments II**
Experiments and Interpretation of Experimental Results By P. Brüesch

66 **Phonons: Theory and Experiments III**
Phenomena Related to Phonons
By P. Brüesch

67 **Two-Dimensional Systems: Physics and New Devices**
Editors: G. Bauer, F. Kuchar, and H. Heinrich

68 **Phonon Scattering in Condensed Matter V**
Editors: A. C. Anderson and J. P. Wolfe

69 **Nonlinearity in Condensed Matter**
Editors: A. R. Bishop, D. K. Campbell, P. Kumar, and S. E. Trullinger

70 **From Hamiltonians to Phase Diagrams**
The Electronic and Statistical-Mechanical Theory of sp-Bonded Metals and Alloys By J. Hafner

71 **High Magnetic Fields in Semiconductor Physics**
Editor: G. Landwehr

72 **One-Dimensional Conductors**
By S. Kagoshima, H. Nagasawa, and T. Sambongi

73 **Quantum Solid-State Physics**
Editors: S. V. Vonsovsky and M. I. Katsnelson

74 **Quantum Monte Carlo Methods in Equilibrium and Nonequilibrium Systems** Editor: M. Suzuki

75 **Electronic Structure and Optical Properties of Semiconductors** 2nd Edition
By M. L. Cohen and J. R. Chelikowsky

76 **Electronic Properties of Conjugated Polymers**
Editors: H. Kuzmany, M. Mehring, and S. Roth

77 **Fermi Surface Effects**
Editors: J. Kondo and A. Yoshimori

78 **Group Theory and Its Applications in Physics**
2nd Edition
By T. Inui, Y. Tanabe, and Y. Onodera

79 **Elementary Excitations in Quantum Fluids**
Editors: K. Ohbayashi and M. Watabe

80 **Monte Carlo Simulation in Statistical Physics**
An Introduction 3rd Edition
By K. Binder and D. W. Heermann

81 **Core-Level Spectroscopy in Condensed Systems**
Editors: J. Kanamori and A. Kotani

82 **Photoelectron Spectroscopy**
Principle and Applications 2nd Edition
By S. Hüfner

83 **Physics and Technology of Submicron Structures**
Editors: H. Heinrich, G. Bauer, and F. Kuchar

84 **Beyond the Crystalline State** An Emerging Perspective By G. Venkataraman, D. Sahoo, and V. Balakrishnan

85 **The Quantum Hall Effects**
Fractional and Integral 2nd Edition
By T. Chakraborty and P. Pietiläinen

86 **The Quantum Statistics of Dynamic Processes**
By E. Fick and G. Sauermann

87 **High Magnetic Fields in Semiconductor Physics II**
Transport and Optics Editor: G. Landwehr

88 **Organic Superconductors** 2nd Edition
By T. Ishiguro, K. Yamaji, and G. Saito

89 **Strong Correlation and Superconductivity**
Editors: H. Fukuyama, S. Maekawa, and A. P. Malozemoff

Springer and the environment

At Springer we firmly believe that an international science publisher has a special obligation to the environment, and our corporate policies consistently reflect this conviction.

We also expect our business partners – paper mills, printers, packaging manufacturers, etc. – to commit themselves to using materials and production processes that do not harm the environment. The paper in this book is made from low- or no-chlorine pulp and is acid free, in conformance with international standards for paper permanency.

Printing: Mercedesdruck, Berlin
Binding: Buchbinderei Lüderitz & Bauer, Berlin